What Could Possibly
Go Wrong . . .

JEREMY CLARKSON

PENGUIN BOOKS

PENGUIN BOOKS

UK | USA | Canada | Ireland | Australia
India | New Zealand | South Africa

Penguin Books is part of the Penguin Random House group of companies
whose addresses can be found at global.penguinrandomhouse.com.

Penguin
Random House
UK

First published by Michael Joseph 2014
Published in Penguin Books 2015
001

Copyright © Jeremy Clarkson, 2015

The moral right of the author has been asserted

Typeset by Jouve (UK), Milton Keynes
Printed in Great Britain by Clays Ltd, St Ives plc

A CIP catalogue record for this book is available from the British Library

ISBN: 978–1–405–91937–1

www.greenpenguin.co.uk

MIX
Paper from
responsible sources
FSC   FSC® C018179
www.fsc.org

Penguin Random House is committed to a
sustainable future for our business, our readers
and our planet. This book is made from Forest
Stewardship Council® certified paper.

The contents of this book first appeared in Jeremy Clarkson's *Sunday Times* column. Read more about the world according to Clarkson every week in *The Sunday Times*.

Contents

For pity's sake, Fritz, please stop fiddling
MINI Countryman Cooper S ALL4   1

No nasty surprises in this gooey confection
Audi A7 Sportback 3.0 TDI quattro SE   5

Oh yes, take me now, Lady Marmalade
Citroën DS3 Racing   9

It's hardly British but learn to haggle
Mitsubishi Outlander 2.2 DI-D GX4, 7 seats   14

Try this moose suit for size, Mr Top Gun
Saab 9-3 SportWagon Aero TiD 180PS   18

Titter ye not, it's built for the clown about town
Nissan Juke 1.6 DIG-T Tekna   22

Those yurt dwellers have got it right
Land Rover Freelander 2 eD4 HSE 2WD   26

Little Luigi's turbo boost
Fiat 500 0.9 TwinAir Lounge   30

I don't fancy Helga von Gargoyle . . . Can't think why
Porsche Panamera 3.6 V6 PDK   34

Damn it, Spock, we can't shake off Arthur Daley
Jaguar XJ 5.0 Supercharged Supersport LWB 4dr   38

Bruce's bonzer duck-billed koala
Ford Falcon FPV Boss 335 GT   42

Botox and a bikini wax and I'm ready to roll
Jensen Interceptor S   46

Oh, barman, my pint of pitbull has gone all
    warm and fluffy
Ford Focus Titanium 1.6 Ecoboost   50

Pointless but fun – what a good wheeze
Renault Wind Roadster GT Line 1.6 VVT   54

Prepare your moobs for a workout
Aston Martin Virage   58

The old duffer trots out in boy-racer colours
Skoda Faiba vRS1.4 TSI DSG   62

What's the Swedish-Chinese for I can't see?
Volvo V60 T5 R-Design   66

I love you now I'm all grown up, Helga
Porsche 911 GTS   70

Oh, miss, you turn me into a raging despot
Mercedes CLS 63 AMG   74

From 0 to 40 winks in the blink of an eye
BMW 640i SE convertible   78

Oh, Shrek, squeeze me till it hurts
Nissan GT-R   82

A world first – the Ferrari 4 × what for?
Ferrari FF   86

Work harder, boy, or it will be you in here
VW Jetta 2.0 TDI Sport   90

Too tame for the special flair service
Audi RS 3   94

An asthmatic accountant in lumberjack clothing
Mazda CX-7   98

Someone please check I haven't left my spleen back there
BAC Mono   102

I thought it looked humdrum. But wow!
Honda Accord Type S   106

You vill never handle zis torture
Mercedes-Benz G 350 Bluetec   110

Strip out all the tricks and it's still a wizard
Audi A6 SE 3.0 TDI   114

Open up them pearly gates . . .
Lamborghini Gallardo LP570-4 Spyder Performante   118

Oh, grunting frump, you looked so fine on the catwalk
Jaguar XF 2.2 Diesel Premium Luxury   122

Now we're flying
Mercedes-Benz SLS Roadster   126

The topless tease luring men to ridicule
VW Golf Cabriolet GT   130

I'm sold, Mrs Beckham – I want your baby
Range Rover Evoque Prestige SD4 auto   134

I say, chaps, who needs a fourth wheel?
Morgan Three Wheeler   138

Beach beauties love my bucking bronto
Lamborghini Aventador LP 700-4   142

Hop in, Charles, it's a Luddite's dream
Mercedes C 63 AMG coupé Black Series   146

It's no cruiser but it can doggy-paddle
Jeep Grand Cherokee 3.0 CRD V6 Overland   150

Uh-oh, some fool's hit the panic button
Chevrolet Orlando 1.8 LTZ   154

Simply no use for taking the kids to see Granny
Audi R8 GT   158

Amazing where bottle tops and string will get you
Hyundai i40 1.7 CRDi 136PS Style   162

Bong! I won't let you go until you love me
BMW M5   166

A heart transplant sexes up Wayne's pet moose
Bentley Continental GT V8   170

The arms race is over and Vera Lynn has won
Aston Martin DBS Carbon Edition   174

Good doggy – let's give the bark plugs a workout
Suzuki Swift Sport 1.6   178

Look what oi got, Farmer Giles: diamanté wellies
Jeep Wrangler 2.8 CRD Sahara Auto 4-door   182

Powered by beetroot, the hand-me-down that
    keeps Russia rolling
Lada Riva   186

The yummiest of ingredients but the soufflé's gone flat
Porsche 911 Carrera   190

I ran into an EU busybody and didn't feel a thing
BMW 640d (with M Sport package)   195

Blimey, you've got this mouse to roar, Fritz
Volkswagen High Up!   199

Styled for mercenaries. Driven by mummy
Ford Kuga 2.0 TDCi Titanium X PowerShift   203

Simply the best, but so bashful buying one is *verboten*
BMW 328i Modern   207

Click away, paparazzi, I've got nice clean Y-fronts
Audi A8 3.0 TFSI   211

Get a grip – it's only a Roller
Rolls-Royce Phantom II   215

I know about your frilly knickers, Butch
Mercedes SLK 55 AMG   219

Fritz calls it a soft-roader. I call him soft in the head
Audi Q3 2.0 TDI quattro SE S tronic   223

Cheer up – Napoleon got shorty shrift too
Mini Cooper S roadster   227

That funny noise is just Einstein hiding under the bonnet
Ford Focus 1.0 EcoBoost 125PS Titanium   231

Gosh, never thought I'd dump Kate Moss so fast
Citroën DS5 DSport HDi 160 automatic   235

Squeeze in, Queenie, there's space next to Tom Cruise
Kia Cee-d '2' 1.6 GDI   239

The wife's away, so come check out my electric extremity
Mercedes-Benz ML 350 BlueTec 4Matic Sport   243

If I go back to Africa, will you take it away again?
Porsche 911 Carrera S cabriolet   247

Oh, Miss Ennis, let's sprint to seventh heaven
Ferrari 458 Spider   252

Yikes! The plumber's van has put a leak in my wallet
Citroën Berlingo   256

Gary the ram raider cracks Fermat's last theorem
Vauxhall Astra VXR 2.0i Turbo   260

Kiss goodbye to your no-claims – Mr Fender-bender
      has a new toy
Peugeot 208 1.2 VTi Allure   264

The nip and tuck doesn't fool anyone, Grandma
Jaguar XKR-S   268

Wuthering werewolves, a beast made for the moors
Lexus LFA   272

It's certainly cheap . . . but I can't find cheerful
Skoda Octavia vRS   276

Ooh, it feels good to wear my superhero outfit again
Toyota GT86   280

OK, Sister Maria, try tailgating me now
Audi S6 4.0 TFSI quattro   284

It's Sunday, the sun is out – let's go commando
Ferrari California 30   288

Yo, bruv, check out da Poundland Bentley
Chrysler 300C Executive   292

Out with the flower power, in with the toothbrush
      moustache
VW Beetle 1.4 TSI Sport   296

You can keep your schnapps, Heidi – I'll have cider
      with Rosie
Mercedes A 250 AMG   300

A real stinker from Silvio, the lav attendant
Chrysler Ypsilon   304

Ask nicely and it'll probably cook you dinner
      underwater
BMW M135i   308

Contents   xiii

The pretty panzer parks on Jurgen's golf links
Volvo V40 D4 SE Nav   312

I ordered a full English but ended up with
    bubble and squeak
Aston Martin Vanquish   316

The cocaine chintz has been kept in check
Range Rover Vogue SDV8 4.4L V8 Vogue   320

Thanks, guys, from the heart of my bottom
Audi RS 4 Avant 4.2 FSI quattro   324

Just like Anne Boleyn, there's no magic
    with the head off
Volkswagen Golf GTI cabriolet 2.0 TSi   328

Come on, caravanners, see if it will
    tackle the quicksand
Hyundai Santa-Fe Premium 7-seat   332

No one can reinvent the wheel quite like you, Fritz
VW Golf 1.4 TSI ACT GT   336

Great at a shooting party – for gangsters
Mercedes CLS63 AMG Shooting Brake   340

Yippee! It's OK to be a Bentley boy again
Bentley Continental GT Speed   344

Thrusters on, Iron Man, this'll cut through the
    congestion
Audi R8 5.2 FSI quattro S tronic   348

They'll be flying off the shelves at Poundland
Porsche 911 Carrera 4S   352

So awful I wouldn't even give it to my son
Alfa Romeo MiTo 875cc TwinAir Distinctive   356

Off to save the planet with my African queen
BMW 528i Touring SE (1999, T-reg) 360

Oh, I hate the noise you make in 'wounded cow' mode
*Toyota Corolla GX (aka the Auris but GX
model not sold in UK)* 364

That puts paid to my theory on the ascent of manual
Aston Martin Vantage V12 roadster 368

Oh, how you'll giggle while strangling that polar bear
Ford Fiesta ST 1.6T EcoBoost 372

Another bad dream in a caravan of horrors
Honda CR-V 2.2 I-D TEC EX 376

Ooh, you make me go weak at the knees . . . and the
hips and the spine
Jaguar F-Type S 380

Mirror, signal, skedaddle – Mr Bump's been
turbocharged
Peugeot 208 GTi 384

Not now, Cato – keep turning the egg whisk
while I push
MG6 Magnette 1.9 DTi-Tech 388

No grid girls, no red trousers – it's formula school run
Mazda CX-5 2WD SE-L 392

Where does Farmer Giles eat his pork pie?
Range Rover Sport SDV6 Autobiography 396

They only make one car. But it's a nice colour
Porsche Cayman S with PDK 400

Say the magic word and the howling banshee turns
sultry sorceress
McLaren 12C Spider 404

Take the doors off and put them back on?
     That'll be £24,000, sir
BMW M6 Gran Coupé   408

Thunderbird and Mustang have gone, so what'll
     we call it, chaps?
Vauxhall Adam   412

Ha! They'll never catch me now I'm the invisible man
VW Golf GTI 2.0 TSI Performance Pack   416

Coo! A baby thunderclap from Merc's OMG division
Mercedes-Benz A45 AMG   420

From the nation that brought you Le Mans . . . A tent
     with wheels
Citroën DS3 cabrio DSport   424

The fun begins once you've arm-wrestled Mary
     Poppins for control
Audi RS 5 cabriolet quattro 4.2 FSI   428

Gliding gently into the parking slot reserved for losers
Peugeot 2008   432

Where the hell did they hide the 'keeping up with
     Italians' button?
Jaguar F-type   436

Go and play with your flow chart, Comrade Killjoy,
     while I floor it
Audi RS 6 Avant   439

Who lent Scrooge the ninja costume?
Lexus IS 300h F Sport   443

Crikey, the Terminator has joined the *Carry On* team
Mercedes-Benz SLS AMG Black Series   447

Grab her lead and forget all about the mess on the floor
Alfa Romeo 4C   451

Goodbye, Dino. It's the age of the mosquito
McLaren P1   455

Watch out, pedestrians, I'm packing lasers
Mercedes-Benz S 500 L AMG Line   459

I can see the mankini peeking out over your waistband
BMW 435i M Sport coupé   463

The crisp-baked crust hides a splodge of soggy dough
Kia Pro_Cee'd GT Tech   467

A menace to cyclists, cars, even low-flying aircraft
Audi SQ5 3.0 BiTDI quattro   471

I'm sorry, Comrade. No Iron Curtain, no deal
Dacia Sandero Access 1.2   475

You're off by a country mile with this soggy
    pudding, Subaru
Subaru Forester 2.0 Lineartronic XT   479

You can't play bumper cars, but the bouncy
    castle's brilliant
Volvo V40 T5 R-Design Lux   483

Drives on water and raises Lazarus in 4.1 seconds
Aston Martin Vanquish Volante   487

For pity's sake, Fritz, please stop fiddling

MINI Countryman Cooper S ALL4

After much careful consideration over the festive season, I've decided that God is almost certainly a German. He created the world and festooned it with all sorts of unusual creations, none of which he liked very much. So then he killed them off and started again. Then he didn't like that lot either, so he turned all the dinosaurs into birds and gave one of the apes opposable thumbs.

Geologically, he's never satisfied. Originally, he placed Scotland in the south Pacific, but he obviously thought the feng shui was wrong, so he moved it to a spot in the middle of what we now call the Atlantic ocean. Then he didn't think the world should have Scotland at all, so he buried it under what has now become South America.

And then he thought that actually England looked a bit lonely sticking out of the top of France, so he dug up Scotland again and placed it on the top of Northumberland, like a jaunty, lopsided hat. And then he decided that England shouldn't really be joined to France any more, so he created the English Channel.

Today, he's decided that the Himalayas should be a bit taller and that there really is no point to Greece, or any of those silly low-lying islands in the middle of the Pacific. And he's realized that the polar bear is so ugly and vicious that it has no place in his toy box.

He fiddles with the weather, too. At first, he thought it should be a hot and steamy planet but then he thought that, actually, it ought to be extremely cold. He's still fiddling today, which is driving all the eco-loonies insane. Just as they think it's getting hotter, the whole of Europe gets covered in snow.

Germans are the same. Give them a country and they want the one next door as well.

There is an upside to this, though. When a German creates something excellent, he does not go home to celebrate with a glass of beer. No. He goes straight back to his office so that he can set about making improvements. In Germany, being better than everyone else isn't good enough. You have to be better than yourself.

They even do this with their wine. Having created the liquid perfection that is Niersteiner Gutes Domtal, they went back to the drawing board and decided that the only way to make a better wine would be to add flecks of gold leaf. So they did. How brilliant's that? Wine that glitters under the lights. Stunning.

Things are very different in Britain. Prince Charles, for instance, thinks the world would be a better place if all progress had stopped in about 1952. And every planning department is run by people who want Britain to look like the front of a Dorset chocolate box. If God were English, your route to work would be blocked every morning by a brontosaurus.

Red telephone boxes were a prime example of this. They were useless and smelt of urine, and you could die of hypothermia before the pips even began. But there was a huge furore when someone – probably a German – suggested they should be updated. Change? Here? In Britain? Are you mad? We are a nation that puts *The Two Ronnies* on every Christmas, even though one of them is dead.

This attitude really doesn't work and it especially doesn't work in the car industry. When the first Range Rover came along in 1970, everyone could see that it was very excellent indeed. So the team responsible for designing it was sent home and the model soldiered on, with almost no changes at all, until 1994. By which time it was a relic.

There's a similar problem with the Land Rover. The car you

buy today is pretty much the same as the car you could have bought after the war. Can you imagine BMW doing that? Designing a car and then keeping it in production for sixty years? It's inconceivable.

But when it comes to resting on your laurels, the crown must go to Alec Issigonis. He made the Mini, which in the late 1950s was an inspired design, and then he decided to leave it alone for ever. Occasionally someone would nail a bit of wood to the side, and they once changed the radiator grille, but, fundamentally, it just kept on rolling down the production line, powered by an engine that could trace its roots back to a time when Scotland was off the coast of South Africa. It would still be soldiering on today, had BMW not arrived on the scene and said, 'For you, Tommy, the warhorse is over.'

Unfortunately, the Germans' obsession with self-improvement is now starting to get a bit silly, because in addition to their original Mini, and the various derivations of that, we now have the convertible, which is fine, and the Clubman, which is fine too, providing you are impervious to its looks and don't want to see out of the back. But sadly we also now have the Countryman. And that's not fine at all.

First of all, it has four doors, seating inside for five and a large boot. This has been achieved by making the car much bigger. So it's not really a Mini any more, is it? At 13½ feet in length, it's a third longer than the Issigonis original and should really be called the Maxi. Or maybe the Twinset.

There's another problem, though. BMW's first effort looked good, and still does, whereas the Countryman looks absolutely stupid. It's like a Mini that's been putting on weight for a part. It doesn't look cool or interesting or practical. It looks fat.

Of course, you might not care about how it looks or what it's called. Fine. But I bet you will care about the cramp it gives you when you drive it in traffic. It's the second Mini on the trot that has done this to me, come to think of it, and you will definitely care about how easy it is to stall, and how hard it is to get going again thanks to the stop-start eco-gadgetry that shuts down the engine whenever you're stationary.

To make matters worse, it's not especially nice to drive. The ride's not bad but the steering is nervous, the dashboard is bonkers and you feel like you're sitting on it rather than inside it. I arrived everywhere late, exasperated, looking silly and with a lightning bolt of pain in my left shin.

On the upside, the Countryman is available with four-wheel drive. It's a simple system that would be flummoxed by the weather we had recently but would get you up and down a farm track easily enough. The thing is, though, that the model I tested – a Cooper S – costs more than £22,000. And that makes it about £3,000 more than the similarly powered Skoda Yeti.

Don't be a snob about this, because the Mini isn't a Mini and the Skoda isn't really a Skoda. It's just a Volkswagen. More pointedly, the Mini is terrible and the Yeti is surprisingly good.

And on that note, can I just wish you all a very happy and exciting new year.

2 January 2011

No nasty surprises in this gooey confection

Audi A7 Sportback 3.0 TDI quattro SE

Well, there we are. 'Public transport' was a very interesting social experiment, but after the debacle of last year, it's probably in everyone's best interests if we all agree it simply doesn't work.

Let us take the trains as a prime example. As we know, they all grind to a halt whenever it is too warm or too cold or too autumnal, but of course the problem is much bigger than that. A railway locomotive is extremely expensive. I don't know how much it would cost to buy one, but I'm guessing that it would be several hundred pounds.

Then you have the rolling stock – and I do know that each carriage costs more than £1m – and the mile upon mile of track that need to be linked and monitored and governed. Just maintaining it costs £2 billion a year. The upshot of all this is very simple. Divide the total cost of the railway network by the number of people who want to use it, and the average price of a ticket should be about £4m. Soon, if what we're hearing is correct, it will be.

I realize, of course, that in theory a high-speed train linking the north and the south of Britain is a fine idea. But since it needs to cross at least five Tory constituencies, it will never happen. And nobody would be able to afford to go on it, even if it did.

Then we have air travel. In principle this should work quite well, but the concept has unfortunately been hijacked by busybodies who now insist on taking photographs of your gentleman sausage and confiscating your toiletries every time you want to go somewhere. This doesn't work. And it doubly doesn't work when planes are now grounded by everything from a bit of weather to a volcanic burp near the Arctic circle.

This leaves us with buses, and oh dear. They really don't work at all because they are simply too full of diseases and knives. No, really. The next time a bus goes by, have a look inside, and I guarantee that the passenger – there's never more than one – will not be the sort of person you would allow within 500 feet of your front door.

I accept that in rural areas the elderly and the infirm need to get to the post office, but why send a supertanker round to their village five times a day? Nobody needs to go shopping that often. Why not send a small Transit van round once a week? Or, better still, why not give those who cannot drive an internet and let them do it all online?

So, we're all agreed that whether you want a pint of milk from the shops or a holiday in the south of France, the car is better, safer, cheaper, faster, more comfortable and less annoying to others. Plus, nobody pats your breasts before you set off and you don't emerge at the other end of the journey with deep vein thrombosis, diphtheria, a knife in your eye and no luggage.

Of course, there are many annoying things about using the roads. Interfering governments have decided, for instance, that the amount of tax you pay should depend on the composition of the gas coming from your tailpipe. This means that cars will soon have to have two motors. One to move you about and one to assist on hills. Even Ferrari is going down the stupid hybrid road.

Then there are the speed limits. For some reason our government thinks that motorway travel should be undertaken at no more than 70 mph, because that was a safe speed when your dad's Ford Anglia had drum brakes. I know. Ridiculous. But there we are.

There are countless other problems, too, but despite everything, the car is still good. The car still works. There is still no alternative. It's just a question of deciding which one to buy.

There was a time when Audi made cars only for German cement salesmen, but in recent years it has decided to make a car for absolutely everyone in the world. There's the Q5, the Q7, the R8, the A1, the A3, the A4, the A5, the A6, the A8 and now the A7.

I'll let you into a little secret at this point. They are all the same. Oh, they may look a bit different, and some are bigger than others, but in essence they are all made from the same components.

Think of it this way. Cakes, buns, Yorkshire pudding and pancakes all look and taste different but they're all made from the same thing. That's how it is with Audis. Flour and eggs mixed up in different ways to create twenty-one different cars.

At first you think Audi may have actually tried something radical with the A7 because it has an all-new platform. But then you learn that this platform will be used in the next A6. It's the same story with the engines and the four-wheel-drive system and all of the interior fixtures and fittings.

However, the engineers can make a difference by fiddling with the steering and the suspension setup, and I must say that in the A7 they have. This does not feel like an Audi. It feels better. The ride is beautifully judged, the handling is lovely and the steering is spot-on. It's not a sports cake but it's not a Yorkshire pudding either. It's just right, in fact, for the fiftysomething chap who wants a stylish hatchback that doesn't break his spine every time he goes over a catseye. Although, when I say stylish . . . it isn't, really. The back looks as though it's melted and the front is just sort of Audi-ish. Mind you, it must be said that it does have an enormous boot and loads of room in the nicely trimmed cabin for four. Not five, though. There is no centre rear seatbelt.

Apart from this oversight – which is bound to have been the result of a marketing meeting at which someone stood up und said, 'Zer is no such sing as ein sporty car mitt five seats' – the only problem is the positioning of the accelerator pedal. The car may be able to keep going when the weather would rather it didn't, thanks to four-wheel drive. But if you are wearing the sort of shoes that enabled you to get through the snow to the car in the first place, you will end up pressing the brake pedal every time you want to go faster.

Despite this, and the melted rear, the A7 struck me as a good car. A bit heavy, perhaps, but good nevertheless. Until I checked out the prices. A top model dressed up to the nines will set you back a massive £91,500. The model I drove, a 3-litre turbodiesel with a seven-speed double-clutch gearbox and four-wheel drive, is the best part of £50,000. And I'm sorry, but it simply doesn't feel worth this much.

Yes, it's big and striking and practical and – we're told – extremely safe, but underneath, it's just eggs and flour. And for £50,000 you could do better. The Mercedes CLS springs to mind. So, too, does the Jaguar XJ.

It's nice to have the choice, though. Because that's what you didn't ever get with the failed experiment that was public transport.

9 January 2011

Oh yes, take me now, Lady Marmalade

Citroën DS3 Racing

Over the past few years, the sort of people who find recycling exciting have predicted the end of the internal combustion engine and said that 2011 will herald the bright new dawn of silent, zero-emission electric motoring, where no one dies and town centres actually look like the models architects make when applying for planning permission.

There's no doubt, of course, that many car manufacturers are working hard on hybrids – which are normal, petrol-engine cars that have a second, electric motor to keep the rule makers in Brussels happy. But pure electric cars? I don't see their Blu-ray/VHS/Sky moment until someone commercializes a hydrogen-based system for recharging the batteries. And that's not going to happen in 2011. Or 2012. Or any time in the foreseeable future.

What I do see happening in 2011 is car makers peeping from behind the terrifying double-dip curtain of financial uncertainty and presenting us with a flurry of machinery that will keep the disciples of internal combustion as happy as if they'd landed the role of 'chauffeur' in a French porn film.

Aston Martin, for instance, will present two new cars this year. One, called the Cygnet, is a 1.3-litre version of the Toyota iQ. Engineered solely to keep the average fuel consumption figures of the Aston range down – and therefore the Euro law makers happy – it will be treated as a joke. The other will not.

It's the One-77, which is made from carbon fibre, has a

hand-built V12 and boasts a top speed of 220mph or more, making it by far the fastest production car Aston has made. The only trouble is that it will cost £1.2m, which is a lot.

Lamborghini is also planning a limited-edition, mega-money car for 2011 and, because it doesn't have to worry about average fuel consumption figures – Lambo is owned by Volkswagen, which makes the Polo – it will be a replacement for the Murciélago.

That, however, will be overshadowed by the new McLaren. Named MP4-12C – or OCD for short – it will have a twin-turbocharged 3.8-litre V8 that develops 592 horse-power. So, with a price tag of just £168,500 – about half what McLaren charged for its last car and around a fifth of the price of the one before that – it will be less expensive and more power-ful than the Ferrari 458. I can't imagine it will be better but who knows.

You want more evidence that the economy's recovering and batteries are on the back burner? Well, there's going to be a con-vertible version of the epic Mercedes SLS and a long-wheelbase option for the Rolls-Royce Ghost. Then we have the new Por-sche 911, which will be exactly the same as every other 911, and a hardcore derivative of the car most Formula One drivers use when their sponsors and engine suppliers aren't looking – the Nissan GT-R.

In the real world, BMW is working on yet another version of the Mini – it'll be a two-door coupé. There will also be a handsome-looking new 6-series and a £40,000 M version of what BMW calls the 135 coupé, even though it's actually a sal-oon.

Never mind the muddle, though; this is one of the cars I'm most looking forward to driving, partly because I reckon the standard car is already the best model in the BMW line-up and partly because, with a twin-turbo, 335-horsepower straight six and a straightforward front-engine, rear-drive, no-styling, no-nonsense approach, it will be a genuine successor to the simple M cars of old.

I'm also looking forward to the new Mercedes SLK, although I'm a bit alarmed that industry insiders are saying it'll be a more hardcore experience than the 'soft' outgoing model. Having owned an SLK 55, which I sold because it was way too uncomfortable, I am a bit worried that the new car might not have any suspension at all.

Strangely, though, in this sea of wholesome goodness, the car I've been anticipating with the most eagerness is Citroën's DS3 Racing. I realize that this is like booking a table at the Wolseley in London and then looking forward most of all to the bread rolls. But the fact is this: when the sun is shining, I like a simple two-seater convertible most of all, but when it isn't – and this is Britain, after all – the type of car that I most enjoy driving is a hot hatchback. And the DS3 Racing is about as hot as hot hatches get right now.

You might argue that the optional paint job, with checks on the roof and all sorts of slogans and symbols that would only make sense if they were splashed on the deck of a Nimitz class aircraft carrier, is a bit stupid. But I disagree. They're a laugh. I even like the warning above the petrol filler cap. 'Caution. Attention', it says. And why not?

Inside, it's just as bonkers, with a bright orange dash, a carbon-fibre steering wheel and epic seats that would be more at home, you feel, in an F-22 Raptor.

After a period in which car makers have looked backwards for inspiration – I'm thinking of the new Beetle, the new Mini, the new Chevrolet Camaro, the new Ford Mustang and the new Fiat 500 – it comes as a refreshing change to find Citroën has decided to face the other way, while taking inspiration from both the US navy and Airfix. Mind you, I suppose that if Citroën had looked backwards, we'd have ended up with a reborn 2CV. And no one this side of the *Guardian* wants that.

I realize, of course, that looks and style are a matter of taste and that some of you may find the Racing garish and idiotic. But without wishing to sound childish, this is my review and I really like it.

However, it would all be for nothing if its body were writing cheques its engine could not cash. Well, let's get one thing straight from the off. It's not, as the name would suggest, a racing car. It's just a DS3 with a few racing-style bits and bobs added into the mix. That said, it's powered by a 204-horsepower version of the turbocharged 1.6-litre used by BMW until recently in the Mini, so it'll do 146mph. And thanks to a lower ride height, a wider track and firmer dampers than the basic DS3, it handles crisply, too.

Yes, there's a fair bit of torque steer and I will admit that the Renault Clio 200 Cup is a tad more dynamic. But the Citroën is more comfortable and less noisy and, of course, every time you see your reflection in a shop window, you will feel like you're on the bridge of the *USS Dwight D Eisenhower*. Whereas when you see a reflection of yourself in a Renault, it's just a reminder that you will soon break down.

In short, I loved the DS3 Racing as much as I thought I would. I loved driving it. I loved looking at it. I love the feeling now that it's parked outside my house and I can use it for a trip to town this afternoon. It is a car that's excellent to drive but, more importantly, it's a car that makes me feel happy. And, of course, because it's a hot hatch, you get all the fun as well as a big boot, folding rear seats and space inside for five.

Issues? Well, the adjustment on the seat is so crude that you either drive sitting bolt upright or flat on your back, and I must say, for a whopping £23,100, I would have expected a few more toys. When you are paying BMW 3-series money for a small Citroën, the least you would expect is satnav.

The worst thing, though, is that to bypass costly legal tests, Citroën has declared the Racing a 'low-volume' car and will make only 1,000. Just 200 will come to Britain. On the upside, there is a loophole in the law that allows Citroën to make a modification to the engine that no one will notice and that lets the company make 1,000 more. Get your name down early, but don't be surprised to find you're behind me in the queue.

16 January 2011

It's hardly British but learn to haggle

Mitsubishi Outlander 2.2 DI-D GX4, 7 seats

Alarming news from among the potted plants at your local plate-glass car dealership. It seems that six out of ten people who buy a new set of wheels these days don't bother to haggle over the price.

I should explain that I'm one of them. Mostly, if I'm honest, it's because I have to pay the full sticker price or the *Daily Mail* will run a story saying that I'm on the take and cannot be trusted. But you do not have the *Mail* breathing down your neck every time you eat food or go to the lavatory, and so you really should try to beat down the man with the cheap suit and the boy-band hair.

If you pay cash, even a Ferrari salesman will give you free door mats. Whereas with something like Citroën, he'll probably give you a 100 per cent discount, £1,000 cashback, 0 per cent finance for 300 years and an evening with his girlfriend and one of her better-looking friends.

I realize, of course, that you are not an Egyptian market trader and that you find haggling completely revolting. You don't try for a discount when you are buying a stamp or a box of corn-flakes, so why would you try for one when you are buying a car? It would be ghastly. If you are English, you would rather vomit on a salesman than negotiate with him face to face over money. But come on. The whole process of buying a car is so unpleasant, a bit of toing and froing over price is nothing.

You've already dealt with the balloons. This is just one example of what the car dealer thinks of you. He reckons that you are so moronic that if he hangs a few colourful balloons outside his showroom, you will think there is some kind of 'do' on, so you'll be unable to drive by.

Then there's the decor. A car showroom, even the fancy ones on Park Lane in London, has all the visual appeal of a railway station's lavatories. You want to get out as soon as possible. But you can't because the man with boy-band hair is on his way over with the handshake of a dead haddock and a silly earring. And he's got lots of impertinent questions about where you live and what you do and your credit rating. As a general rule, it should also be noted he knows less about the cars he's selling than you do about the moons of Jupiter.

Then there's the worst bit. When he grabs a form and steps outside to tell you what the car you wish to part exchange is worth. In short, it's worth about an eighth of what you thought. This is because the man from Take That has found a scratch, and it's grey and grey's not very popular at the moment. Except for the fact that 75 per cent of all cars sold in Britain are one shade of grey or another.

Then, of course, it's time to sit down and choose some options for the car you're buying. And this is a terrifying place to be because when you are spending £25,000 on the car, £200 sounds like nothing. So yes, you decide you'll spend £200 on a DVD player and another £200 on metallic paint and another £200 on a sunroof, and pretty soon you notice Jason Orange has grown a third leg. By the time you've finished, the money you're spending would be enough to clear up a medium-sized oil spill.

All of this, however, assumes you've been able to choose what sort of car you'd like to buy in the first place. Obviously, some people are swayed by balloons, or dealers would stop using them as a marketing tool. And many simply buy an updated version of what they have now. But some people insist on buying the car that best suits their needs. This is like being thrown, naked, into an acacia tree. You're going to end up thrashing around for a while. And then you are going to become dead.

Let's say, for instance, that you have a family. Many people do.

So you'd think it might be a good idea to buy something practical. Obviously, you cannot have a Citroën Picasso or a Renault Scénic because nothing says you've given up on life quite so succinctly as a mini MPV.

Then you decide that the mini MPV would be all right if it had some Tonka toy styling, a raised ride height and perhaps four-wheel drive. Four-wheel drive implies that you go hunting for bears at the weekend, and besides, it will be useful should the snow come back.

So, you want lots of space, four-wheel drive and chunky styling. That's narrowed your choice down to pretty well every single car maker in the world. And to make matters even more complicated, many of the cars that appear to be different . . . aren't. Take the Citroën Cross-Dresser, for example, or the Peugeot 4007. Underneath, they are Mitsubishi Outlanders. They're even built by Mitsubishi. So which do you pick?

Well, if you are suffering from rabies, forget the French offerings and go for the recently updated Japanese original. There are many symptoms of this debilitating ailment – agony and frothing at the mouth are two – but so is an extreme thirst. And on this front the Outlander scores well because it comes, in the front alone, with no fewer than five cupholders.

What's more, in the back, there's seating for five on two rows of seats. Though a word of warning here. Anyone volunteering to sit in the boot should remove their head and legs first.

Mitsubishi says there's another reason for picking its offering. In the blurb, it claims the Outlander has a distinctive 'jet fighter' grille. Well, I've studied the front end for quite some time, and I don't think this is quite correct, mainly because jet fighters don't have grilles.

Perhaps the best reason for choosing the Itchy Pussy is because, unlike the rivals from Peugeot and Citroën, its 2.2-litre diesel engine comes with variable valve technology. That means fewer emissions, better power and more miles to the gallon. Absolutely, but it also means a very narrow power band and the consequent need to change gear every one and a half seconds. There's even a light on the dash instructing you to shift up, constantly.

Other problems? Well, it's boring to look at, boring to sit in and extremely boring to drive. It feels like the suspension and steering are made from cardboard. Apart from a few joke cars from the former Soviet Union, I cannot think of any other car that feels quite so inert.

Of course, if you are not an enthusiastic driver, this will not matter. You will be far more interested in the promise of great reliability, a genuinely good satellite navigation system and all those cupholders, in case you are bitten by a French dog.

But really, are you better off with this, or the Peugeot, or the Citroën, or the Land Rover Freelander, or the Nissan Kumquat or Honda CR-V, or a Ford or a Jeep or a Volkswagen? The answer, with cars of this type, is very simple. Since they are all largely the same, simply telephone the dealers, ask for their best price and buy whichever is the cheapest.

If you end up with the Outlander, it's not the end of the world. But don't expect the earth to move, either.

23 January 2011

Try this moose suit for size, Mr Top Gun

Saab 9-3 SportWagon Aero TtiD 180PS

In the days of the cold war we knew we had four minutes to respond to the Soviet threat and we developed the hugely powerful English Electric Lightning fighter to deal with that. But up in the frozen north, Sweden had its face pressed against the Iron Curtain and needed even faster reactions. Which is why it came up with the Saab Viggen.

This was the most powerful single-engined fighter in the world. For a while it held the international speed record and it remains the only fighter to get a missile lock on an SR-71 Blackbird spy plane. It also packed the most powerful cannon and a very advanced radar. But it was a bit more than brute force and a big fist.

Because Sweden covered the West's northern flank against the Soviet Union, the Scandawegians reckoned that if the balloon went up, their airfields would be destroyed in short order. So, when a Viggen's nose wheel hit the deck, reverse thrust was triggered instantly, allowing the plane to stop in little more than 500 yards. This meant it could be operated from roads, frozen lakes, even school sports pitches. It was also extremely economical.

Unfortunately, the Swedish government refused to sell military hardware to any country it considered to be undemocratic. Which meant that the Swedish air force had to buy every Viggen that rolled off the production line. And that's why, for a while, it was the fourth-largest air force in the world.

Still, at least there was one accounting upside, because here in Britain everyone thought that if they bought a Saab car, they were actually getting a Viggen with a tax disc. That still holds true today. But actually this hasn't been entirely accurate for some time. And not only because the Viggen's engine was made by Volvo.

In the early days, it's true, the aircraft designers were employed to work on the car's aerodynamics, but that stopped years back. The car is not a jet. It's a Vauxhall Vectra in a moose suit.

Oh, Saab is still banging on about the aircraft connection. It fits a button that turns off all the dashboard lights at night, so you can feel like a night fighter pilot. But you don't really. You just feel as if you might be running out of petrol.

Other features? Well, Saab says, 'A wide range of functions can be pre-set according to personal preference.' Sounds good. But one of the things listed is the clock. Yes. You can set it to whatever time you like! And another is the air-conditioning system. Wow! So it has a heater that can deliver a range of temperatures.

It seems, then, that I was dissing it unfairly when I said it was just a Vauxhall Vectra in an antler suit. In fact, it's a Vauxhall Vectra with a heater and a clock. And a diesel engine that produces no torque at all. Technically, this isn't possible. But somehow Saab seems to have managed it.

If you dribble up to a roundabout in second gear at 5 mph, spot a gap and put your foot down, you will roll into the gap you spotted, still doing 5 mph, only now the van driver you pulled out in front of is leaning on his horn, mouthing obscenities and wondering why you don't get a bloody move on.

Once you're moving, and provided you keep it in the right gear, the power is not too bad. But when the turbocharging is on song, the steering wheel does protest mightily, writhing about as though it's in physical pain. And guess how much you're expected to pay for all this. Yes, £29,000. That's more than BMW asks for the 318 diesel estate.

To make matters worse, there was recently a great disturbance in Saab's force. General Motors had bought half the company in

1989 and the rest in 2000, but realized last year it didn't want it any more. The production lines stopped and for a while it looked as though the company would be gone. But then it was rescued by a Dutch outfit that makes the Spyker supercar.

In many ways this is a bit like Mr Patel from your local corner shop deciding to buy Harrods. It sounds terribly romantic, but if you're going to take on the big boys, you need to have deep pockets. A billion won't cut it. Toyota probably spends that on pot plants.

But here's the thing. I do not want Saab to go. I'm glad that in Britain 6,000 architects decided to buy one last year and I hope that number continues to grow. Which is why I have a tip for the new company.

The 9-3 is old. It has a nasty engine. And, while I acknowledge the standard fitment of both an adjustable heater and a clock, it is also quite expensive. But it does have one feature that sets it aside from almost every other car on the market. It's comfortable.

Today all car makers have got it into their heads that, despite the traffic and the price of fuel and the war on speeding, what motorists want is sportiness. A hard ride. Nervous steering. Bucket seats. Big power. There was a time when Volvo sold itself on safety and VW on reliability and Mercedes on quality. Not any more. Now, they all make racing cars.

Before a new model goes on sale it is taken to the Nürburgring, where final tweaks are made to the suspension to make sure that it can get round the 14-mile track as fast as possible. This is fine, of course, if you live in the Eifel mountains and you use the Ring on the way to work. But it's not fine at all if you live in Esher and your office is in Leatherhead. And it's also not fine if you ever encounter a pothole or have a bad back.

I know that people in a focus group will tell the inquisitors in the polo-neck jumpers that they would like their next car to be 'sporty' because that's the motoring dream and has been since

Christopher Plummer roared away from the battle of Britain in his zesty MG. But in reality, sportiness is a pain in the backside.

Recently, I bought a new sofa because it looked good. Sharp. Modern. Crisp. It's an aesthetic masterpiece, but after a hard day at work, when I just want to slob out in front of the television, I'd be better off sitting on the floor.

At my age I crave comfort, and that's why I have enjoyed my week with the Saab so enormously. It's dreary to drive and underpinned by one of the worst car platforms in modern history, but the seats are superb, and the suspension is capable of keeping the pothole bomb blasts to nothing more than a shudder.

Plus. And this is the really good bit. As I cruised about, with the adjustable heater providing me with just the right amount of heat, and the clock telling me precisely the right time, everyone else – apart from the occasional van driver – was looking at me and thinking, Ooh, look. It's Chuck Yeager.

30 January 2011

Titter ye not, it's built for the clown about town

Nissan Juke 1.6 DIG-T Tekna

To this day, I remain baffled by the Ford Scorpio because at some point someone must have walked into an important board meeting and said, 'Well, everyone. This is what it's going to look like.'

Why did no one present say, 'Are you joking?' or, 'Have you gone mad?' or, 'Take some gardening leave, you imbecile'? They obviously just sat there thinking, Yes, we have had cars in the past that were designed to look like sharks and cars that were designed to look like big cats. So why should we not now have a car that looks like a wide-mouthed frog?

It's strange. I know who designed almost every single car in recent times. I know who did the Lamborghini Countach, VW Golf and Volvo 850. I know several people who claim to have done the Aston Martin DB9. But nobody in all my years has ever put their hand up and said, 'Yes. It was me. I did the Scorpio.'

I bet you would have a similar struggle if you set out to find the man who did the Toyota Yaris Verso – the only car ever made that is five times taller than it is long. I pulled up alongside one yesterday and studied the driver for some time. Do you realize, I wondered, how utterly ridiculous you look in that?

Then there's the Pontiac Aztek, which was unusual in that it managed to look wrong from every single angle. Normally, even the most hopeless designer gets one tiny feature right by accident – the rear tail-lights or the C pillar, for instance. Even the Triumph TR7 had a nice steering wheel. But the Aztek looked like one of those cardboard cities you find beneath underpasses in Mexico.

And let's not forget the SsangYong Rodius. Plainly, they set out to build a coupé and then decided at the last minute that what they actually wanted was a removal van. And then, when those two concepts had been nailed together in the most unholy merger since Caligula fell in love with his horse, they realized that the only wheels they could afford were the size of Smarties.

It's easy, when you look at a SsangYong, to imagine that the designer simply doesn't know what he's doing. But that ain't necessarily so. Remember the Musso? That was as awful to behold as a frostbitten penis and yet, amazingly, it was styled by the same man who designed that old warhorse the Aston Martin Vantage and the Bentley Continental R.

The problem is that there's a language to car design. Some of the language is written down. Ideally, the wheels should be half the height of the car, for example. But mostly, it's a dark art. All I know is that the car must look like it's capable of great speed, or else it looks wrong.

Look at the kink at the bottom of every BMW's rearmost pillar. The one between the back window and the back door. It's got a little kink and that makes the car look like it's pushing forwards, straining at the leash. BMW is also very good at making the body look like it's been stretched to fit over the wheels. As if there's barely enough skin to contain all the muscle.

This doesn't just apply to sporty cars, either. Look at the new Vauxhall Astra. It's a handsome thing because it's all straight lines and sharp angles. There's a whiff of the fast patrol boat. And that gives a sense of howling turbochargers and sea spray – even if the engine under the bonnet is a miserable diesel.

This brings me on to Nissan. A few years ago, it decided to try to make a car that didn't look fast. The company reckoned that in a world of road rage, traffic and simmering rage, it would be good to have a car that was friendly and unthreatening. So it produced the Micra.

I hated that car. It had the sort of face you wanted to punch.

And because it was 'happy', it was bought by the sort of people who were never in much of a hurry. I'd love to know how much of my life has been stolen by Nissan and its Micra experiment. One day, I may send it a bill.

But in the meantime, the company has changed tack again and come up with the Juke. It's not ugly by any means but it is, without any question or shadow of doubt, the stupidest-looking machine to see the light of day since the Ronco Buttoneer.

What were they thinking of? Why, for instance, are its rear wheel arches bigger than those you would find on a modern tractor, even though the wheels are the size of Polo mints? And why are the front lights mounted on top of the bonnet? It's all completely ridiculous.

I first encountered it at Heathrow airport early one Monday morning. The office said it would leave a car for me in the valet parking bay and so there it was, sitting among the Maseratis and the Mercs, like a big comedy hat at a funeral.

At first, I assumed it was some kind of electric car, and that filled me with horror and dread as a busy week lay ahead and I really didn't have the time to spend eight hours a day looking for somewhere to charge it up and then another eight hours drinking coffee while the batteries replenished themselves with juicy electricity. Made from burning Russian gas.

Happily, as I turned the key, I was greeted with the welcome sound of internal combustion. So why, I wondered, have they made it look so mad? Perhaps, I thought, it's a four-wheel-drive crossover vehicle. Well, for sure, there is an all-wheel-drive version but the model I had was based on a front-drive Micra.

So maybe, then, it has the silly body because it's somehow capable of doubling up as a bus. Nope. It has seating for just five and a boot that is surprisingly small.

Then I noticed something odd. In the middle of the dash is quite the most baffling onboard computer I've ever seen. It tells you every single thing you don't need to know, including, wait for it, how much g you are experiencing at any given moment. So this idiotic high-riding car with its small wheels, street lighting and arches from a Massey Ferguson thinks it's a jet fighter.

It really isn't. Yes, the engine's a turbocharged 1.6 that produces 187 horsepower, but it doesn't ever feel fast. Or exciting in any way. I'm not suggesting that it is nasty to drive or that it kept crashing into trees, but it's not good, either. It is just some car.

And that means I'm stuck. Normally I can tell what sort of person might be interested in a particular car and I try to tailor my conclusion to meet their specific requirements. But I've trawled my memory banks and I can't remember ever meeting anyone who might be interested in buying a car that looks absolutely stupid.

The best I can come up with, therefore, is this: if you just want a normal five-seat hatchback, buy a Golf or a Ford Focus. If, on the other hand, you want a normal five-seat hatchback but you enjoy people pointing at you and laughing, then the Juke is ideal.

6 February 2011

Those yurt dwellers have got it right

Land Rover Freelander 2 eD4 HSE 2WD

The phone rings. It's a friend who's just crashed his Jag and is thinking of spending the insurance cash on a new Range Rover. I explain that, all things considered, it's probably the best car in the world, but advise against buying one brand new. First, I say, the initial depreciation can be alarming and second, I am aware the battery on new models goes flat rather too easily.

I therefore advise him to buy the last of the old diesels from the second-hand market and am rather surprised by what he says in reply.

He explains that he lives in a part of the world where middle-aged women pour paint on friends if they are caught buying eggs from a battery farm. Come election time, you could be forgiven for thinking, as you see the posts in people's gardens, that there is only one party, and it's not blue, red or yellow. This is north Oxford. This is where the ultimate status symbol is a wicker trolley on the back of your bicycle and where everyone secretly wants to live in a yurt. As a result, my friend doesn't want to buy the old model. He wants the new one because it's more eco-friendly.

Hmmm. Although he doesn't realize it, he has a point. It is far more eco-friendly to buy a car built just 50 miles away, even if it is a massive off-roader with a turbocharged V8, than it is to buy a Toyota Prius, the components of which have covered half a million miles before they are nailed into the vague shape of a car and shipped to your front door.

However, as eco people are not very bright, I fear my friend's neighbours may not see it this way. And I'm absolutely certain that his argument about the new car being more eco-friendly than the old one won't wash even a tiny bit. In north Oxford a Range Rover of any sort is the devil.

I'm regularly told by people there that cars caused the hole in the ozone layer, usually when they are getting something from their trendy old fridge, or applying some deodorant. The other day, someone even blamed the motor industry for deforestation, even though the only car company still making its cars from wood is Morgan. And I hardly think a cottage industry making seventeen units a year in Malvern can be blamed for all the logging in southeast Asia.

However, because there is so much claptrap floating about in the ether, a company such as Land Rover must feel like it's under siege. And that's before we get to the rather more important question of fuel consumption. I had a supercharged Range Rover on loan recently and in one week of normal motoring it gulped down £250 worth of fuel. That is catastrophic.

As a result, it must be extremely tempting for Land Rover's marketing department to do something stupid . . .

It is, of course, extremely important that I approach every single car that is reviewed on these pages with an open mind and no preconceived ideas of what might lie in store. However, because it's so much more fun to write about a car that is rubbish than one that is OK, I do occasionally book test drives in cars that are likely to be awful.

And that brings me to the new Freelander 2 eD4 – the first car in Land Rover's long and important history to drag itself into the market using only its front legs. I can see the logic, of course. Better fuel consumption and more ecoism.

But, I'm sorry, the notion of a front-wheel-drive Land Rover is idiotic. It's as daft as Tarmac launching a new scent. Or Spear & Jackson moving into the lingerie market.

There's more. Because when all is said and done, a front-wheel-drive Freelander is simply a very expensive and hard-to-park alternative to, say, a Ford Focus. They have the same number of seats and don't be fooled into thinking the Land Rover is better able to withstand a barrage of everyday bumps and scrapes. It looks that way thanks to a trick of the stylist's pen. But it isn't. And because it's so tall, your elderly dog will struggle to get into the boot. So you'll have to pick her up and that will make your hands all dirty.

As a result of all this, I approached the Freelander wearing the cruel smile of an SS officer who'd been given some pliers, a dungeon and a freshly downed Tommy airman to play with. I was going to torture it. Ridicule it. And then rip it to shreds.

Unfortunately, it's a bloody good car. First of all, the chintzy bits and bobs that ruin the look of the modern Range Rover look rather good on the baby of the Land Rover range. It may only be a hatchback on stilts but it looks expensive. Regal almost.

And although it may be hard to load an elderly dog, those stilts do make you feel imperious as you drive along. There are many 'soft roaders' on the market these days, but none offers such a commanding view as the Freelander.

Inside, many of the features are lifted directly from the Range Rover, which can cost nearly three times as much, so again, you don't feel like you're driving around in something from the pick'n'mix counter at the pound store.

However, the best thing about this car is the way it drives. The removal of the four-wheel-drive system has resulted in a weight reduction of 75kg and you can feel this as you bumble about. I'm not going to suggest for a moment that it feels sporty, but it does feel agile. The steering in particular is delightful and the ride is sublime. Driving this car is like lying in the bath. It's brilliant.

Of course, it's not going to get as far into the woods as the four-wheel-drive version, but if you needed to go into the woods, you wouldn't have bought it in the first place. However, that said, because of the ground clearance, it will get you further in tricky conditions or bad weather than a normal five-seat hatchback.

The only drawback I could find in the whole package was the engine. It has slightly less power but more torque than the previous 2.2-litre Freelander engine and that's fine. You get quite a big punch when you put your foot down. But while I have no complaints about the performance, this is certainly not one of those cars where passengers say, 'Is this really a diesel?' In fact, as they sit there, vibrating, they may ask what you are using instead of fuel. Pebbles? It's like a powerplate with a tax disc.

It's so unrefined when it starts that after a while I disengaged the system that cuts the engine when you stop at the lights and starts it again when you put your foot on the clutch. This may save half a thimbleful of fuel but it drove me mad.

Because of this roughness, the car cannot have a five-star rating. However, it does get four. Which, is four more than I was hoping to award. The fact is, though, that the cost of fuel and the blinkered prejudice found in the nation's mental yurt-heads has resulted in something that's pretty damn good.

13 February 2011

Little Luigi's turbo boost

Fiat 500 0.9 TwinAir Lounge

I spent most of last week playing with the new McLaren MP4-12C and I must say that, in a technical, mathematical, common-sense, add-up-the-numbers sort of way, it is extremely impressive. Plainly, it has been designed for the serious business of going fast. And yet there are no histrionics at all. In fact, in road mode, it rides and sounds like an S-Class Mercedes. It's also beautifully made, so, unquestionably, this is a car that you could use every day.

As a result, even though it's a bit more Ron Dennis than Ron Jeremy, it is certainly the best car ever to wear a McLaren badge. It's definitely better than the old F1, which I hated. And it's definitely better than the more recent SLR, which had a switch masquerading as a brake pedal – you either went through the windscreen, or you didn't slow down at all.

It may even be better than the Ferrari 458, which is not something I thought I'd be saying any time soon. And yet I don't yearn to own one.

It was the same story with the Bugatti Veyron. Yes, it was a masterpiece, a composite and magnesium firestorm of brilliance, perseverance, engineering persistence and planet-stopping power. But at no time did I ever think, Crikey. I'd love to have one of these on my drive.

I experienced much the same sort of thing at Heston Blumenthal's new restaurant in London the other day. He makes food in the same way that McLaren and Bugatti make cars. The duck is stripped down to a molecular level, treated with exotic gases and then reassembled before being cooked by a team of men who are dressed up like the guards in a Bond villain's lair. Even the ice cream is made with a sewing machine.

The results are simply spectacular. Without any question or shadow of doubt, Heston's rhubarb mousse is the second nicest thing I've ever put in my mouth, and although the texture of the duck fat was a bit like a quilted anorak that's been left in the rain, it tasted astonishing. It was a duck plus. A super-duck. A duck Veyron.

And yet, while I admire Heston's skill and respect his knowledge of food preparation, I'm not sitting here yearning for the day when I can sample his wares again. Did I like it? Yes, very much. Am I glad I've tried it? You're damn right I am. But will the day ever come when nothing but a plate of his bone marrow will do? I doubt it.

I think it's because, in our complicated lives, we yearn only for the simple. An evening in front of the telly. A nice sit-down. A game of cards. At a drinks party, I can find myself talking to a fascinating and beautiful woman who's just written a book about something interesting and clever. But what I yearn for is to be in the pub with my mates.

This is especially true of food. When I am struck with a sudden craving, it's always for something simple: a chicken sandwich, an apple, some tongue or, more usually, a pot of crab spread. It's never a truffle in a rich jus made from a koala's ears.

The same can be said of cars. I like the Mercedes SLS, the Jaguar XKR, the BMW M3 and the Ferrari 458 very much. But the car I yearn to own most of all is the Citroën DS3 Racing that I wrote about on these pages last month. And close behind is the little Fiat 500.

Of course, you are familiar with the 500. Your estate agent's daughter probably has one. And unless you are James May, the chances are you like it very much. You like the cheekiness and the way it's both retro and very modern at the same time.

It gets better, because while the Fiat is very similar in concept to the Mini – they're both fashion statements first and cars second – it is much cheaper. And as a little bit of icing on the cake, here is a car that doesn't have to be grey or silver like 75 per cent of all the other cars on the road. It can be powder blue or egg-yolk yellow or child's lipstick red. You can even cover it in stickers. And you should.

In short, the little Fiat is a joyous machine that makes you smile, but the car I'm talking about today is different. And better. It's the new TwinAir, so called because it has an engine quite unlike anything else we've ever seen before.

First of all, there are only two cylinders, which is not a revolutionary idea. The original Fiat 500 was similarly equipped. However, in the new version, there is no camshaft. Instead, the exhaust valves drive the inlet valves using hydraulics and electronics, and that sounds like the greatest ever solution to a problem that doesn't exist. But the end result is spectacular.

First of all, there's the noise. Remember that sound you got when you put a lollipop stick in the spokes of your bicycle wheel? It's that. Only amplified. It is one of the best engine noises I've ever heard. It's nearly as good as a Merlin.

And then there's the grunt. Yes, it may be tiny – just 875cc – but it is turbocharged so you get 85 horsepower. That means you can cruise down the motorway with ease. And it takes off from the lights like it's being kicked into action by Toby Flood.

There's more. Because there is much less friction in a two-cylinder engine than there is in a four, it is incredibly efficient, which means it produces less carbon dioxide than a fat man on one of Boris Johnson's rent-a-bikes. As a result, you don't have to pay the London congestion charge.

Certainly, this little car is ten times more environmentally friendly than the Toyota Prius because it's smaller and it's made from fewer parts and Fiat doesn't have to plunder the Canadian countryside and cause acid rain to make its batteries. With this little car, everybody wins.

Especially the oil companies, because unfortunately, the TwinAir is not what you'd call economical. It could be, if you drove it sensibly, and if you press the eco button on the dash, it probably is. But you won't deploy the eco button. And you won't drive it sensibly because it's impossible. It's as impossible as expecting a puppy to sit still.

I've had the car for a week and because I've enjoyed the noise it makes so much, I've averaged just 38 mpg. I got more from the hot Fiat 500 Abarth. And to make the economy argument even less palatable, the TwinAir costs around £1,000 more than a similarly specced model that has twice as many cylinders. So it's not cheap to buy and, unless you have the will power of a donkey, it's not cheap to run, either.

And it doesn't matter because, as you sail through central London, flicking V-signs at the congestion camera and beating bikes off the lights and revelling in that fantastic noise, you really won't care. The 500 is a great little car. And now you can have it with what is almost certainly the best engine . . . in the world.

20 February 2011

I don't fancy Helga von Gargoyle . . . Can't think why

Porsche Panamera 3.6 V6 PDK

I don't like marzipan. I'm aware that it is categorized as a food-stuff and that you are supposed to put it in your mouth and move it about and swallow it, but honestly, I'd rather lick the back of a dog.

I don't like kidney beans, either. Or Piers Morgan. I know that he has a nose and a liver and all the other things that qualify him to be classified biologically as a member of the human race, but he grates, and I find myself gloating in an unkind way over the true fact that his new television show now has fewer viewers than *Kerry Katona: The Next Chapter*.

We all have likes and dislikes and it's often hard to find rational explanations. I don't like whisky, for instance, and I can't understand why. Everybody else I know likes it, but as far as I'm concerned each sip is a nose-busting reminder that in the morning I shall wake in a puddle of sick with a headache. It's the same story with calvados.

And Surrey. I have many friends who live in its dingly dells. I even work there one day a week, but each time I visit, I am consumed with an irrational need to leave again as soon as possible. It's strange.

Nearly as strange as my unbridled hatred of Marks & Spencer. It is a proud boast that in all of my life I have never bought anything from M&S, even though I am aware that its clothing is well made and its sandwiches are nutritious and delicious.

I think I don't like it because of the flooring or because I have it in my head that everyone in the queue for the tills is going to be a magistrate with firm home-county views on youth crime and bad language. But I can't be sure because I've never been through the door.

Strangest of all, though, is my dislike of Porsches. I joke that my hatred of the 911 stems from the simple fact that both Richard Hammond and James May have one. But this, if I'm honest, isn't it. I haven't liked these arse-engined Hitler-mobiles since way before Hammond was even born. This is annoying because in many ways a Carrera 2S would be almost completely perfect for the life I lead and the driving I do.

It is easy, of course, to say that I prefer the excitement of cars from Italy, but in truth, and on a wet Tuesday morning in February, a Lamborghini Gallardo or even the magnificent Ferrari 458 would drive you absolutely bonkers. They are pantomime dames: fun when you are in the mood, but the booming and the palaver would quickly drive you insane when you were not.

The Porsche is not like that. You can drive a 911 in the same way as you drive a Ford Mondeo – quietly, to the shops, where its relatively small size means it is easier to find a parking space. What's more, provided you avoid the silly high-performance versions that have scaffolding in the back instead of seats, you can take the kids along, too.

And not only can it handle family duties, it is electrifyingly good fun to drive when the road is empty and it's just you and the sun is shining and you fancy getting a move on. Plus, it is built by people who are German, which means it is likely to be fifteen times more reliable than a Swiss pacemaker.

So. It's a great car. A brilliant car. A perfect car for the man who wants everything. And compared with all the competition, it's not even that expensive. But I don't like it. Wouldn't have one in a million years.

It's not like I have a problem with the Porsche badge. I loved the old 928 and the 944, and I may be the only man alive who has publicly professed a fondness for the 924 – even though it was propelled along by the engine from a Volkswagen van and was, as a result, slightly slower than continental drift.

However, I do have a problem with the current offerings because the Boxster is a palindrome, the Coxster is stupid, the Cayenne looks like a 911 with elephantiasis of the underparts and then, sitting on top of this festering pile of aesthetic dreadfulness, we find the Panamera.

You can see what they were trying to do. To make a big, comfortable four-seater with a family resemblance to the 911. The spine of Porsche's reason for being. This might, just, have been possible had they employed a stylist who actually knew what he was doing. But, unfortunately, it seems they chose instead to give the job to a committee made up of the man who did the Ford Scorpio, the man who did the Pontiac Aztek, Ray Charles and some lunatics. It is the ugliest car on the road today.

This is annoying because, underneath, it's not that bad to drive. I recently tried the four-wheel-drive turbocharged V8 and, my God, it was fast. And a nice place to sit as well. Partly because of the tall centre console and partly because from behind the wheel you can't see it.

If only, I thought, they'd restyle the ruddy thing, this would be a cracking car. A genuine rival to the Aston Martin Rapide and the wonderful but fragile Maserati Quattroporte.

Instead, however, Porsche has chosen to make its gargoyle slower. I can't quite understand the logic behind this move because I cannot imagine that anyone has spent the past couple of years thinking, Hmmm. Yes. I'd buy one of those Panameras if only the damn thing wasn't so fast.

Of course, the new model is much more economical and far cheaper than the V8, but in base trim with rear-wheel drive and a six-speed manual gearbox, it's still £62,783. And that kind of money buys an awful lot of 5-series. Still, if you don't like BMWs and Audis and Mercs, and you are impervious to bad design, you might be interested to know more.

So, you get a 3.6-litre V6 that develops 295 horsepower, and that's enough to make the car move about. You can also have a seven-speed double-clutch gearbox that is just like all the other modern double-clutch boxes – ponderous and dim-witted at low speeds in town. Like the Mercedes SLS, this is not a car that can be used to exploit gaps in traffic.

There's another problem, too. It's enormous. To get through the width restrictions on Hammersmith bridge, ooh, you have to breathe in.

But the worst aspects are the interior fixtures and fittings. The gear lever feels like one of those toys McDonald's gives your children when you buy a Happy Meal, and the electric window motors sound as if they've been asked to move a mountain. I know it's a well-made car, but it doesn't come across that way.

And it really isn't exciting or special to drive. You pay a premium to have a Porsche and you should be rewarded with a 'feel' that you don't get in mass-produced cars. In the V6 Panamera, though, you don't. Yes, the centre console is still delightful and the driving position is perfect, but the ride and the acceleration and the steering – they're flat. They're inert. They're Korean, almost.

Still, all this means I can at least be rational in my conclusion. I don't like it because it's not a very good car.

27 February 2011

Damn it, Spock, we can't shake off Arthur Daley

Jaguar XJ 5.0 Supercharged Supersport LWB 4dr

In the far reaches of your satellite television's hinterland, way out past *Kerry Katona: The Next Chapter* and Piers Morgan talking to someone you've never heard of about a movie you don't want to see, it is possible to find a channel that's showing *Minder*. I recommend it because it just might be the best television show of all time.

Today, when you watch it on CabSat Freeview 757, it's like a history lesson. You cannot believe that there were ever that many parking spaces in London or that the traffic was ever that light. They'd go from Terry's flat in Fulham to the Winchester in Notting Hill in about three minutes. But only after Terry had slept with some ladies and punched a foreigner in the middle of his fez. You could then. It was allowed.

I was such a big fan of *Minder*, I had my wedding reception at a place called the Winchester. I even hired 'Dave' to be the barman. And then got very cross when guests called him Glynn and asked what it was like to have starred in *Zulu*. He's not Corporal Allen. He's Dave and it's his job to get you a large VAT.

We think of *Dad's Army* as a classic, and it was, but *Minder* was tighter. *Minder* was written to an even higher standard. And the characters were just perfect. I saw Patrick Malahide the other day pretending to be a hotshot CIA spy and I just kept pointing at the screen and shouting, 'It's Detective Sergeant Chisholm.'

It's the same story with Dennis Waterman. He's still about, cropping up on TV from time to time, trying to convince us he's not an ex-boxer with a Ford Capri parked outside. But it's all hopeless. In my mind, he's Terry McCann. And he always will be.

It wasn't just the characters that became etched in our minds, either. It was the props. The hat. The coat. Terry's bomber jacket. And, of course, the cars. Because of Arthur Daley, I've never quite trusted anyone with a Jag. I like people with Jags. They are usually interesting, but I wouldn't leave them alone with my silver.

In my mind, even today, and purely because of *Minder*, the Jag driver is always having a 'spot of bother' with the taxman. He's always asking if he can crash at yours because of a 'misunderstanding' with the mortgage company. I like to think that most of the people in prison today for crimes such as art forgery have an XJS in a barn somewhere. Robbers have Vauxhalls. Rogues have brogues and a Jag.

That's why the new XJ worries me, because when you step into that extraordinary cabin, you do not even catch a whiff of Arthur Daley's ghost. There is blue lighting in the door pockets. The glove box is lined with purple velvet. And when you select Dynamic mode, the dials glow red. It's like being in one of those bars in central London where visiting businessmen go to meet ladies.

I like it. It's a fantastic, futuristic place to sit. But there's no man with pointy ears in the passenger seat, and where you expect to see NCC-1701 on the steering wheel there's a leaping cat instead. It feels strange. Like taking off a page 3 girl's clothes and finding that underneath she's Yootha Joyce.

The exterior is weird, too. Again, I think it's very bold and brave of Jaguar to make it look so different from anything that's worn its badge before. I think it's very striking. But it's also a bit odd. And you obviously do, too, because since this car was launched six months ago, I have not seen a single one on the road.

Last week, Bertone, the Italian styling house, showed off its designs for a new Jag, and they were right. Its car was sleek. And the new XJ? It's many things, but sleek isn't one of them.

Then there's the question of interior space. Tricky one this.

Because, in a Jag, you are supposed to sit low down, with your buttocks kissing the catseyes. You're supposed to feel cocooned, too, as if you're in an Elizabethan pub. But that won't do these days. If Jaguar wants to capture market share from Mercedes, it must convince the chauffeurs who ferry Posh and Ant around London that their car is at least as spacious in the back as an S-class.

So once again, Jaguar has ditched tradition, ditched the beams and the horse brasses and gone for space. In the long-wheelbase version – £3,000 extra – there's tons of it, to stretch out and watch the world slide by through the big glass roof panel while listening to the 1,200-watt stereo until your ears bleed. You even get climate control in this new car, rather than a wood-burning stove.

But will you want to be in the back? The answer's yes, if it's a diesel. That's built for economy and it does a fine job. But if you have the supercharged V8, the answer is a big emphatic 'I'd rather get in the back of Brian Blessed'.

On paper, this engine doesn't look like it will pass muster. You get just 503 horsepower, and these days German cars use that much to operate the automatic parking brake. But you also need to look at how much the XJ weighs. Because, thanks to an all-aluminium construction, it is even lighter than Porsche's Panamera 4.8 V8 Turbo. In a strong wind, you'd be advised to fit mooring ropes to stop it blowing away.

And you don't just feel this lack of weight when you accelerate or when you stop or when you look at the petrol gauge. No. You feel it all the time, through the seat of your pants and, more especially, the steering. This is not like a sports car to drive. It is a sports car.

Sadly, to achieve this flickability, the suspension is a little harder than you might expect. It's a problem that affects all Jags today. A hard ride is the only reason I don't own an XKR. But that said, at no point would you ever call the XJ uncomfortable. Or noisy. Or nasty in any way. It is absolutely bloody brilliant.

Taken on face value, it is the only car that marries the raw driver appeal of a Maserati Quattroporte with the space and luxury of a Mercedes S-class. By rights, the centre of London should be chock-full of nothing else. But it isn't . . .

There's a very good reason for this. We don't buy cars by the numbers. Nobody ever test drives all the models that might seem suitable.

We may pore over the options list of whatever model we've chosen, kidding ourselves that we really need parking sensors. But it's all haphazard. We don't buy with our heads or our hearts. It's just gut instinct. That looks nice. I can afford it, just. So I'll have it.

And that's where the Jag falls down. It meets all the emotional challenges and the numbers stack up, too. However, the bounders and the cads want a Jag, but not a Jag with a purple glove box. And the people who do want a purple glove box don't want to be tarred with the Arthur Daley label.

It is, then, a magnificent car. A brilliant car. But sadly, *Minder* means half the world won't buy it because it's a Jag. And the other half won't buy it because it's not a Jag.

6 March 2011

Bruce's bonzer duck-billed koala

Ford Falcon FPV Boss 335 GT

On the face of it Australia is much like any other modern, developed nation. But for a number of reasons it isn't, and chief among those reasons is the koala: you may not know this but it spends almost all of its life off its face on dope and then, whenever it feels frightened, it catches chlamydia. You do not find this sort of thing going on with any other creature in any other part of the world.

Then you have the kangaroo. The red variety can travel at 40 mph, which is fast enough to give a G-Wiz a run for its money. But no kangaroo of any sort can back up. They have no reverse gear at all. Plus, all female kangaroos are permanently pregnant.

Life is very different for the female Sydney funnel-web spider. She has to spend her whole life in the burrow and is not allowed out until she dies of old age. The males, meanwhile, like to roar around Sydney at night, swimming in people's pools, hiding in children's shoes and eating anyone who gets too close.

I like Australia, but almost everything you find down under is unique. The duck-billed platypus, for instance. Surely the strangest animal ever to leap from the fumes of God's chemistry set.

On land it is four-wheel drive, but underwater it becomes front-wheel drive and uses its rear legs for steering. So it's an amphibious fork-lift truck, with a beak.

We see the same sort of thing in sport. Elsewhere in the world, you have American football or proper football. Whereas down under there is Aussie Rules, which is strange, because from what I can gather there aren't any rules at all. Apart from no poofters, obviously. The game itself is part soccer, part rugby and part basketball, but what sets it aside from all three is that each side consists of about 17,000 players, all of whom wear rather unattractive skin-tight vests.

Another notable thing in Australia is a fanatical approach to health and safety. There are more speed cameras than people, and if you wish to go snorkelling you must dress up in a giant nylon all-in-one. This means no part of your skin, including hands, feet and face, is in contact with the water, and so you cannot get stung by a box jellyfish.

In some ways it is a wise precaution. But I'm sorry – splashing about in an acrylic submarine rather spoils the point of snorkelling. And it's not as if the box jellyfish is unique to Oz. The little critters are everywhere, and no other nation makes you get into a condom just in case.

It gets worse. After my snorkelling expedition, I tried to rent a jet ski. But a state law meant that I had to sit down, in the blazing sunshine, and take a written exam. What's to learn? There's a throttle and that's it. My daughter was riding a jet ski at the age of five. An idiot could do it. I pointed all this out to the blond surfer dude who was running the course, but it was as if he'd been programmed: safety is everything.

Not on a jet ski, it isn't. Fun is everything. Whizzing about and trying to splash your mates is everything. Getting knocked off by a big wave is everything. If you want to be safe on a jet ski, get off it.

So there you are, in a country where they drive on the same side of the road as us, speak the same language and have the same head of state. And the same summer weather, if my recent trip is anything to go by. But it's not the same at all.

And it especially isn't the same when you look at cars. In every other country in the world people may like the brand of car they drive, but not so much that they would punch someone in the face for driving something different.

To the average Aussie there are two brands. Ford and Holden. And even if you are a solicitor and you drive an Audi, you are instinctively in one of these two groups. I was going to say it's like the Catholics and the Protestants in Northern Ireland, but it isn't. It's more ingrained than that.

There is such fanaticism, in fact, and loyalty, that both Ford and General Motors make cars specifically for the Australian market. We're talking about a country of just 22 million people – most of whom are in Earls Court. That would be like making a car specifically for Romania. It wouldn't happen.

That said, it is not expensive to engineer a car for the Australian market. Certainly, you don't have to employ a stylist. The current crop of Holdens aren't too bad, in a meaty, knuckle-dragging sort of way, but the Fords . . . Oh dear. And it was always thus, even back in the days of Mad Max and his Interceptor.

I tried the new Ford Falcon FPV Boss 335 GT when I was over there and it struck me that someone had spent a few quid on the engine and then nothing at all on anything else. I'm told this is how it should be when you are upside down. Which you will be if you try to make it go round a corner.

The old Falcon V8 was a bit of a problem child because the turbocharged V6 model was faster and nice to drive. So Ford has teamed up with Prodrive – the famous Aussie motor racing house in Oxfordshire – to create the new one.

It was a big ask because the base engine comes from a Mustang and it has asthma. To try to insert a bit of ephedrine, it is now fitted with a supercharger and an intercooler, which means you get lots of grunt. So much that every time you set off you 'lay a couple of darkies'. People cheer when you leave a skid mark like this in Australia.

So it's quite gruntsome in a straight line, but it is too big, too soft and too heavy to be remotely good at anything else. And inside, you get the impression everything is made from Cellophane. My snorkelling suit felt more robust.

By rights the FPV should not exist. It's pointless. But then so is the koala, and we'd all be a bit sad if we woke up one morning to find that the last button-nosed little stoner had fallen out of his tree.

10 April 2011

Botox and a bikini wax and I'm ready to roll

Jensen Interceptor S

It's the coolest name ever given to a car. 'I'll pick you up at eight. I'll be in the Interceptor.' Imagine being able to say that. Or even: 'Darling. About tonight. Shall we take the Interceptor?' It sends a shiver down your spine. Maserati is a good name. Thunderbird is even better. But Interceptor? That's the best of them all.

Of course, it wouldn't be quite so good if the name were writing cheques the body of the car couldn't cash. You can't be called Clint Thrust if you have a chest like a teaspoon and limbs from the canvas of Laurence Lowry.

Happily, the Interceptor looked magnificent. It was big, with a body styled by Carrozzeria Touring of Italy that included a thrusting bonnet, an unusual wraparound back window and gills. It was distinguished. It was fantastic. It was one of the best-looking cars ever made.

Sadly, I never drove one of the originals and it's hard to find out what they were like since the only person I know who had one was Eric Morecambe and, unfortunately, he's no longer with us. From what I can gather, though, the driving experience was 'absolutely awful'.

The engine was a Chrysler 6.2 V8, which turned money into noise but produced very little by way of power in the process. Which was probably a good thing since the enormous live rear axle wasn't really attached to the car in a way you'd call finished. And to make matters worse, Jensen would simply pop down to the local supplier whenever it needed a steering rack and come back with whatever was on the shelves. Some Jensens, by all accounts, were accidently sold to customers with steering designed for the Triumph Stag.

As a result of all this, the Interceptor sits in the bargain basement bin of the nation's classic car market. While you are now expected to pay hundreds of thousands for an Aston DB5 or an E-type Jag, a decent Interceptor can be yours for around £5,000.

You may think it would be worth it, just so you could offer to take people out in your Interceptor. But I should imagine you'll get there two days after you set off, covered in soot. Reliability wasn't a Jensen strong point and things won't have improved with time. So that's that, then. Memory Lane this morning has turned out to be a dead end.

Except it hasn't, because the car we're talking about isn't really a Jensen Interceptor. It started out in life that way, but then an Oxfordshire-based company called Jensen International Automotive came along and gave it the full bikini wax and Botox treatment. The car you see is an Interceptor S and it's absolutely brilliant.

First, the original American engine has been thrown away and replaced with another one. It's the rather good all-aluminium 6.2-litre V8 from a Corvette. You get a Corvette gearbox as well, but from there backwards, things get a bit more complicated. The live rear axle, the leaf springs and all the rest of the Roman technology is replaced with a fully independent setup, with some bits coming from the old Jaguar XJS – the limited-slip diff, for example – and some from the new company.

The front AP Racing brakes have six-pot callipers with ventilated and grooved discs. The tyres are low profile. The dampers are adjustable. And . . . are you dribbling yet?

I was. So I climbed inside and it all got better and better. It's a faithful reproduction of the Jensen original with white-on-black Smiths-style dials, quilted seats and, best of all, a push-button radio that offers you the choice of 5 Live with a slight crackle. Or just the crackle. There is, however, a discreet iPod connection in the glove box.

It's extraordinary when you sit in a car from this period how light and airy the interior feels. Because the pillars are there simply to support the roof, rather than to absorb the impact of hitting a bridge, they are thin and spidery. It feels like you are sitting in a glass bubble, and that makes you feel like you're on show. Which, if you are driving an Interceptor, is exactly where you want to be.

Be warned, though. So much of you is visible as you drive along that you need to think about what you're wearing. I suggest a g-suit of some kind. Or, if you don't have one, a black polo neck and some Jason King-style sunglasses.

I genuinely felt, as I set off for the first time, that this could be the perfect car. Olde worlde style with modern dynamics. A Georgian house with central heating.

Sadly, it was not to be. The problem is that if the new company changed every single feature, the end result would be classified as a new car, and would need to face all the modern safety and emissions tests. So some of the period features remain. The wipers, for instance, which move back and forth nicely. But remove not a single drop of moisture.

Then there's the wind noise. The fact is that, back in the Seventies, cars had rain gutters and 'that'll do' was the guiding principle of all West Midlands panel beaters. So at 70 mph on the motorway, even in the new car, it sounds like you're wing-walking.

The biggest problem, though, is the steering. They've been forced to keep the original rack, and no matter how many adjustments you make to the geometry, you're not going to get round the fact that it was designed by a man who wanted most of all to go on strike. As a result, it's heavy, there's little self-centring and it's so low-geared you need a lot of arm-twirling just to move three degrees off the straight ahead.

Right. That's the bad stuff out of the way. Now let's get on to the good bits. Starting with the heater. Oh, the joy of being able to have warm feet and a cool face. Climate control is all very well but give me the simple Jensen setup any day.

And then there's the sheer speed of the thing. Put your foot down and, with a lot of angry lion noises, the bonnet rears up and in 4.5 seconds you're doing 60 mph. This comes as a big surprise to the following Audi driver, who's desperately dealing with life in your jet wash, while not quite believing his eyes. Flat out, you could be doing 155 mph – or more. That's 155 in a car made in 1975.

It gets better, because although the steering feels a bit Victorian, there's no question that there's plenty of bite in the bends and none of the understeer I was expecting. It's not a sports car. But it gets close.

On top of all this, there's the ride. We have become used in modern cars to the fat, low-profile tyres loved by stylists transmitting every single ripple and ridge directly to our bottoms. But in the Jensen, it's like floating. It's as comfortable as a modern-day Bentley.

But it costs less. Despite the sheer amount of work that's gone into this car, it will cost you – including the donor vehicle – from just £107,000.

So, to conclude. What we have here is one of the most beautiful cars ever made, stripped down to the bare metal, repainted, retrimmed and fitted with just enough mechanical components to make it handle properly, ride even better and go like a bastard. That's a properly good idea.

Best of all, though, for the first time you can say, 'I'll pick you up at eight. I'll be in the Interceptor.' And you'll actually be there on time.

17 April 2011

Oh, barman, my pint of pitbull has gone all warm and fluffy

Ford Focus Titanium 1.6 Ecoboost

In the olden days most people who needed a family car bought a simple hatchback. A Ford Escort, perhaps, or, if they were feeling racy, a Volkswagen Golf. Not any more. Now, ordinary won't do. Simple is dreary. The Hush Puppy has been ousted to make way for the clown shoe: the SUV. The crossover. The funky little retro bomb with scorpions on the bonnet and chequerboard door mirrors.

I wonder why. When I finish work today I shall potter over to the house and pour myself a glass of Château Léoube, the pink sort. It's what I drink before supper. And during supper. And afterwards as well, usually. Wine is simple. It's easy. And a bottle contains exactly the right amount for a single evening.

Beer is good, too. The Ford Escort of beverages. On a really hot day, when you've been busy outside, you don't think, Oooh. What I really need now to quench my thirst is a banana daiquiri. You always want a beer: not the sort James May likes, with twigs in it, but a Peroni, in a glass with condensation dripping down the outside.

One of the things I hate most in life is when people come round to my house and ask for a gin and tonic. That's four ingredients I must go and find. The gin, the tonic, the ice and the lemon. We never have a lemon in the house: why would you? And if we do, invariably it was picked before the Boer war and has the texture and juiciness of a marble.

And then there's the ice tray, which either contains no ice at all, or it does but it's one big lump that will not, even with the assistance of a hammer, come out of the container. To get round this we recently installed a fridge that dispenses ice at the touch of a button. In theory this is brilliant, but what happens in fact is that you hold the glass under the nozzle and you get water instead. So you start again, and now it delivers enough ice to keep a Spanish trawler at sea for several months.

The first time this happened I attempted to clear up the mess with a vacuum cleaner, and now I have a message for you all: do not do this. Because Henry burped a bit and then broke.

Seriously, going to someone's house and asking for a gin and tonic is like asking for a shepherd's pie. And you wouldn't do that because you know it'd be a nuisance. Especially if your wife has asked for a vodka and cranberry juice, with just a hint of lime. We don't have any cranberry juice, or lime. Or vodka usually. Because the kids' friends have drunk it all.

It's always best, for convenience, then, to keep your drinking preferences simple. Yes, on a lazy Sunday morning it's possible to spend an hour or so making a super-complicated Bloody Mary, but, no, in a packed City pub it is not acceptable to shout from the back, 'Four Pimm's, please, with all the trimmings.'

And now, in a link so tenuous a spider would call it flimsy, we shall move on to the modern-day equivalent of a pint of stout. The new Ford Focus. The car you didn't buy because you fancied a Fiat Harvey Wallbanger or a Citroën Shirley Temple instead.

Ten years ago I bought a Ford Focus, and I still have it today. Occasionally I use it, and I am always amazed by what a joyous thing it is to drive. Thanks to independent rear suspension, an expensive solution to a problem no one in the world has ever noticed, it is an attack dog in the corners. The engine's good, too. And, as we know, it can carry just as many people and dogs as a big 4x4. More, in fact, because our ancient labrador can no longer leap up into the Range Rover, whereas she can get into the back of the Ford.

The best thing about the Focus, though, is the amount of times it's broken down. Have a guess. No. Because in ten years it hasn't gone wrong once. Every single thing still works. It's a five-star car, that.

But what of the new model? Well, there's no getting away from the fact that it's a looker. With its black-painted sills, it appears sleek and slinky, more like a coupé than a family hatchback.

Inside, it's a button-fest. There are millions of them on every flat surface; so many, in fact, that it takes several hours to find the one that starts the engine. I don't mind this, especially as they are laid out in the same pattern that Porsche uses on the Panamera, and anyway they are necessary because this car has a lot of features.

Even the base models come with blue teeth and voice activation for the main controls. This doesn't ever work, of course – like the iPod connector – but there are hours of fun to be had on long journeys asking the dashboard to do one thing and then wondering what it will do instead.

Problems? Yes. One. It's all very well fitting wipers that move about in an interesting and unusual way, but they don't clear even a small portion of the screen. John Prescott could hide in the blind spots.

Underneath all this, though, it's a festival of electrotechnology designed to keep you pointing in the right direction. There's a torque-vectoring system – wake up at the back – that gently brakes the inside wheel in a corner to prevent understeer and, as an option, a device that applies all the brakes very firmly indeed if you are about to crash into the car in front.

This kind of thing – and there's tons more besides – is all very clever and demonstrates very clearly that the people at Ford who live with their mums and play with laptops could walk off with the geek of the year award at next year's Mr Nerd competition. But when you put a layer of electronics between the driver and the road, some driving purity must be lost.

In short, this new car may be safer and more economical than the Focus I bought. But as a driver's car it's not in the same league.

And that, of course, means the Focus has lost its USP. I bought one because it was demonstrably better to drive than anything else. But why would you buy the new one rather than a – much cheaper – Hyundai or Kia?

The only thing I can think of is the engine that was fitted to my demonstrator model. This variant is called the Ecoboost, but don't be fooled into thinking it runs on lentils and has an output that's measured in flower power. It's a turbocharged 1.6 that, even in sixth gear, provides a genuinely surprising chunk of grunt. And what's even more surprising is that it's very economical as well.

However, even here there's a problem, because today you'd have to be a swivel-eyed lunatic to buy a car that runs on petrol. Or a billionaire. And if you're a billionaire you won't be interested, I'm guessing, in a Focus.

This means, then, that if you buy the Ford, you'll specify a diesel. Nothing wrong with that. Ford's diesel engines are fine. Not brilliant. But OK. And there's the problem. You end up with an OK car with an OK engine. And is that what you want?

In short, there's no big reason for not buying the new Focus. But there's no big reason for buying one, either. Especially when other pints of beer are available for a lot less.

1 May 2011

Pointless but fun – what a good wheeze

Renault Wind Roadster GT Line 1.6 VVT

I've been cycling. It was a charity bike ride and I completed the five-mile course in a little over two hours. Everyone overtook me. Partly this was because the uphill stretches were extremely difficult and partly because on the downhill stretches I daren't build up speed because I was absolutely convinced the front wheel was about to come off.

This would have caused the forks to dig, suddenly, into the road and as a result I'd have been catapulted over the handlebars, landing at high speed, on the tarmac, on my face. I don't like my face very much, but I do need it for talking and seeing where I'm going and so on.

And no, I wasn't wearing a helmet. I don't wear a helmet for skiing, either. Or on building sites. Helmets make the wearer look foolish. So I decided, after a long discussion with my lungs and my quads, that it'd be better if I simply went slowly. This, then, is what I did and actually it was fine.

What, however, was not fine was getting my bicycle and my wife's to the start line. Needless to say, someone else fixed the bike rack – which we seem to own for no obvious reason – to the back of our Volvo and loaded the bikes onto it.

This worried me for two reasons. First, I'd have to drive half-way across England in a Volvo with two bicycles stuck to the back, which is like a scene from a Ski yogurt commercial, and second, after the ride was over, I'd have to fit the bikes into the rack myself for the journey home. And I am to this sort of thing what the Duchess of Kent is to spot-welding.

Still, I figured that since it was a charity bike ride, the finish line would be awash with eager, Lycra-buttocked weird beards who'd take pity on the petrosexual and help out.

But there was a problem. After dropping my wife and her bike off at the start point for fit people, I was setting the satnav for the slobs' start point when, KERPOW!, a man who was 700 years old reversed into the bike rack, knocking it clean off the car.

Have you ever tried to assemble such a thing? It is impossible. It makes no sense. You clip some straps behind the tailgate and then it just sort of rests on the bumper. To my eye that looked all wrong.

But since time was pressing, I moved on and examined the procedure for attaching my bike. This made even less sense, since all that prevented it from falling off and bouncing through the windscreen of the car behind, decapitating everyone inside, were two of those twisty things you use for doing up freezer bags.

I pushed and heaved and got chain oil on my face until eventually I decided I had to set off. That's when I discovered two very sturdy-looking straps with big military-style clasps on the end. These didn't seem to be important, though, so I left them dangling.

By driving very gingerly, I made it to the start line for fatties without beheading anyone. But now, in my mind, a big question mark hangs over the safety of things you attach to a car. Not just bike racks but roof boxes as well.

Ever fitted one of those? Of course not. Because lifting them into position will break your back, and dropping it, which you will, will remove all the paint from your car. Better, and cheaper, to buy whatever it is you were thinking of putting in your roof box when you get to wherever it is you're going.

And another thing. A roof box is shaped just like a cruise missile, and if it becomes detached it will do as much damage. But have you seen the nuts and bolts they provide for affixing the box to your car? You'd be better off sticking it in place with chewing gum.

Let me give you a word of warning. If you come up behind a car with a box on the roof and some bikes attached to the back, keep your distance. Because they will have been fastened to the vehicle by someone who likes roof boxes and bicycling. Not Isambard Kingdom Brunel.

Then you have snow chains. They are supposed to keep you moving when conditions are atrocious. But what they actually do is sever all your fingers and, if your wife is in the car, make you divorced.

Best, then, to buy a car that suits your lifestyle rather than a car to which various things must be attached in order for it to fit the bill.

Not the Renault Wind, then – the first car to be named after the effects of indigestion. Designed by a Frenchman, based on the humble Clio hatchback and built in Slovenia, it sounds perhaps the most stupid car in all of modern history.

It gets worse, because although it is a two-seater convertible, it was plainly not built to be light and sporty in the mould of an MG or a Lotus Elan or a Mazda MX-5. No. With its electric flip'n'over roof and its tiny little engine, it's more of a city-centre pose-mobile, a Christian Louboutin shoe with a tax disc. I'm surprised, frankly, that its undersides aren't red.

Of course, it's normal at this point for the petrolhead to scoff, to suggest that the Wind's body is writing cheques its engine can't cash. That it'd be burnt off at the lights by a pedestrian. But I'm not going to do that because, many years ago, I used to own a Honda CR-X. And what you have here is the modern-day equivalent. I like it.

Yes, the 1.2-litre turbo engine's a bit too small, but the non-turbo 1.6 isn't bad at all. Go for this option and you get 131 bhp, which is enough to let you exploit the Botty Burp's really rather excellent chassis. All quick(ish) Renaults feel lively in a Lucozade, good-for-you sort of way and this is no different. On country lanes it was – despite a ridiculously large steering wheel that makes the bigger driver feel cramped – fun, and, better still, it's not so fast that your passenger complains after five seconds about having the roof down.

In terms of practicality – well, you're not going to get a bike in the boot, but because the roof folds in such a clever way, not as much space is robbed as you might imagine.

And to top it all off, prices start at £15,205. The range-topping 1.6 GT Line is only £17,010, which means it's considerably less than the 1.6-litre Ford Focus I drove last week. Given the amount of equipment you get as standard, that looks good value.

However, there is just one chink in the armour. It feels as cheap as it is. This is not a car designed, I suspect, to be passed on to the next generation. The dash, the switchgear, the levers – everything you see and touch feels brittle. As though everything will last about as long as – well, without wishing to be too lavatorial – a fart.

8 May 2011

Prepare your moobs for a workout

Aston Martin Virage

If you want to spend a lot of money on a house, there is a very large list of options. It could be in France or Florida. It could be old or new. It could be nestling on a bed of gravel in the Cotswolds or surrounded by a turquoise moat in Alderley Edge.

It's the same story with restaurants and art and furniture and holidays. Money buys you choice. Unless you are planning on buying a car. Because when you are rich enough to take a seat at the top table in petrol heaven, there's no choice at all.

You want a big, fast BMW. It'll be uncomfortable. You want a big, fast Mercedes. It'll be uncomfortable. You want a Ferrari or a Maserati or a Porsche. Uncomfortable. Uncomfortable. Uncomfortable. All of them have suspension made from concrete and tyres with the give of an African warlord.

The fact of the matter is this: with the exception of footballers and tennis players, most people who have enormous lumps of money to spend on a car are in their forties or fifties. They've done their time sleeping on chairs after parties and kipping on the floor because they can't afford a taxi home. What they want, at all times, is to be comfy, to have a nice sit-down, to relax.

But the car makers have got it into their heads that this simply doesn't apply when you're coming home from work. No. The car makers think that you want to feel every ripple and every catseye. They think you want seats with the cushioning of a kitchen chair. They think that you want to feel at all times like you're going for a lap record at the Nürburgring.

If the designers at AMG made sofas, they'd be fashioned from gravel and would come with spikes.

The new BMW range of luxury carpet: 'Made from Lego bricks and upturned plugs.' And the new Maserati bath: 'Instead of water, or ass's milk, we allow our customers to soak away the strains of the day, up to their necks in sulphuric acid.'

It's bonkers, and you can see what's going on. Car makers want their cars to be liked by *Autocar* magazine, and what *Autocar* likes is fast lap times. Engineers like fast lap times, too, because that shows they are better engineers than the idiots at BMW, whose cars are slower. It's all just one big peeing competition, with you and me playing the part of the suckers with the chequebooks.

This brings me to Aston Martin. In the beginning it made the DB9, and we saw that it was pretty good. It rode as if it had been designed for the road, not the Nürburgring. But instead of making the new, smaller car – the Vantage – in the same mould, Aston's boffins decided it should be more uncomfortable. And then, when they fitted that with a bigger, V12 engine, they decided that it should be more uncomfortable still. And then along came the DBS, in which they did away with the suspension altogether and fitted steel girders.

You would imagine that this would cause the marketing department at Jaguar to think, Aha. Now that Aston Martin has decided to make a range of racing cars, we see an opportunity, so let us soften the supercharged XKR. Because there are many middle-aged men with moobs and very wide bottoms who might like such a thing.'

'Fraid not. Jaguar decided that anything Aston could do, it could do better. So the current XK rides around on suspension seemingly made from a blend of granite and chest freezers. Run over a pothole in that car and you shatter.

As a result of all this, I had high hopes for the Aston Martin Virage. It was billed as a cheaper, more comfortable version of the DBS. All the style. All the speed. All the lovely interior detailing. But none of the rock-hard, racetrack, carbon-fibre nonsense that no one either needs or wants.

Well, it may have an automatic gearbox but it's still a bitch. You can tell when you run over a white line whether the paint was gloss or emulsion. You know when you run over a pheasant whether it was a cock or a hen. And you can't just feel the suspension refusing to budge when it encounters a bit of gravel; you can hear it, too. Raging away with a series of clumps and bangs.

It is a huge missed opportunity. It could have been the only expensive car currently on sale designed for people who actually exist. But it is just as uncomfortable as all the others.

In almost every other way, however, it's better. With new sills and a new front spoiler, it looks even more beautiful than the DBS. It looks more beautiful than the most beautiful thing you can think of. Especially in deep, dark, last-vestige-of-the-day navy blue. And doubly especially if you go for the convertible version.

What's more, it's £25,000 cheaper than the DBS and, really, it's hard to see why, since the two cars have the same 6-litre V12 engine. It may have been mildly detuned in the Virage, but you still get 490 horsepower, and that's enough to get you from rest to the wrong side of the national speed limit in 4.6 seconds. Provided you are in the right gear – and the auto box can be a bit dim-witted sometimes – this is a very, very fast car.

It's even fast at stopping, thanks to carbon ceramic brakes, and, of course, because the suspension and the tyres are so hard-core, it is utterly thrilling to hustle. You've never actually seen an Aston being hustled, of course, but if that's your bag, the Virage is the best of them all. After the V12 Vantage, perhaps.

Drawbacks? Well, behind the wheel it is a bit cramped, and the price you pay for all that design elegance is that the buttons are quite hard to find. And even harder to press if you are on a bumpy road at the time.

The worst thing, though, is the new satnav. Unlike the old system, which only told you where you'd been, this one only tells you to slow down. Constantly, with a series of bongs. If it even thinks there could be a speed camera nearby, off it goes, yelling and panicking.

It may well be, of course, that there's a button for turning this feature off, but finding that would mean reading the instruction manual. And that's not going to happen. I'm a man.

What's more, when I told it I wanted it to go to London, its next question was, 'What house number?'

We all need the same thing from a satnav system, so why do all car makers give us a choice about how the screen looks or what sort of voice we want? Choices mean submenus, and submenus are for people who live at home with their mums. Submenu people are the only people on earth who don't actually need satnav because they never go further than the fridge.

So the Virage is a missed opportunity in this respect, too. And yet, I'm afraid I'm completely in love with it. It's a hard car, and a hard car to operate, and there are those who say that the wheels are coming off Aston's previously untarnished brand kudos. But get into a Virage in the morning and I guarantee you will feel good. Better than if you were getting into almost anything else.

At the raggedy edge, a Ferrari 458 is more rewarding to drive and a Mercedes SLS is more fun. But both those cars are a bit flamboyant. And that's where the Aston scores. It isn't.

15 May 2011

The old duffer trots out in boy-racer colours

Skoda Faiba vRS1.4 TSI DSG

In his first year in office the transport secretary, Philip Hammond, announced that he would scrap the M4 bus lane, stop funding speed cameras and raise the motorway limit to 80. What he should have done next is gone home and started a well-earned retirement. But, sadly, when you are the transport secretary you are expected to go to work every day. And, of course, when someone is at work they are duty-bound to do stuff and think of things. This is fine if you are a doctor or a telephone repair man but when you are transport secretary it's hard to think of things that make any sense.

This is a problem when you are invited to speak on the *Today* programme. You can't very well sit there and say you've not thought of any ideas, because people will think you have been lazy. So you have to come up with something. And that's what Hammond did recently. He took a deep breath and said he was going to get the police to clamp down on boy racers.

Of course, this was an excellent thing to say because the people who listen to the *Today* programme do not have gel in their hair, or acne. Or an electric-blue Citroën with a huge exhaust pipe and no suspension.

Radio 4 people think that boy racers sit in the social mix between rapists and Hitler. So they will have leapt up from their Shackletons wingbacks, delirious with joy that Mr Hammond was finally going to make their life on the road a little less terrifying.

Sadly, however, if you examine the details of Hammond's half-formed excursion into the world of middle England tub-thumping you see that it doesn't make any sense at all. For instance, he says he's going to get the police to clamp down on the lunatic fringe, to which I say this: what police?

The last time I saw a jam sandwich patrolling the motorway, it was a Ford Granada. Today you get Highways Agency traffic officers and the odd plod-dog van, but actual police? They're all at the station, learning how to climb ladders.

Then we get to what Hammond thinks constitutes boy racing: tailgating and undertaking.

Quite what he has against undertakers, I don't know. In my experience they drive very carefully. Unless, of course, he means people who overtake on the left. In which case he's just plain wrong.

These days I undertake other cars as a matter of course. And I'm fifty-one, which means I'm not much of a boy. The problem is that in the olden days everyone on the road had at least a rudimentary grasp of lane discipline. But today – how can I put this without sounding as though I'm from the *Daily Mail*? – many of Britain's motorists learnt the art of driving in more exotic parts. And they simply have no idea, as they trundle up the M40 at 50 mph in their £200 Toyota Camry, that they should keep left.

You can flash your lights, indicate, make hand gestures, huff, puff and die of a heart attack but it will make no difference. They don't realize they're doing anything wrong.

That's why I glide by on the left. And if I am stopped by one of Hammond's non-existent policemen, I shall explain that if I had the space to undertake, then the person around whom I drove must have had the space to pull over. He should therefore be prosecuted for driving without due care and attention.

Then there's the issue of tailgating. This is done exclusively by people in Audis with Montblanc pens, Breitling watches, Oakley sunglasses, those shirts with horses on them and a fondness for squash. I don't know what you'd call people such as this – 'awful' springs to mind – but they're not boy racers.

So when Hammond says that he will be targeting undertakers

and tailgaters, he's actually targeting the victims of the middle-lane hogs, and people who play squash. Unless he really is talking about people who drive you to the church when you're dead. In which case it truly is time for him to stop thinking of things and doing stuff.

Actual boy racers, I should imagine, are now getting very irritated because they'll have seen the picture of the car I'm reviewing this morning, with its white roof and its big wheels, and they'll be thinking, Get on with it, you imbecile.

So get on with it I shall. It's a Skoda Fabia vRS, and the last version of this car was OK. I liked it a lot, even though it waded into battle with a diesel engine. And that's a bit like competing in a 100-metre running race while wearing wellies. The new one has a 1.4-litre petrol engine that is supercharged and turbocharged. The result is 178 bhp, and the result of that is 0 to 60 in a little over seven seconds and a top speed of 139 mph. Or 140 if you buy the aerodynamically cleaner estate version.

Weirdly, the people at Skoda have sent me a comparison chart, which shows that in terms of performance the vRS is a little slower than the Clio Renaultsport 200 and the Vauxhall Corsa VXR. They've also sent me a laminated card saying that the No. 1 key feature of their car is that it has a three-point seatbelt. It's almost as though they don't want me to like it.

And that's fortunate, because I don't. There are some things, though, that are rather good. I like the styling especially. I'm not sure why, but it reminds me of a bemused and slightly cross second world war squadron leader. And I like the way it has a white roof.

But most of all I like the price. It's £16,265 and, although Skoda doesn't provide figures to show this, it is way cheaper than every one of its rivals. Even if you fit the useless satnav and blue teeth and climate control, it's still £1,000 less than the Volkswagen Polo GTI. And that's especially odd, because underneath it's exactly the same car. Same engine. Same everything.

So what's it like to drive? Well, the seats are comfy and the ride is surprisingly pliant, given that it's running on wafer-thin low-profile tyres. But there's a problem. This is a turbocharged and supercharged hot hatchback, so it should make you want to drive like you are on fire. It should encourage you to pass every other road user on whatever side takes your fancy and never brake for corners. Hot hatchbacks are supposed to fizz but the vRS doesn't.

The double-clutch flappy-paddle gearbox is reluctant to change, and the steering is too low-geared. Couple this with the noisiest tyres in Christendom and what you mostly want to do in this car is slow down for a bit of peace and quiet. It is horribly noisy.

And, while I don't mind the interior, I must say it's a bit gloomy. Perhaps that's why the vRS looks like a bemused squadron leader. Because it's not really a hot hatch, so 'why the bloody hell has someone painted me the colour of an Opal Fruit?'.

You are better off with a Fiat 500 or a Mini or a Citroën DS3 or a Twingo Renaultsport 133. These are the real boy-racer cars. The Skoda looks like it might be a laugh but actually it isn't.

22 May 2011

What's the Swedish-Chinese for I can't see?

Volvo V60 T5 R-Design

Many years ago, I came up with a solution to drink-driving and because no one has thought to make it law, pubs are currently closing down at the rate of twenty-nine a week.

At present, we are told that if we are going out for a drink, we should use public transport, but this is not possible in the shires because there isn't any. And if I were to call for a taxi at 11 p.m., it would not arrive until mid-September.

So, we bumpkins are told that if we are going out we should designate a driver, who must sit there, all night, staring into his Britvic, willing his heart to stop beating. Not drinking in a pub full of people who are is like being the only sane man in a lunatic asylum. Death is preferable.

My plan, then, was very simple, and completely workable. Whenever a driver feels a bit tipsy, he or she must clip a flashing green light to the roof of their car before setting off. Once in place, they would be limited to 10 mph, a speed at which they could not possibly be a danger to themselves or anyone else.

Besides, pedestrians and drivers coming the other way would see the green light and think, Uh oh, this bloke's had a few. I'd better give him a wide berth.

Of course, anyone found to be drink-driving without a light on the roof of their car, or exceeding the 10 mph limp-home limit would face the consequences. Which would be execution.

There are many upsides to this idea: no one would ever wake up in the morning and wonder where the bloody hell they'd left their car; you would never have to use the hateful last bus; and in pubs, the lonely squeak of a barman polishing his glasses would be replaced by the joyful buffoonery of people having a nice time.

Everyone wins, except, of course, for your local minicab firm, whose drivers would be forced to sell their horrible, sick-stained Toyotas and get a proper job that doesn't involve quite so much leching.

Anyway, I've now come up with another plan that, frankly, is even better. It's this. Occasionally in life, all of us face an emergency that means we have to break the speed limits, and at the moment there is no system in force that allows us to be let off. Wife in labour? Child's head stuck in railings? Mother had a stroke? Doesn't matter. You still get three points and this is simply not fair.

Policemanists and ambulance drivers are allowed to drive fast in an emergency, so why not us too? You might think they are trained for this sort of thing and we're not but the fact is, many aren't. Constable Plod, whizzing about in his diesel Astra – he's no more qualified to do 90 than Princess Anne.

Of course, I recognize that there are many scoundrels out there who would claim that every journey they make is an emergency. To stop this, everyone would simply download a free app that, when deployed, tells a central police computer that they are about to set off on a journey where speed is imperative. And this can only be used, say, once a year. You therefore wouldn't dare waste it on something trivial.

The only problem with this scheme is that today it's virtually impossible to make super-speedy progress on the motorway because the outside lane is a permanent home for the sanctimonious, the belligerent and the stupid.

The sanctimonious won't let you past because they can't see why anyone should drive fast in these days of global warming; the belligerent won't let you past because it would suggest you are better than them; and the stupid don't know you're there. Usually because they are in a van. And they knocked the door mirror off in the yard at a builder's merchant last week.

I was in a big hurry on the M40 last week and could not believe how many people just sat in the outside lane. But then nor could I believe what happened when they finally pulled over and I tried to get past.

I was in a Volvo V60 T5, and those of us who remember those epic Touring Car races from the early Nineties know what that means. T5 means, Yes, I'm in a Volvo and, yes, there's a Georgian tallboy in the back, but underneath my tweed suit I'm wearing a crotchless leather G-string and I have a death tattoo on my back, and I am bloody well coming past.

A Volvo T5 is a Cotswold tea shoppe where they serenade the customers with a medley of hits from Wayne County & the Electric Chairs. It's a Sex Pistol in a twin set, anarchy in the Home Counties. And the model I was driving came with the optional R-Design package, which includes bigger wheels and stiffer suspension. So, when I put my foot down to overtake the van that had finally pulled over, I was expecting an explosion of power and a surge of acceleration that bordered on the insane. But it never came.

Unlike previous T5s, this does not have a five-cylinder engine. It's a turbocharged four, which means that the offbeat strum has gone. But so too has the lunacy. When you caress the throttle pedal, you can feel what seems like a big muscle tensing and you think that all is well, but when you really go for it, especially if you are in sixth gear at the time, nothing happens.

Later, on the lovely road between Banbury and Rugby, it was the same story. The car would float deliciously round a corner – it handles and rides very well indeed – but when I accelerated onto the straight? The tumescence was gone. Frankly, you may as well save a few quid and buy the diesel.

Or something else entirely. There are many good things about the V60. It is extremely comfortable, for a kick-off. And like all Volvos, it was plainly designed by someone who has a family. That's why you can have raised seat bolsters – effectively, child booster cushions – in the back. Touches like that are what makes the XC90 the school-run king.

Load it with the safety options and it will also be festooned with warning lights that illuminate whenever the car feels you may be in peril. You get a warning if a car is in your blind spot. You get another if you stray out of lane. And if you get too close to the car in front, the dash lights up like a Pink Floyd gig. Should it suspect you are about to hit a pedestrian, it will actually apply the brakes on your behalf.

This all sounds very noble and Volvoey, but there's a very good reason why you need to be warned of impending doom. The V60 is a hard car to see out of. Because of the swooping and rather attractive bodywork, coupled with small windows, the all-round visibility is quite poor. And because of the sloping roofline, the boot isn't as big as you might imagine.

I can't quite work out how they got it so wrong. Maybe there's a language problem between the Swedish engineers and the new Chinese owner. I can't imagine there are many translators who can manage that combination.

But whatever, anyone after a performance car would be better off with the equivalent BMW 3-series, and anyone who just wants to lug around dogs and chests of drawers would be better off with . . . well, with what? It's a good question.

Just recently, we have seen a raft of rather good-looking estate cars come onto the market. The Vauxhall Insignia and the Honda Accord stand out in particular. Boring choices, yes. But good, in these draconian times, for occasionally driving through the motoring rule book without being noticed.

29 May 2011

I love you now I'm all grown up, Helga

Porsche 911 GTS

I'd pretty much decided over the past year or so that I couldn't abide Sebastian Vettel. All that finger-pointing when he won a race. And the hair. And the way he blamed his team-mate for the crash last year. Ghastly jumped-up little German prig.

But last weekend, there I was, enjoying a plate of scrambled egg in Monaco, when I looked up to see the man himself, running towards me like he'd just crossed a desert and I had the keys to a fridge full of cold beer. We chatted about his forthcoming appearance on *Top Gear* and he was utterly charming; delightful.

The day before, I'd bumped into Mark Webber and he was charming too. I reminded him that the first time we met, he'd been employed by Ford to chauffeur fat drunks in dinner jackets from a hotel to the Goodwood Festival of Speed. 'Hey,' he said. 'Don't mock. I got eighty quid a day for doing that.'

In my brief visit to the principality, I met lots of people involved in Formula One. And they were all much the same. Michael Schumacher. Nick Fry. Martin Brundle. Christian Horner. Rubens Barrichello. All of them made the Duke of Cambridge look like a lout.

And then we get to Bernie Ecclestone. It was late, and I was wandering about the harbour, wondering whose party I was going to gatecrash next, when down the ramp of what appeared to be a floating city bounded the octogenarian. Without wishing to sound like Piers Morgan, he was all smiles, and after dispensing a good deal of bonhomie, he invited me for a drink on Flavio Briatore's boat. You won't believe this, but he turned out to be charming too. Well, I think he was charming. Flav doesn't bother much with consonants. He just sort of makes a noise when it's his turn to speak, but he did a lot of smiling and gave me a lot of wine.

So, behind the sponsorship and the nonsense and the back-biting, I have to report that the silly world of F1 is rammed full of people you'd like very much to have round for dinner.

Unfortunately, the people F1 attracts to its showcase Monaco event are not quite so charming. Let's deal with the men first. There are two kinds, as I see it. There are those who have the money and they are all very greasy. And then there are those who ride around on the big shots' backs, like oxpecker birds, picking at their fleas.

This is a mutually beneficial arrangement because the rhinos get to be surrounded by acolytes who agree with everything they say and laugh at their jokes until they are told to stop. And the oxpeckers scratch out a living by selling the rhinos superyacht insurance and hideous watches.

Occasionally, I would be grabbed by an oxpecker and made to meet his rhino and there is no small-talk manual in the world that covers this sort of encounter. The rhino has no clue who I am – he has someone to watch television for him – and the oxpecker is not really allowed to speak. And you can't ask the rhino what he does for a living because you know full well he sells guns and arranges for people to be murdered. Besides, to prevent you from asking any questions at all, he spends the entire time in your company yawning. Billionaires yawn almost all the time.

I'm told that on one of the really big boats, there was a young man who is employed to sit around all day, getting a tan and stay-ing fit. His job? He's the owner's heart donor.

Then you have the women. Mostly, they are prostitutes. I sus-pect that if you were so minded, you could come home with a veritable smorgasbord of sexually transmitted diseases. But not the billionaires. They have someone to make love for them.

If you were to drop an atom bomb on Monte Carlo during

the grand prix weekend, you'd mourn the loss of the sport's inner circle. But on the plus side, with the outer circle gone as well, there would be a measurable improvement in the planet's quality of life.

Of course, you might imagine that if you were to drop an atom bomb on Monte Carlo at any time, you'd achieve the same result. But I'm afraid not. The billionaires don't actually live there. They employ a man to go into their apartments once in a while to make phone calls and switch the lights off and on, so the tax authorities think they do.

All things considered, then, I was very pleased to leave Monaco to come home and watch the race on television. But I was not at all pleased to discover what car was waiting for me at the airport. A Porsche 911 GTS.

This is a reviewer's nightmare. It's like asking a restaurant critic to write about a McDonald's burger that has exactly the same ingredients as all the others but in a slightly different arrangement. Some colleagues of mine recently worked out that there are currently 153 different options available across the twenty-strong 911 range and that, as a result, there are 9.6 trillion mildly different permutations of what is basically the same bloody car.

There is, however, one thing that sets the new GTS apart. The price. If you were to buy a standard Carrera S and equip it to the same level as the new model, it would cost around £95,000. But this car – including a few extras – is just £81,968. And thrown in for free is the much better-looking wide body from the Turbo and a bit of black paint here and there.

I suspect there's a good reason for this unusual act of generosity. Next year we will see the arrival of a new 911 – which will be the same as all the others since Hitler first came up with the idea – and they need to get rid of all the parts before the production switchover. What you are buying, then, is not a new car. It's the last version of the old one.

I'm told by enthusiasts of the breed that it is also possibly the best. They like the look, the rear-drive simplicity, the value and the Alcantara steering wheel. They say that it combines all the best things from the massive Porsche option list in one unbeatable package and that everyone should have one immediately.

My eyelids are starting to droop. Because if there's one thing I hate more than writing about a Porsche 911, it's driving one. I feel like such a plonker. Fifty years old. What am I saying? It's one of two things, actually. I'm an enthusiastic motorist (in which case, give me a wide berth at parties) or I'm having a terrible midlife wobble (in which case, give me a wide berth at parties).

Plus, I've never really liked the way a 911 feels. I've always quietly respected Porsche's attempts to marry thrill-a-minute driving with everyday usability, but I've always thought that it was chasing an impossible dream. The two things are mutually exclusive. To be fun, a car must be a bit mad. And the 911 isn't.

So why did I enjoy my time with the GTS so much? And why did I also enjoy the GT3 version that I drove onto these pages not so long ago? The car hasn't changed – at all – which means I have.

And that's probably true. Yes, a Lamborghini or a Vauxhall VXR or a Mitsubishi Evo are all fantastically insane and I love them for that. But now I'm past fifty, I don't really want flames coming out of the exhaust any more, and a ride that cripples my back.

You don't drive a GTS. You dance with it. It is a beautiful experience, actually, and yet there are no histrionics. The satnav and the iPod connectivity all make sense. And it's not huge or loud or uncomfortable. It's as lovely as Sebastian Vettel, in fact.

So bear that in mind when you see a middle-aged man driving a Porsche. He's not having a midlife crisis. He's just grown up.

5 June 2011

Oh, miss, you turn me into a raging despot

Mercedes CLS 63 AMG

I wonder if we realize just how fast the age of electronic communication is taking over our lives, and shaping them and ruining them.

Unless you are a slipper and sherry enthusiast, you will be aware of a computer game called *Call of Duty.* The idea is simple. You run about shooting people in the face with a selection of large weapons. And then, if you believe the nonsense, you go out for a pizza and are overwhelmed by a sudden need to stamp on a tramp.

Of course, you can play by yourself or with friends. But, staggeringly, you can also play against unseen people in Canada or Israel or Siberia. It is incredible. And all the people you're trying to kill are being operated by unseen tramp-stampers in sitting rooms and shops and offices all over the world. If you have a microphone, you can even speak to them as you play, whooping whenever you fire a 12-gauge shotgun directly into their testes.

Unfortunately, like nearly everything powered by ones and noughts, it sounds brilliant but it doesn't quite work. You start the game. It tells you it's searching for other people in the world. It finds some. It does some electronic wizardry. And then it says the connection has been lost. So you go through the process again. And then again. And then again. And then you have a game of Scrabble instead.

We see the same thing with wireless routers. Wonderful. A must-have accessory. But, as I've said many times before, they work 10 per cent of the time and you spend the other 90 per cent of your life with your head in a cupboard, on the phone to a man in India.

The problem is, of course, that electronics companies always want to be first with a new idea. So the idea makes it onto the market before it's completely ready. This is why nothing electronic ever quite works.

Satellite navigation is a prime example of this. In the early days it was hopeless and would try to send you through Leicester Square, which was pedestrianized by William Pitt. The system in my last car refused to acknowledge there was such a thing as the M40. And we were constantly reading stories about people who'd obeyed the electronic voice of reason and ended up in a river, with a crab in their nose. But that was probably their fault for being idiotic.

Today you'd imagine that all of the mapping issues had been resolved, and to a certain extent they have. But the back-room boys – the sort of chap who wears a black T-shirt, lives with his mum and doesn't wash terribly often – are always shoehorning new submenus into the setup. And those are being rushed out as well.

Take the traffic warning technology. The idea is that the map informs you of hold-ups ahead so that you can plan a route around them. Very clever. It cuts congestion, saves fuel, spares your temper and keeps the polar bears happy.

But the system in the Mercedes CLS that I've been driving for the past week is forever getting its northbound and its southbound muddled up. Which means I spend an hour dribbling along a country lane, with my door mirrors in the blackberry bushes, avoiding a queue that's going the other way.

What's more, a stern-sounding woman interrupts Chris Evans to say in a weird voice that there is a queue ahead. She even gives you the average speed in the queue and adjusts your estimated time of arrival accordingly.

Because of one of her warnings last week, I realized that I would not make it to the restaurant I'd booked before it shut. So

I called to cancel the reservation, phoned home to disappoint the children and plodded onwards towards the jam. WHICH WASN'T BLOODY THERE.

There's more. One of the features provided is a list of all the restaurants in the area. It asks what sort of food you want and then takes you to the nearest eatery that is equipped to help. The trouble is that people only ever want Chinese, French, Italian or Indian. But it would be racist to limit the list to just four options. So, to keep everyone happy, it comes up with every single country in the world. If you have a modern Mercedes and you live in the highlands of Scotland, do please enter 'Balkan' and let me know what on earth it comes up with.

I was also amused by the other things it will help you find. Many are useful. Hospitals, police stations and so on. It will even help you locate the nearest mosque, which is clearly important if the sun is going down and you are a Muslim.

But then the black T-shirt brigade obviously thought, Uh-oh. We can't list just mosques, because it looks as though we are favouring the children of Muhammad over those who support other teams. So, it will also find the nearest synagogue. But – black mark here – it does not seem to think that Methodist chapels are worthy of a mention.

Also, it could not find the Devils Dyke pub on Devils Dyke Road, just north of Brighton. And it will not let you enter a seven-character postcode. It ended up making me very angry. The command-and-control system in my old Mercedes is very good. This new one? It's so clever, it's actually a drooling vegetable.

And then we have the phone system. Until very recently Mercedes fitted an actual telephone that was hardwired into the car. What that did was work. Now the nerds in the back room have decided that Bluetooth is good enough. It isn't. People speaking on Bluetooth sound like deep-sea divers, and that's when they're both in an anechoic chamber. Communicating with someone in a car on Bluetooth is like trying to communicate with a corpse.

Electronically, then, the Mercedes CLS has taken a couple of steps forwards and about five in the other direction.

It's the same story with the shape. The original CLS is said to have been designed by a young stylist who wanted to see how a Jaguar would look if it were made by Mercedes. It was weird, but undeniably attractive.

From the front the new one is even better, but, as with other new Mercs, there's a styling detail over the rear wheelarch that simply doesn't work at all. Styling details need to be there for a reason – a hump in the bonnet hints at great power beneath, for instance – but this one is just fatuous. I pretty much hate it.

I also hate the gearbox. The old seven-speed auto has been replaced with the double-clutch flappy-paddle system found in the SLS. It's electronic, so it works well, except when you are in town going slowly. Then it's jerky and unwilling to respond when you want to exploit a gap in the traffic.

The rest of the car, though. Wow. It's been festooned inside and out with lots of neat bits of jewellery that stop just short of being blingy and, in the case of the CLS I've been driving, add £27,000 to the £80,000 price tag.

And then there's the engine. It's AMG's new twin-turbo V8 – with, on my test car, a performance upgrade to 550 bhp and 590 lb ft – and it's much more muted than the old 6.2. Under big acceleration you still get some machinegun noises from the tail-pipes, but it's quieter, more civilized. I'd go so far as to call the driving experience imperious. When slower drivers see this coming, they get out of the way in a big hurry. You feel a bit like Idi Amin. Or was that just me?

Overall, however, I think that some of the original CLS's appeal has been lost. And, as a result, if I wanted to buy a big, stylish four-door saloon, I'd just walk past this and go for the Maserati Quattroporte.

12 June 2011

From 0 to 40 winks in the blink of an eye

BMW 640i SE convertible

Have you actually stopped for a moment and looked – really looked – at the new BMW 5-series estate? All things considered, I would say that this is one of the most handsome cars ever made. It has all the BMW hallmarks: the body seems to have been stretched to the limits simply to cover the wheels, and there's the traditional Hofmeister lean-forward kink in the rear pillar. It's a tiny design detail that makes the car look as though it's going a thousand miles an hour even when it's in a golf club car park.

And yet the car doesn't look old-fashioned. There's something about the shape of the bonnet that makes it look as though it may be visiting us from the future.

It's the same story on the inside. All cars of this type, if we're honest, feel and look pretty much the same from behind the wheel. But not the Beemer. It's all very minimalist and unusual. As if you've accidentally plonked yourself down in a Bang & Olufsen catalogue. And the 5-series estate is not alone. The new(ish) Z4 moons me with its beauty when I see one go by, and then there's the limited-edition 1-series M coupé. It's not a coupé, actually; it's a saloon. Actually, it's not even that. It looks like the box in which your washing machine was delivered, only to make sure you don't throw it away by mistake, it has enormous wheelarches.

I love this car because what it says when you look at it is this: I am very fast and I don't need to shout about it. It is very fast, too. Faster than a Cayman R. Faster than a Lotus Evora. Faster than you would believe possible. The 1-series M coupé is very probably my favourite car on the market right now. Or it may be the M3. I'm not sure. But it's one of those two.

I can't believe I just said those things. I've spent the past ten years laughing at the man who styled BMWs and the idiots who drove them. BMWs? They were nothing more than expensive and very hideous mounting brackets for your stupid personalized plate. Montblanc pens with windscreen wipers.

I was a Merc man. But Mercs these days are getting a bit too chintzy for my taste. And so are Jags. And the Lexus is still stuck at the back, in its hairy sports jacket, wanting to smoke its pipe. So now I've switched my hero worship from United to City. I've become a Beemer man.

I should say at this point that I don't like all BMWs. The X1 is very terrible, and so is the genital wart that is the X3. Then you have the X6, which is idiotic, and the little 120 diesel, which seems to be a bit boring. But the rest? Mmmm.

That's why I almost skipped with delight to the door of the car of the new 640i convertible, and I was looking forward to driving it very much.

We see some of the current BMW good and the bad in the styling. Roof down, it's excellent, but when you press the button to put the roof up, it seems a slightly tipsy man comes and erects a tent with which he's not completely familiar above your head. Why the vertical rear window? It looks stupid.

There's more silliness, too, in the name. There was a time when the numbers on the back told you the model and the engine size. So a 325 was a 3-series with a 2.5-litre engine, and a 750 was a 7-series with a 5-litre engine. Very sensible. Very German. But now the logic has gone, so the 640 I was testing was a 6-series with a turbocharged 3-litre engine.

It's a new power plant that develops many horsepowers and big wads of torque. And it's allied to an eight-speed gearbox that is fitted solely so the man who invented it can go to gearbox conventions and tell other gearbox enthusiasts that his box has more cogs than theirs. It shifts well enough, but eight speeds? In a car with this much torque? As the XJS proved all those years ago, three is all you need.

And three would suit this car far more because it is the laziest machine I've driven since the old Rolls-Royce Silver Shadow. It is not built for point-and-squirt hammer-time trips to the race-track. Shock and awe? Lock and snore, more like. It even comes with a TV screen of such vastness, it's unfair to call it a television. This is a home cinema.

This, then, is built to move you and your family, provided the children are both amputees, around in the greatest possible comfort. It is a car designed for Houston dentists. A fatboy car.

We should rejoice at the news. Too many expensive cars these days are ruined by low-profile tyres and ebony-hard suspension. Too many are honed on the Nürburgring. So, while they work on a track, where they will never be driven, they are bone-shakingly horrid on the road, where they will.

The BMW is soft. Squidgy. Comfy. The steering wheel is connected to the front wheels with a big soft bag full of melted chocolate, and the noise it makes at speed is the hum of a gardener who's happy in his work.

I'm sure that if you were really determined, it could be made to whiz about at 155 mph in a fog of tyre smoke. But that would be like trying to make a teenager tidy its room. Possible but, ultimately, too much effort is required.

You don't drive the 640. You waft along. It's one of those cars that's extremely happy to plod up the motorway at 65 mph, unless you aren't concentrating, in which case it will slot some Barry White into the CD player and drop down to about 3 mph.

At the lights, the engine is so lazy that when you come to a halt it can't even be bothered to keep going and falls asleep. The idea is that when you take your foot from the brake pedal it wakes up. And that sounds very green and polar-bear-friendly. But you wait till you're trying to pull onto a fast and busy main road. It can be a bit hairy, spotting a gap and then thinking, Oh Christ. The engine's nodded off.

Even the satnav is lazy. After you've given it the first four characters of a postcode, it can't be bothered to listen any more and finds what it thinks is roughly the right road to nearly where you're going. In some cars these days the satnav will suggest alternative routes to miss jams that lie ahead. Not the BMW. It can't be bothered. It likes jams. It can have a nap.

I know a great many people who would absolutely love this car. People who have a few bob and want to park something a bit tasty on the drive, but who actually can't abide driving and aren't very good at it. The 640 would be perfect, because it isn't really a car. It's a bed.

When you climb into most BMWs, you feel as if you should be wearing racing bootees and a suit made from Nomex. In this one, the correct attire would be a pair of pyjamas.

19 June 2011

Oh, Shrek, squeeze me till it hurts

Nissan GT-R

Are you a serious car enthusiast? I mean, really serious? Do you drive round every corner as fast as the laws of physics will allow? Do you open the taps whenever you can to revel in the intoxicating, mesmerizing power of internal combustion? Does G-force tickle your G-spot? Do you talk about torque at parties? Are cars, for you, the light and the life and the meaning of everything? Right. Well why don't you have a Nissan GT-R, then?

The GT-R is not designed to impress other people. There is no hand-stitched leather and no monogrammed luggage. It's a Nissan, too – a Morphy Richards in a world where Dolce & Gabbana rules. Does it look good? No. Will it turn heads? No. But only because no one's neck muscles can move that fast.

The GT-R is designed to examine carefully the scientific laws that govern movement and then systematically to break them. It is designed to go faster than you ever thought possible, possess more grip than is physically allowed, change gear more quickly than you can blink, and stop with such ferocity that you can actually feel your face coming off. No style. Just engineering.

It is made in a hermetically sealed factory, which is climatically controlled to ensure all the components are in the same state of thermal expansion when they go together. The tyres – this says a lot – are filled with nitrogen because the normal air used in humdrum cars such as Ferraris and Aston Martins is too unpredictable. It expands and contracts appreciably according to tyre temperature. Nitrogen does not.

Then you have the wheels. They are knurled to stop the tyres coming adrift during cornering. Does a Ferrari have that? No. It's not necessary.

The new 2011 GT-R is built along exactly the same lines but now there's more power, more grip, more downforce and even more speed. It's still not designed to impress your passengers. It's designed to hurt them.

Let's begin with a standing-start full-bore acceleration run. You put the gearbox in race mode, and then you hold down the traction control button for a moment, put your left foot on the brake and mash your right foot into the carpet. When the revs have settled, and the 523-horsepower 3.8-litre twin-turbocharged engine is screaming its head off, you take your foot off the brake.

What happens next is extraordinary. There is no wheelspin. The clutch does not slip. One second you are stationary, and the next you are doing 100 mph. Imagine sitting in a deckchair in your garden on a summer's day. It's quiet and peaceful and you are enjoying the birdsong. Then you are hit from behind by a Boeing 747. That's what the acceleration feels like in a GT-R. Absolutely unbefrigginglievable.

Of course, there are lightweight, low riding track-day cars that, on paper, can get from 0 to 60 just as quickly. But they don't get from 0 to 5 with anything like the savagery. And they run out of puff at 100. The GT-R does not. It just keeps on going, and when you get to the red line, you pull the paddle and instantly – not something that could be said of the previous model – the next gear is engaged. I have never experienced anything quite like it, if I'm honest. It's wild. It's relentless. It's intoxicating. It's amazing.

Certainly, you should never use the launch control in this car unless you are bracing your head against the headrest at the time. Because if you're not, the whiplash could put you in hospital.

It's the same story in the corners, where the steering wheel becomes nothing more than a handle to hold on to so as to prevent yourself from being flung out of the seat. I should like very much to see an X-ray photograph of someone's heart when they are cornering a GT-R, because one thing's for sure. The G-force is so severe, there's no way it would be heart-shaped.

I was desperate after just a couple of laps of the Top Gear test track to turn off the traction control. This would let the car slide, which would a) be more fun and b) reduce the pressure on my neck. But it's not wise to turn off the safety features, because if you spin a GT-R, you will break many complicated components.

So, the GT-R is very good. But I know what you're thinking. You're thinking that you'd still much prefer a Ferrari 599 or Aston Martin DBS. If that's the case, I'm obviously not getting the message across. The Nissan will eat cars like this. Chew them up; spit them out. Bring whatever you like to the party. The GT-R will blitz it. It blitzes everything. It recently blitzed the Nürburgring in seven minutes and twenty-four seconds. Do you know anything else with number plates that could get round as fast? Because I don't.

Of course, a Ferrari is a much nicer thing to own and to behold and to touch, but when it comes to the business of driving, or going from point A to point B as fast as possible, no Ferrari would see which way the GT-R went. Ferrari is Manchester United. The GT-R is Barcelona.

There are, however, some problems. First of all, it is extremely ugly. It tries to be unshowy in the same way as a bouncer tries to be unshowy when he slips into a dinner jacket. You can always see the tattoos and the neck like a birthday cake, so you know. You know with the GT-R, too, because of the scoops and the exhaust tailpipes, which are even fatter than before.

Inside, it's worse. I can see what Nissan has tried to do. Keep it simple. But the slab of carbon fibre on the centre console is embarrassing, and the central command unit, which shows you the state of all the components and how many g you generated in the last bend? No. It's all a bit too fast and furious for my taste. A bit too Jason Statham.

I wish Nissan had had the guts to truly hide its light under a bushel. As it did with the old Skyline. Not to pretend.

But, that said, the GT-R is a proper four-seater and it has a boot into which you could fit many things. It is also surprisingly quiet and remarkably comfortable, even on a traditional pot-holed British road. Of course, there's a harshness to the feel, a sound that hints at the racetrack, but there's no volume. And I like that.

I also like the price. Yes, it's rocketed up by about £10,000 to £69,950, and that's a lot for a Nissan. But it's much less than half what you'd have to pay for a slower, less electrifying Ferrari 458. And it's not as though the salesman can mug you with a list of options, because I've been on the online configurator and there aren't any.

The new GT-R is demonstrably better than the old one. It's faster, and the gearbox is a significant improvement. This means it's demonstrably better than what was a benchmark. Yes, it's an ugly son of a bitch, and there are some stupid gimmicks, but this car is a genuine phenomenon.

You're interested in cars. You love driving. You like engineering. You have to have a GT-R. It's that simple.

26 June 2011

A world first – the Ferrari 4 × what for?

Ferrari FF

It was a normal Saturday morning and the roads were jammed with DIY enthusiasts on family trips to the local hardware store. This sort of scene is bad news if you're in a hurry, because the sort of person who erects shelves himself is not going to drive to the shop at more than 4 mph and waste the money he's saved.

Saturday morning is now, for me, the worst time on the roads. They're a cocktail of the mean, the elderly and the frightened. Nobody's quite sure where they're going, and no one can concentrate because the kids in the back are explaining sulkily that they'd rather shoot space aliens than traipse around B&Q looking for self-tapping screws.

Happily, however, my jaunt to the Midlands last weekend wasn't so bad because I was driving the new Ferrari FF. And all you do when a Hyundai or a Peugeot gets in the way is pull the left-hand gear-shifter paddle a couple of times and press the accelerator down – suddenly it isn't in your way any more. In the FF you could easily overtake an Australian road train before you'd got out of your own drive.

The engine is a 6.3-litre direct injection V12 that develops a stratospheric 651 horsepower and 504 torques. And what that means is a car that gets from 0 to 62 mph in 3.7 seconds and then onwards, propelled seemingly by its own seismic shockwave, to a dizzying 208 mph.

Of course, there are other cars that can go this quickly, but none of them feels like a Ferrari. The paddles, for instance: when you pull them, they feel as if they aren't actually connected to anything, which in reality, of course, is a fact. They aren't.

They work the gearbox in the same way as the light switch in your kitchen works the bulb. Only a little bit faster.

Then there's the steering. As is the way in all modern Ferraris, it is disconcertingly light. Turning the wheel requires as much effort as dusting a polished work surface. So you imagine that there can't possibly be any feel. But there is. You know all the time exactly what those front wheels are doing, how much grip is left and what you should be doing with the throttle as a result. Ferraris these days are like Vietnamese masseuses. Soft, but acupuncture accurate. And they handle – there's no other word – beautifully.

They are like other cars in the same way as a Mac is like a PC. In other words, they are not like other cars at all. And just as a Mac has no right-click – which drives me insane – Ferrari have irritations, too. All of the controls are on the steering wheel. Indicators, lights, wipers, the horn, the starter button, radio tuning, volume, gear-shifting and the traction control switch. The lot.

The idea is that you never have to take your hands from where they should be, and in a meeting that makes sense. It makes sense in a grand prix, too, but on the Fosse Way what it means is that you indicate left when you want to go right because the wheel is upside down at the time you hit the button. And you turn the wipers on to say hello to friends going the other way.

It's all a bit bonkers but it does add to the sense of occasion, as do the leather and the sense of space and the howl from the engine. A modern-day Ferrari feels very, very special.

However, even though the new FF feels this way, it is not like any Ferrari I've driven before. Chiefly because it's the first Fezza to be fitted with four-wheel drive.

There's a conventional way of doing this. You take drive from the engine to a centrally mounted transfer box, which then distributes power to front and rear axles. Naturally, Ferrari decided not to do this. It says that if you send the power down shafts below the engine to the front axle, the engine must sit up high, which is bad for the handling and bad for the styling, too.

So, instead, the FF sends its power to a rear-mounted gearbox and then to the rear wheels. But then, if sensors detect that those wheels do not have enough grip, two small two-speed gearboxes being driven by the crankshaft at the front of the engine are engaged and the front wheels are asked to join the party.

It's an idea that was first tried by Ferrari in the Eighties, but back then the electronics necessary simply weren't available. I'm surprised they're available now. It sounds an almost fantastically complicated solution, and I wonder if it really will work.

I drove the FF pretty hard and at no time did I sense the front wheels were being driven. And even if they were, they are never allowed to take more than 30 per cent of the engine's power. And none at all if you go past 130 mph. At this sort of speed the front-mounted two-speed gearboxes can't cope and shut down. The words Heath and Robinson keep springing to mind here. Followed by a simple question. Has anyone at Ferrari ever driven a Nissan GT-R?

Ferrari says that the system will be of enormous benefit to those who wish to take their Ferrari skiing. And that may be so. But, as the owner of any Bentley Continental or even Range Rover will testify, you can drive as many wheels as you like; if you don't fit snow tyres your journey will end in Moûtiers.

Of course, you may be wondering where on earth you would put your skiing clobber in a Ferrari. Aha. Well, not only is this the company's first four-wheel-drive car, it's also its first hatchback. It has fold-down rear seats, and you can fit more in the boot when they are up than you can in a Renault Scénic.

So, a very special, very fast car, with a dollop of practicality and a four-wheel-drive system that may not add much. But it doesn't take anything away, either.

However, there are a couple of problems. First of all, the FF is as big as a medium-sized US state. On a normal British B-road you really do flinch when you pass traffic going the other way, and in a town it's next to useless.

But worse than that is the styling. It looks fantastic from the front and wonderful, too, from the side, but I'm afraid its back-side is hopeless. Bland. Dreary. Kia does a better job. You may think you could live with this but I guarantee that when the time came to sell, you'd struggle to find a like-minded soul. As a result, I expect the FF to drop in value like a hammer falling from a tower block. Be aware of that before you run amok with the comically expensive and lengthy options list. For instance, don't have the boot lined with leather (£1,728) if you really are going to take your FF to St Moritz.

Overall, I liked this car. It was exciting when I was in the mood, but strangely docile and comfortable when I wasn't. I like the way it feels different from everything else on the road, and I'm not bothered about the four-wheel-drive system either way. But that rear end? They really should have had a look at Pippa Middleton before they made merry with the ballpoints.

10 July 2011

Work harder, boy, or it will be you in here

VW Jetta 2.0 TDI Sport

I suppose I ought to come clean. The cars I review on these pages every Sunday are sometimes nothing like the cars you can actually buy. Every car company runs a fleet of press demonstrators, which motoring journalists can borrow for a week. We imagine, of course, that the cars on these fleets are plucked at random from the production lines, in the same way as a famous restaurant reviewer expects that the food he's eating is exactly the same as the food everyone else is eating.* But I fear that, sometimes, they are not.

Many years ago, when cars were judged only on acceleration times, Austin Rover made all sorts of wild claims about how its new Maestro turbo could get from a standstill to 60 mph in six seconds. And indeed the press-fleet cars supplied for testing could do just that. But only once. Because then they'd blow up, causing everyone to wonder if the wastegate valves hadn't been welded slightly shut.

There were also tales about car makers stripping down cars that would be going to the press, and then rebuilding them, very carefully and at huge expense, by hand. And I must say that the Ferraris that turn up for performance testing always seem to be noticeably faster than the cars that are supplied for photographic purposes. That could be my imagination, though (he said, aware of the laws of libel).

But even if the car supplied for testing really does come from

---

* *This isn't necessarily so. Normal people are not, as a matter of course, served food containing quite so many bodily fluids.*

the production line, the experience is still a bit skewed. First of all, it's delivered, fully taxed and insured, in an extremely clean state with a tankful of free fuel. That makes the reviewer feel very gooey about life in general and his job in particular. And when the tank is empty, a man comes with another car and takes the first one away. It's all completely unrealistic, really.

And so are the cars. Most press-fleet managers will ensure that every demonstrator is fitted with every single optional extra. They will argue that this gives the reviewer a chance to sample all that's available. Yes. But a layer of exciting buttons and knobs can also mask the dreariness of the product underneath. In the same way as a spicy sauce masks the fact that the curry you've ordered is full of dead cats.

So, in short, I spend half my life driving around in a £15,000 car that's been hand-built at a cost of £250,000 and has been supplied with eighty quid's worth of free fuel, free insurance, free tax and a range of optional extras that are worth twice what most people would pay for the entire car.

Not this week, though, because Volkswagen supplied a new Jetta in what can only be described as hire-car spec. I assumed that this was because the company was so proud of the actual car, it didn't want to spoil the experience with lots of unnecessary electronic or cosmetic flimflam. I was wrong. Because this is the dreariest, most depressing car ever made in all of human history.

I am not saying this because it is an ordinary car of a type people buy. I am saying this because it really is the four-wheeled equivalent of drizzle.

In front of the gear lever are five switches, and in a standard car all of them are blanked off with plastic shrouds – little reminders that you didn't work hard enough at school and that life's not going as well as you'd hoped. If only you'd clinched that last deal, you could have bought the £440 parking-sensor pack. Then you'd have only four blanked-off buttons.

It's much the same story with the central control system. Push the button marked 'Media' and a message flashes up saying, 'No medium found.' This is not a reference to Doris Stokes. It's another gentle dig, another reminder that you couldn't afford to fit an iPod connection. That your whole life is going down the khazi.

So you push the button marked 'Nav' and you get another baleful message saying that no navigation disc has been found. You couldn't afford it, could you? You only got three Bs and Exeter said no. You ended up at a glorified poly and your life's gone downhill from there.

It must have done for you to have ended up in a Jetta. As we know, it's a Golf with a boot on the back instead of a hatchback, and what's the point of that, exactly? There was a time, when *Terry and June* was on television and people doffed their caps to aldermen, that a saloon was perceived to be more upmarket than a hatch.

There are also places in the world, in Africa mainly, where a saloon marks you out as someone special. But here? Now? No. We have come to realize that a saloon is just a hatchback that's less practical and more boring to behold.

The Jetta is extremely boring to look at. It's boring to think about. This is the sort of car you would buy not realizing that you already had one. It is catastrophically dull. As dull as being dead.

It is not, however, dull or boring to drive. No. It is absolutely awful. First of all there's the suspension, which is plainly tuned to work only on a billiard table. On a road it transmits news of every crease, ripple and pebble directly to your spine, and, to make matters worse, the seats appear to have been fashioned from ebony. They are rock hard.

The backrest, which is even less forgiving than the squab, seems to have just two positions. Bolt upright and fully reclined. Only once can I remember ever being so uncomfortable, and that's when a doctor was examining my colon with what felt like the blunt end of a road cone.

Then there's the air-conditioning. Or, rather, there isn't. VW calls it semi-automatic air-con, and I'm sorry, but there's no such thing. It doesn't work. And as for the trip computer, it told me about oil temperature, which isn't interesting, or it was a compass. And that's not interesting either.

So what of the engine? Well, you've a choice of a 1.4-litre petrol, which comes in two states of tune, or a brace of diesels. I opted for the 2-litre, which, by diesel standards anyway, was reasonably quiet and refined. It was also reasonably powerful, clean and economical.

I must be similarly kind about the quality. The interior does appear to be well screwed together, but then the Jetta is made in Mexico – and Mexico, as I have recently learnt, is a byword for industrious attention to detail. On a personal note, I'd far rather have a VW built by Pablo in Central America than a VW built by some sloppy German who just wants to spend the day sleeping and being kidnapped.

Apparently, Volkswagen wants to sell 3,000 Jettas in the UK this year, which in the big scheme of things does not sound an ambitious target. But I cannot think of even three people who would be happy to live for more than a few seconds with this hateful, dreary, badly equipped, uncomfortable, forgettable piece of motoring-induced euthanasia.

Better alternatives include the Golf, the Passat, every other car ever made, walking, hopping and being stabbed.

17 July 2011

Too tame for the special flair service

Audi RS 3

God made a bit of a mistake when he was designing women. He made the birth canal so narrow that babies have to be born when they are nowhere near ready for life in the outside world.

A newborn horse can run about and feed itself five seconds after emerging from the back of its mum. And it's the same story with dogs. My labrador gave birth to a litter of nine puppies while asleep, firing them out like a Thai hooker fires ping pong balls. And within moments they were up and about, being doggish.

A human baby, though, is not capable of anything. For week after interminable week, it can't sit up, crawl, speak or operate even rudimentary electronic equipment, and sees absolutely nothing wrong with sitting in a puddle of its own excrement. Babies are useless. Stupid, mewling, puking noise trumpets that ruin life for anyone within half a mile.

Unless, of course, the baby is yours, in which case you rejoice in its ability to grip your finger with its tiny little hand and are keen to take it on as many aeroplanes as possible. Why is this? Why is your baby so perfect and wonderful when everyone else's is as irritating as a microlight on a peaceful summer's evening?

We see the same problem with literature. There is a book called *Versailles: The View from Sweden*. It's excellent in a game of charades, but as a light read I should imagine it's not excellent at all. Unless you could get a copy signed by the author. Then you'd love it.

I have a copy of Monty Python's *Big Red Book* that is signed by all of the Pythons and as a result it is my most treasured possession. The one thing I would rescue if my house were to catch fire.

We see this specialness in other things too. There's a little spot just below the village of Keld in the Yorkshire Dales. It has grass and some trees and a bit of sky, all the ingredients you would find on a roundabout in Milton Keynes. And yet the spot I'm talking about is special and a roundabout is not. Why? Dunno. It just is.

Then there's the iPhone. As soon as I was shown some pictures and discovered you could enlarge them by moving your fingers apart on the screen, I had to have one. I treat it with great care and become very defensive when BlackBerry enthusiasts are critical.

It is just some wires and a bit of plastic. It's not signed by Steve Jobs; I did not give birth to it and took no part in its creation. It is not unique and yet, to me, it is special. And that, in my view, is what makes the difference between a product that you want and a product that you need.

Specialness is particularly important when it comes to cars. Recently, on *Top Gear,* I drove something called the Eagle Speedster. It was a modern take on the old Jaguar E-type and in many ways it was a bit rubbish. There were no airbags or antilock brakes, and while the lowered, more steeply raked windscreen meant the car looked good, I couldn't see where I was going. And yet, despite the shortfalls, it is the most special car I've driven. Do I need it? No. Do I want it? Yes. More than my left leg.

And now let us spool forwards to the Nissan Pixo. This is the cheapest new car on sale in Britain today and in many ways it is excellent. You do get power steering and antilock braking and I have no doubt that it will be a faithful and reliable servant for many years. And yet, despite all this, I want one about as much as I want a bout of herpes.

There is a similar issue with the new McLaren MP4-12C. It is a superb piece of engineering and, my God, it's fast. But the excitement and joy and specialness that you get from a Ferrari or a Lamborghini is missing. You sense that it's the brainchild not of a man called Horacio or Ferruccio or Enzo, but of a man called Ron.

So what about the Audi RS 3? A roundabout? Or my special place in Swaledale?

Well, fans say that because it has a turbocharged five-cylinder engine and four-wheel drive, it harks back to the original quattro, which, in second-generation 20-valve guise, was one of the most special cars ever made. So it has good genes.

Good manners, too. Unlike the original quattro, the engine is not mounted several yards in front of the front axle, which means that the catastrophic understeer of yesteryear is gone. You just get normal understeer, which is dreary but not fatal.

The power's good, though. And so's the speed. And so is the noise and so is the seven-speed dual-clutch flappy-paddle gearbox. Provided you are at the Nürburgring and your family's life depends on your lap time. If, however, you are not at the Nürburgring and your family is not being held hostage, you may find it a bit irritating.

It's a problem with all these gearboxes. They're good when you are travelling fast, but in town they jerk. You don't get the creep of a normal automatic or the slip from a clutch pedal in a normal manual. I realize, of course, that flappy paddles mean better emissions, which is good news for polar bears, but for smooth driving, they're pretty hopeless.

Now we must address comfort. There isn't much, because, like the gearbox, the suspension is set up for fast lap times. It's not as bad as in some cars but you do need to scour the road ahead carefully so that you don't accidentally run over a pothole.

Inside, you are reminded that while the RS 3 is a new car, it's based on a car that is not new at all. It feels old-fashioned. Boring. And it's time I mentioned this: there's a wee bit too much choice.

An example. Would you like the climate control to deliver

17.5 degrees or 18? And what would your passenger like? I'm sorry, but half-a-degree increments are plainly silly. You either want to be chilled or warmed. Two settings would do.

It's the same story with the satnav. Yes, we like to be able to adjust the scale of the map. But in the Audi you twiddle the knob for 90 minutes and it zooms in from something like 900 metres to the centimetre to 875 metres to the centimetre. That's not necessary.

Neither is the price. It's just shy of £40,000, which is a lot for what, when all is said and done, is a fancy Golf. Yes, I know that it's a limited-edition car and that, as a result, second-hand values will be good, but if I were spending that much on a car, I'd want it to feel and look and be a lot more special.

That's a trick BMW has pulled off very well with the 1-series M coupé. It's about the same size and price as the Audi and delivers the same sort of get-up-and-go. But while you emerge from every trip in the Beemer wearing an enormous grin, you emerge from the Audi smiling only because the trip is over.

24 July 2011

An asthmatic accountant in lumberjack clothing

Mazda CX-7

How you feel when you arrive somewhere in a car is more important than how you feel on the way. On the way, a car is just a tool, but when you get to your destination, and your hosts are waiting for you, and people you know are looking, that's when the true value of your wheels shines through.

I went to a party the other night, and when I arrived, there were twelve Range Rovers already parked in the drive. Mine made it thirteen. This made me feel gooey and part of a club: like I'd read the social circles in which I move well and that I was keeping up. The drive to the party had been normal – a row about why it had taken my wife so long to get ready and some light rain – but the arrival was terrific.

However, then the fourteenth car arrived. It was my friend Alex. He screeched into the yard in a seven-year-old Renault Clio and that looked bad. He knew this. So to make up for the deficiency of the rust bucket, he didn't just pull up and park. He kept his foot on the throttle, turned the wheel slightly and applied the handbrake. This was a cool thing to do, and it more than made up for the brownness of his wheels.

The fact is that, if you exclude the very cheap and the very expensive, all cars feel broadly similar to drive. A BMW and a Merc? Essentially, they are the same. A Renault and a Fiat? Same story. Look at the brakes on your car, then go and look at the brakes on your neighbour's. Both sets are made by the same people in the same factory. And it's a similar story with the power-steering system and the wiper motor and the shock absorbers.

Take the engine in your Mini Cooper S. You may think that it's lovingly hand-crafted by gnarled old men in Oxfordshire and that BMW is fiercely protective of its secrets. Not so. You will find the same engine in the Peugeot RCZ and the Citroën DS3 Racing.

And then there's the Fiat 500. Lovely little car, so cute and chic and perfect. Except that if you peel away the body, it's exactly the same as both the Fiat Panda and the Ford Ka. Do you want a Ford Panda or a Fiat Ka? Why not? They will drive and feel and go just like the 500. But they will not feel even remotely similar when you arrive at your friend's house because your friend will not come out and go, 'Aaah.' As they would if you pulled up in the Fifties throwback.

All of this means that brand image is critical. But even more important than that is the styling. And that brings me on to the Mazda CX-7.

There are now many crossover 4x4 family school-run SUV MPV, whatever you want to call them, vehicles, and for the most part they are all absolutely terrible. Pull up at anybody's house in any one of them and I can pretty much guarantee that no one will open the door. I'd rather be friends publicly with Piers Morgan than friends with someone who has a crossover car.

Crossover cars are for fools. They offer no more space inside than a normal family hatchback but they are perceived by the idiots who buy them to be tougher. Why? They are made from the same grade of steel and the same quality of plastic and they have the same suspension components. You are fooled by the high-riding stance into thinking that they have been built to take on the Kalahari but they have not. All the tall stance means is worse handling and inferior fuel consumption.

The trouble is that crossover cars do look quite good, in a Tonka Toy sort of way. They look better than a Ford Focus. More interesting. They are like accountants underneath but they are wearing lumberjack shirts and Timberlands. It makes them stand out. And just about the best-looking of all of them is the four-wheel-drive Mazda CX-7.

I love the flared arches and the way its window line tapers. If you were in the market for a car like this, you might well see one in your local town and think, Mmmm. That'll do nicely. I'm with you. I liked the old model quite a lot, but the new one has a few issues. Take a deep breath – we're going in . . .

First of all, you would imagine that, being Japanese, it is built to outlast Scotland's mountains. Well, you will be disappointed to note that when you slam the doors, they sort of clang, and that when the electric window goes up, it crashes into the door frame with such a thud you think the glass will break. Oh, and the offside wiper hit the A pillar on every sweep, which was deeply irritating. It was more relaxing to drive in the rain with the wipers turned off, seeing where I was going using nothing but the Force.

Then there's the gear change. First, third and reverse are separated by a millimetre, so every time you set off you don't really know whether you will go forwards, go backwards or stall.

And stall you will, because the turbocharged diesel engine is woeful. They say it will get from 0 to 60 in eleven seconds, which raises the question: sixty what? Certainly not miles per hour. There doesn't even seem to be much in the way of torque, normally a pleasant by-product of diesel motoring. On even the slightest incline, you need to change down. Which normally means engaging reverse by mistake.

Mazda says the CX-7 produces less nitrous oxide than just about any other car made, which is lovely. But that's like saying it produces very few eggs or rice pudding. All anyone cares about these days is carbon dioxide. That's what the tax is based on, and on that front it produces a lot.

And don't think you can get round all these problems by buying a petrol version with an automatic gearbox. Because there's no such thing. It's a diesel manual. Or something else.

I haven't finished with the problems yet. The satnav screen is the size of a Third World postage stamp, the boot is even smaller and there is only enough room in the rear for people who have lost their legs.

Now this might just be acceptable if the car were small. But it's chuffing massive. So big that it would not fit into the parking space in my local underground car park. I even struggled to fit it into a standard London meter bay. And to make matters worse, you cannot see any of the corners from the driver's seat.

So, big on the outside, small on the inside, badly made, ill-conceived, woefully slow, hard to drive and I wish that were an end to it. But no. It's also uncomfortable.

The only good thing, apart from the alluring looks, is the price. Considering the equipment provided as standard, it's not bad. But that, be assured of this, does not make it good value.

It's strange. Because most cars these days are fundamentally the same, I thought the days of the truly bad car were a thing of the past. There are boring cars and ugly cars and slothful cars. But bad? Outside America, I haven't driven one of those for years.

I have now, though. So if you must have a CX-7 for whatever reason, please remember to take a leaf out of my mate's book. Arrive everywhere in it with a handbrake turn.

31 July 2011

Someone please check I haven't left my spleen back there

BAC Mono

The biggest problem with really fast cars these days is that they are way too fast. I was made aware of this recently when driving the Lotus T125, a sort of quasi Formula One racer.

It is capable of accelerating so savagely that when you put your foot down, your head comes off. And it brakes with such ferocity that all your internal organs become detached and your face squelches into the front of your helmet.

Think of a corner that you encounter on the way to work. Any corner you like. And think of the speed you drive round it. Well, if you tried to do that corner at that speed in the Lotus, there would not be enough air passing over the wings to generate downforce. So you'd crash and be killed.

You have to take the corner you have in mind at a speed that is insane. And it's hard to convince your mind that this is possible. You know that if you lift off, you will die. But you also know that if you don't lift off, you will die. It's all very terrifying.

And in a Lotus T125, no one can hear you scream, partly because your larynx has been crushed and partly because you've just trodden on the accelerator pedal and your head's come off again.

Happily, you cannot take this car on the road. But you are allowed to drive about on Britain's highways and byways in a V8-powered Ariel Atom. This is like being licensed to drive a horse that is propelled by a Saturn V rocket. You accelerate. You hit a tree. Your head comes off.

And it's not alone. A modern Ferrari comes with a telephone connection and iPod connectivity and electric windows. So it's like the perfectly reasonable-looking man at the school gates with a bag of sweets. Apparently harmless. But if you make the mistake of getting inside? Well, it's going to be ugly.

I cannot think of one yard of British tarmac where you could sensibly put your foot down in a modern Ferrari. Not one. Because by the time it's gone through second gear, it's broken even our most relaxed speed limit, and by the time you're through third, your head is in the boot.

A lot of people wonder why *Top Gear* films these really fast cars on an airfield. The reason is simple. On a road, almost all of them are borderline idiotic. And that's why I was so pleased to climb on board the BAC Mono this morning. Because it isn't.

BAC is the world's newest car company. I first heard about it last year and I must confess, I smiled. It had based itself in Cheshire and I thought, I see. So, soon we will be treated to the first car made entirely from onyx. I expected it to have gold fixtures and fittings and a stone dog by the door.

It didn't turn out like that at all. To get inside, you remove the steering wheel and then lower yourself into the single seat until you are completely wedged. All you can move are your feet and your hands. It's like you've been tinned. You then pull on your helmet – it would be silly to drive it without one because you might be hit in the face by a bee – and start it up. It all feels very racy. And a bit scary.

This car was designed to look like an F-22 Raptor and it's festooned with all sorts of imagery and branding from the world of motor sport. The F3-spec gearbox is from Hewland. The brakes are from AP Racing. The pushrod suspension is from Sachs. You fear that if you even go near the loud pedal, you will die, terrified and alone.

Its maker claims it can get from 0 to 60 in 2.8 seconds and onwards to a top speed of a billion. So, with much trepidation, you start it up. There's an explosion of noise behind you and the steering wheel comes alive with readouts that you don't understand. You push the neutral button with your left hand and pull a paddle with your right to engage first. There's an almighty clunk. You have just booked an appointment with your executioner.

You engage the clutch. The car moves. You change into second. There's another enormous clunk. The executioner is on his way. So you think you may as well get it over with and open the taps.

What happens next is odd. You know you are moving very quickly indeed but you feel like it's a speed you can handle. Perhaps that's because you are always aware that while the 2.3-litre engine was made by Cosworth, it is basically the same four-cylinder unit Ford uses in its Galaxy. And there is nothing on God's green earth less scary than a people carrier.

Still, there's a corner looming and you know what happens when you try to do one of those in a car that weighs about the same as a hot-water bottle. It goes straight on. So you brake, and you notice straight away that the Mono doesn't pitch forwards. Then you turn the wheel and it doesn't roll, either. It stays level.

In the next corner, you try a little harder and it's the same story. This gives you the confidence to really push and there are no unpleasant surprises at all. Because the engine and the gearbox and you are all in a line, low down, right down the middle of the car, it handles absolutely beautifully. There's a whiff of understeer to let you know that you're getting near to the limit, but a little more power corrects this and you end up cornering like Fangio, in a controlled four-wheel drift.

And because the speed of the thing feels manageable, you can concentrate on what you're doing rather than not dying. With most cars of this type – the Caterham 7 Superlight and the Atom, for example – you need to know what you're doing or they will kill you. But in the Mono, a complete numpty could manage, no problem at all.

There's more good news, too. It is designed so that it can handle speed bumps. It has lights and indicators and there's even a boot that is big enough for your helmet. It's a road car. Of course, if you use it on the road, where there are other people, you will look a bit foolish. But the fact is you can.

Of course, it's not cheap: £79,950 is a lot for a one-seater car that has no radio, windows, satnav or even carpets. But that said, a similarly specced V8 Atom is £146,699.

Sadly, though, there were a few flies in the ointment. First of all, I experienced a small fire. And then the gearbox broke. And then the engine decided it wouldn't work at all below 4000 rpm. All of this was very bad, but in BAC's defence, this was a proto-type, work-in-progress car. Deliveries don't start for a little while so it still has time to, I dunno, move the carbon trim a little fur-ther away from the hot exhaust tail pipe.

If BAC can get it all working properly, it'll be great. The only really fast car that isn't actually too fast.

Neill Briggs, the engineering director of BAC, said, 'The exhaust trim that started smouldering when Jeremy pushed the prototype car to its limits – and watching Jeremy put a car through its paces is an impressive thing – was temporary. It will be replaced in the final version of the car and we're confident that the minor problems he experienced will be sorted out.'

14 August 2011

I thought it looked humdrum. But wow!

Honda Accord Type S

Don't you think it's strange? You buy a BMW one day and you are told that it is the ultimate driving machine, that it is all about balance and grip and immediacy. Whereas the very next day you are told that exactly the same car is all about joy. It was designed and built to be happy and to make you happy as a result. Welcome to the world of advertising.

Volkswagen's advertising agency told us for years that its cars were very reliable. But then the agency decided that actually you don't buy a VW because it's well made; you buy one because it's cheap. Right? So. Has there been a philosophical sea change at the factory in Wolfsburg? Or has there been a meeting in Soho?

In the olden days engineers would tell advertisers what they had made and advertisers would pass the message on. Now it's the other way round. Advertisers tell us what the engineers were thinking. Even when it's plainly obvious they weren't.

Do you really think for a moment the new BMW 5-series was built with 'joy' in mind? It's German. And in Germany the word for 'joy' will almost certainly be 16 miles long and mean, literally, 'the unusual and unexplained phenomenon that occurs in your inner being when someone of your acquaintance accidentally slips on a banana skin'.

All things considered, the current BMW 5-series is possibly the best car on sale today. It is handsome and well made and spacious and economical and comfortable and fast. It is a brilliant driving machine. But it is about as joyful as a technical lecture on the inner workings of a telephone junction box.

Things on the advertising front are particularly difficult for

Mercedes. It knows that its two-seat convertible models are particularly popular among women. This seems to annoy the
marketeers. So with the SL we had Benicio Del Toro hammering
through the desert, and with the SLK we had a good-looking
chap being chased by what appeared to be the god of thunder.
And neither worked. The cars remained very popular with girls.
I shouldn't be surprised if the next ad showed a docker spitting
and scratching his backside. Before we cut to the pack shot: an
enormous scrotum.

The only 'lifestyle' ads that match the car they're promoting
come from Honda. 'Isn't it nice when things just work?' The
message is simple. We don't do fuss. We don't do flimflam. We
are sensible. And that's what Hondas are. Sensible.

Because they are so sensible, my shoulders sagged quite a bit
when I walked out of the house last Monday morning to find
that a brown Accord with a diesel engine was sitting in the drive.
I had many miles to cover that week and, frankly, I didn't fancy
doing any of them in the motoring equivalent of wholemeal
bread. So I loaded up the boot of my Mercedes and took that
instead.

Sadly, the following Monday, the Accord was still there and I
was overcome with guilt. So, with a heavy heart, I climbed inside,
fired up the motor, pointed its sensible, car-shaped nose at the
capital and pressed the accelerator.

What happened next was alarming. We are conditioned to
expect a certain level of response from a diesel engine. It's the
response you get from a fat man in a vest who's spent the afternoon sitting in a deckchair. Not this diesel engine, though . . .

Honda – the last mainstream car maker to get into diesel
engines – brands this top-of-the-range paraffin stove the Type S,
and that means the 2.2-litre turbocharged motor now develops
178 horsepower. That's 30 more than you get from the standard
car and, boy, oh boy, do you feel it. This car may be brown and
as interesting to look at as the periodic table, but it goes like a
scalded cock.

Of course, you may imagine that by upping the power, Honda has sacrificed fuel economy. And you'd be right. It has: 1.9 mpg of it. But you should still be able to get more than 50 mpg, and that, thanks to a massive fuel tank, means you need visit the filling station only once every 650 miles. Let me just say that again. Once every 650 miles. That, all on its own, is a good enough reason for buying this car. But there's more.

The Type S package means an 'aero' body kit – which I couldn't spot – bigger wheels, low-profile tyres and sports suspension. You would imagine, therefore, that you were in for a bone-shaking ride. But you're not.

At the BBC's underground car park in White City there are speed bumps of such severity that in most cars I weave about through the bays rather than drive over them. Even at 1 mph they hurt. But with the Accord they weren't there. I didn't feel them at all. It was as if I was trying to park a hovercraft.

So, it's fast, economical, comfortable . . . and almost unbelievably well made. Slam the door on a Subaru Legacy – another well-made car – and it makes the sound of a shot pheasant hitting the ground on a frosty morning. Slam the door on the Honda Accord and it makes the sound of a pheasant coming in to land . . . after you've missed it. It's almost silent.

There's a similar sensation of quality on the inside. This is a car that doesn't feel assembled. It feels as though it's been hewn from one solid block of steel. It's a Barbour jacket. It's a Scottish mountain. Push a button in a Honda and it feels as if you could push it a billion times and it would still be working. It's the exact opposite, then, of an iPhone.

Right. Now it's time to talk about the drawbacks. Well, while this car is available as an estate, it is not available with an automatic gearbox. And that's odd. Also, it comes with many electronic features that are understandable to only the sort of person that would switch from vodka to sherry more readily than they'd switch to a Honda Accord.

I suppose I ought to point out as well that while the engine delivers all that you could ask, it is not quite as refined as the diesel engine you get in a BMW. And that's it.

The diesel Accord Type S is well priced, considering the amount of equipment it comes with as standard, it's pretty spacious, it's lovely to drive and – I can't remember if I've mentioned this already – you only need take it to the pumps every 650 miles.

Yes, it's boring to look at, but even that can have its advantages. It'll never be vandalized, and I think I'm right in saying that not once in all of human history has an Accord driver been stopped randomly by the police. That's because they know that anyone who bought a car as sensible as this will have the correct paperwork, no alcohol in their bloodstream and no sub-machine gun in the boot.

If it were available with an automatic gearbox, I'd be tempted to give it five stars. But it isn't, so, reluctantly, I won't.

Instead I'll sum it up by saying that it's nice to find something that just works.

21 August 2011

You vill never handle zis torture

Mercedes-Benz G 350 Bluetec

It emerged recently that the least reliable car you can buy is a Range Rover. An extensive study found that on 02-registered cars, there was a 56 per cent chance of a fault developing within a year. My own findings suggest that brand-new models have a battery issue that could put you on the bus in weeks.

To make matters worse, the company's designers seem to be hell-bent on ruining the quiet, restrained, tasteful looks with more and more chintz. The front end now looks like a branch of Ratners, and soon, you get the impression they will fit fake Roman pillars on either side of the driver's door. I suspect they won't be fully happy, though, until the whole car is made from onyx.

I don't doubt for a moment that it is all very lovely if you live in Alderley Edge, but in the rest of the country, where showing-off is considered poor form, it's all just too vulgar and horrid for words. Small wonder, then, that we are starting to see a re-emergence on the streets of the Mercedes G-wagen.

It's been on sale in Britain before but now it's back in two versions. Both are long wheelbase but one is from AMG and therefore has a supercharged V8, and the other is the one I've been driving for the past week, the G 350 Bluetec diesel.

It is extremely handsome. Restrained. Dignified. And cool in a menacing sort of way. If it were a gun, it would be an AK-47. It is, then, the complete opposite of the modern-day Range Rover, which is like a gangsta's diamond-encrusted Colt. Small wonder that in Notting Hill many media types even stopped pedalling for a moment to give the big beast an appreciative nod.

People like looking at this car. It feels, therefore, worth the £81,700 asking price. Driving it, however, is a rather different story.

You may remember that recently on *Top Gear* I brought news of a half-million-pound E-type Jaguar. Built by a company in East Sussex called Eagle, it was the most beautiful man-made thing I'd ever seen. Better than the Humber Bridge. Better than the Riva Aquarama, even.

However, it was nothing like a modern car to drive. Yes, many of the components were brand new, but you couldn't get away from the fact that the basic architecture came from a time when people would travel miles to gawp at a top-loading washing machine.

Then there was the Jensen Interceptor that I reviewed earlier. The idea was brilliant. You had the beautiful Italian styling from the days of the loon pant and the tie-dye T-shirt, but you got a modern engine, modern brakes and modern suspension. Sadly, you did not get antilock braking or airbags or a satnav system. Or wipers that could wipe the windscreen.

I can see why you would be interested in buying an updated Jensen or an Eagle E-type. They are approximately 18,000 times more interesting than the modern-day equivalents from Jaguar or Aston or Mercedes-Benz. But for every point you score on the kudosometer, you will lose one when you run over a man-hole cover. Or into a tree.

The G-wagen is much the same. It was originally designed for the German army in the 1970s, which means that, underneath, it is made from 1970s technology. This means that on roads you know to be perfectly smooth it will pitch and writhe about like one of those bucking broncos you can now rent for children's parties.

It's amazing. I remember driving a G-wagen in the early Eighties and I thought back then that it was extremely refined and that it rode very well. By today's standards, though, it is absolutely woeful.

And the steering is worse. You need a block and tackle to turn the wheel, and even if, by some miracle, you do manage it, the car will stubbornly refuse to actually go round the corner.

In an attempt to make the interior feel modern, the car is sold as standard with things such as cupholders and climate control and a rear-view mirror that dims automatically when it's being blinded by the car behind. But all of these things have been shoehorned into a cockpit that was designed before electricity was invented.

This is particularly noticeable when you try to operate the command and satnav centre. It is very difficult, because the only place it could be fitted was right down at the bottom of the dash, next to your left ankle.

And even if you could read what the buttons do, there is absolutely no chance of pressing the one you want because as you extend your left arm into the footwell, you will run over another piece of grit and the whole car will leap about as if it's been hit by an RPG.

Then there's the driving position. Because people in the army like to be extremely uncomfortable at all times – this is why all British military equipment comes with as many sharp edges as possible – Mercedes decided that the seat should be mounted only 2 inches away from the steering wheel. You drive this car like you sit at a kitchen table.

And yet you pray the journey will never be over because you know that when it is, you will have to get out and close the door. This is not actually possible unless you have just won a competition to find Britain's strongest man. And even if you have, you will still need the silver and bronze medallists to give you a hand. The tailgate is even worse. To open this, you need a JCB. And there's no point because the boot is nowhere near as big as you might have been expecting.

Yes, the engine is modern, and as a result it produces very little by way of oxides of nitrogen – wow! However, it also produces very little power, and certainly not enough for a car that weighs more than Scotland. The result is a top speed of 108 mph, which is what most automotive experts call 'strolling'.

It's hard, really, to think of any good points at all. I liked the fact that it is a proper off-roader with proper off-roading features. I also liked the television sets in the back, but they were a £1,940 option. I liked the reversing camera, too, but that was an extra £460. In fact, my test car was fitted with so many options, the actual price was £94,200.

I will admit that even though this car is made by hand, it appears to be very well screwed together and, yes, the looks are appealing. But do not imagine for a moment that just because it has many modern features and is still being made today that it is a modern car. Really. It's an Austin Seven in a fat suit.

I would, therefore, still choose a Range Rover instead. Yes, it is more likely to drain its battery of juice when you've left it for two minutes outside the newsagent's. And, yes, the new front end appeals only to jackdaws. But at least it can run over a pothole without breaking your back, you can open and close the doors without using heavy lifting gear, there's space for a human being behind the wheel and it's capable of getting from 0 to 60 before you do.

28 August 2011

Strip out all the tricks and it's still a wizard

Audi A6 SE 3.0 TDI

I spent a day last week recording the voice for a new satellite navigation system. This meant sitting in a darkened room saying, 'In 200 metres, turn right. In 200 yards, turn right. In 300 metres, turn right. In 300 yards, turn right. In 400 metres, turn right. In 400 yards, turn right . . .' It wasn't as interesting as it sounds.

It also felt slightly ludicrous, like I was the kettle on the bridge of a nuclear-powered Nimitz-class aircraft carrier: an old-fashioned ingredient in a world that's not old-fashioned at all.

Have you ever stopped and wondered how the satnav system in your car works? It's astonishing. There are twenty-four American military Navstar satellites in space, around 12,500 miles from Earth. At any point on the planet's surface, there is a direct line of sight to at least four of them.

But they're not standing still. They're moving. Which means they are not fixed points as such. So, the little receiver in your poxy Volkswagen has to find them, and they're only the size of wheelie bins, then work out precisely how far away they might be at any given moment. We're talking major algebra here.

And bear in mind that the device must work out how long it takes for the signal to reach an object in space that is moving at several thousand miles an hour. That means a clock that can keep up with the speed of light. Get it wrong by a thousandth of a millionth of a second and your VW will think it's just outside Kiev.

And then, when it has worked out where you are on the surface of Earth, it must compare the information with an onboard road map. And it still isn't finished because you've just asked it to get you from where you are now to a postcode just outside Pontefract. This means it must analyse the 246,000 miles of tarmac in Britain and work out the fastest route. And if it takes more than five seconds, it knows you will be sitting there saying, 'Oh, for God's sake. Come on. You useless piece of junk.'

In the early days of satellite guidance, mistakes were common. The first time I ever used such a system, it tried to direct me through Leicester Square, which had been pedestrianized by the Iceni. And only recently, the systems fitted in BMWs absolutely refused to acknowledge the existence of the M40.

I spent a lot of time thinking, Crikey. This whole thing was designed so the Americans could post a cruise missile through a letter box 7,000 miles away and it can't even find a sensible route from Beaconsfield to London. Now, though, I have to say, mistakes are extremely rare.

Which is why I'm always surprised when an old lady driver tells hospital staff the reason she drove her car off a cliff, or through a river, or into a cave full of wolves, is because the satnav system in her car told her to.

You hear these stories all the time. People who turn left at a level crossing, straight into the path of the four fifty from Paddington, or left at a crossroads that isn't there, and into the saloon bar of the White Horse in Tiverton. I've always assumed that people like this must be stupid. However . . .

Earlier in the summer, while filming in the south of France, I set off in a large convoy of camera cars and crew vans to a pre-determined location. We were being led by a man whom we shall call Rod. Which is a bit annoying for him because I've just remembered that is his actual name.

Anyway, Rod had programmed his portable satnav to where we were going and off he set. Alarm bells began to ring in my car when we turned onto a very small country lane. And they became very loud indeed when the lane became a track. And then it stopped.

Rod was absolutely perplexed. His satnav system was saying

we were just 500 metres from our destination, which may well have been true, but the only way we could have got there on the route it had in mind was if we'd turned ourselves into goats. Many people blamed Rod for this. Me? I blamed the French.

Satnav was fitted to the all-new Audi A6 I was driving last week. But it could do something other than find wheelie bins 12,500 miles away and get you to Pontefract. It could also operate your headlights, altering the shape of the main beam, depending on whether you were on a country or urban road or a motorway, and even switch everything on at junctions so other road users could see . . . that you've apparently gone mad.

And this is just the flake on the tip of the iceberg. Because there is also a device that can spot bikes and suchlike in the blind spots, and another that flashes up a warning message on the windscreen via a head-up display if it thinks you are travelling too close to the car in front. Then you have night vision, which puts a *Blair Witch Project* image of the road ahead on a screen in the dash.

You can even drive this car when you are fast asleep. Citroën was the first car manufacturer in Europe to introduce lane assist, a device that buzzes if it thinks you're drifting out of lane on the motorway. Audi, though, has gone one better. Providing you are travelling at more than 40mph, its system will actually steer you back in line. And if you have the active cruise control switched on, it will even brake on your behalf if there's an obstacle ahead. All that's missing is an alarm clock to wake you up when you arrive at your destination.

Wi-fi? Well, as we know, this doesn't work in a house if the walls are more than 2mm thick, but somehow, Audi has made it work in a car. Which means that actually you could drive down the motorway, catching up on your emails, safe in the knowledge that the steering, braking and navigation are all being taken care of by electronics.

Of course, you might think that this veneer of mostly optional electro-trickery has been fitted to mask the shortcomings of a fairly dreary car. But no. It's lighter than the old A6 and, even though it's shorter, it's more spacious inside. It is also extremely well made and finished beautifully. It is a wonderfully nice place to sit, and thanks to absolutely fantastic seats, comfortable too.

Until you set off. Yes, you can adjust the way the gearbox, the throttle and the suspension behave, but the simple fact is that no matter what settings you select, this new car does not ride quite as well as the BMW 5-series. It doesn't handle as well, either. And the entry-level 2-litre turbodiesel engine is not quite as refined as the unit that BMW uses. But that said, it should be capable of averaging 57 mpg, which is remarkable.

So. Yes or no? Well, I much prefer it to the overstyled Mercedes E class, which was designed mostly to take Carol Vorderman to the airport. And I think for a number of reasons it is better than Jag's XF, but what about the 5-series?

Tricky one. Taken in their base forms, there's no way to split them, really. The Audi matches the 5-series for economy and the Beemer is slightly nicer to drive. They really are Manchester City and Manchester United.

For sure, if you fit a few options to the Audi, you will have something that is mind-bogglingly good. But if you're not careful you could end up spending more than BMW charges for the bigger-engined 530d. And that's better than mind-bogglingly good. All things considered, that's probably the best car in the world right now.

4 September 2011

Open up them pearly gates . . .

Lamborghini Gallardo LP570-4 Spyder Performante

A man died recently. He was called Lieutenant-Commander Peter Twiss and he began his career, quietly, as a tea taster for Brooke Bond. Then, in 1939, he joined the Fleet Air Arm where, as a carrier pilot, he had much success in the Mediterranean, earning a Distinguished Service Cross and bar.

In 1943 he came back to Britain and switched to the twin-engined Mosquito, in which he spent a great deal of time bombing France and generally shooting down German Junkers 88s. Before the war ended, he left for America where he became Tom Cruise, testing naval fighters.

After the war ended he became a test pilot with Fairey Aviation and on 6 October 1954, he took the experimental Fairey Delta 2 on its maiden flight. I had a model kit of one of these as a child and thought it to be the most beautiful thing in the world. For Twiss, though, it was more than just that.

He became convinced that the delicate little jet with its Concorde-style drooping nose could fly faster than anything else in the skies at that time. And so, five months after a US air force Super Sabre had set a new record of 822 mph, Twiss climbed aboard his beloved FD 2 and headed for a course that had been laid out along the south coast near Chichester.

Flying at 38,000 foot, he did indeed break the record. Although actually what he did was smash it. Because the average speed of his two runs was 1,132 mph. Twiss, then, had become the first man ever to beak the 1,000-mph barrier. And for his services he was awarded the OBE and the threat of several lawsuits from market gardeners in Sussex who claimed his sonic boom had smashed all the glass in their greenhouses.

You might imagine that after he retired from test flying, he'd put his feet up and do a spot of gardening. But no. After 4,500 hours in the big blue, in 148 types of aircraft, he appeared at the helm of a Fairey Marine speedboat in the Bond movie *From Russia With Love* and features in *Sink the Bismarck!* at the controls of a Fairey Swordfish torpedo plane. And he developed cruisers. And flew around in gliders and was married five times.

This is what I call a proper life. A full life. The life of a man who could lay on his deathbed and think, Good. I didn't waste too much time in the ninety years I was awarded.

This brings me neatly on to a track called 'Time' from Pink Floyd's *The Dark Side of the Moon*, which is about how we waste and fritter our days away, waiting for someone to show us the way.

This is how most people live. Life is long and there is time to kill today. But, actually, there isn't. Life is desperately short and no matter how much you do, there's always a twang of regret that you didn't do more. No matter how much you see, you always think that if only you'd gone round one more corner and over one more horizon, you'd have seen something else. I bet when Tom Jones is summoned by the grim reaper, he'll think of a girl that he could have slept with but didn't, and that will make him a bit sad. I bet that even Peter Twiss spent at least some of his life wishing he could have shot down one more German bomber and gone 1 mph faster in that Fairey Delta.

I'm in the same boat. God knows, I've travelled over the years, but instead of reflecting on all that I've seen and all that I've done, I will go to my grave thinking, Shit. I never went to Pontefract.

This is why, if you have the wherewithal, it's very important that you go out tomorrow morning and buy a supercar.

I've had my share of them over the years and they are all stupid. Impractical, ruinously expensive, difficult to park and they leave you with dirty fingers every time you open the bonnet to retrieve your (very small) suitcase.

However, that said, supercars are to cars what jet fighters are to the Airbus that took you to Corfu this year. They are built to excite the speed gene that lives in us all; they are designed to release the cocktail of chemicals that is titillated when we are small and we are pushed higher and higher on the swings.

What's more, supercars are the last great division of the global motor industry where engineers and stylists can try out new ideas and new ways of thinking. When you build a car that can travel at two hundred and something miles an hour, you need to make the brakes from exotic materials and think carefully about what effect the air has at those sort of speeds. Anyone can make a saloon. It takes a genius to make a supercar.

At present, there are two standout examples of the breed. The McLaren MP4-12C, which is science and maths, and the Ferrari 458, which is also science and maths. With a bit of Renaissance art thrown in for good measure. Both use brute force to give you the go, the periodic table to give you the stopping power you need and Palo Alto electronics to give you a level of grip in the corners that beggars belief.

And yet, if I were in the market for such a car, I'd buy neither. I'd buy the Lamborghini Gallardo LP570-4 Spyder Performante.

Let me talk you through the name. LP says that the V10 engine is longitudinally positioned in the car; 570 is the metric horsepower that Lamborghini claims it delivers – it's equivalent to 562 bhp. The -4 signifies that it has four-wheel drive. Spyder tells us that it's a convertible, and Performante that it's the performance version of something that's pretty damn fast in the first place.

The extra oomph comes mainly from a raft of weight-saving measures. Carbon fibre, for instance, is used to make the huge engine cover, the door panels, the seats and even the bits that shroud the door mirrors. It sounds, then, like this is another example of science and maths. But it isn't.

Ferrari and McLaren, first of all, are racing teams and Lamborghini isn't. Lamborghini therefore feels no need to give its customers a taste of Formula One, a taste of all that behind-the-scenes trickery. Lambos are designed mainly to make a lot of noise and cause small boys to clutch at their private parts in excitement.

So, while the Ferrari howls and a McLaren hums, the Lambo bellows. And while the racers were styled by aerodynamicists, the Lambo was designed to make people say, 'Wow!' Which it does.

What's more, with most serious supercars, you would never buy a convertible version, because you'd know it wasn't quite as good, dynamically, as the stiffer, more rigid hard top. But since you don't buy a Lambo for the last 0.01 of a g it can generate in the bends, who cares? Best to have no roof, really. That way you can hear the engine more clearly more of the time.

And anyway, it's not like the Performante dawdles. The acceleration is savage, the braking is fierce enough to tear off your face and, unlike most four-wheel drive cars, it does not resort to chronic understeer when you exceed the limit. Plant your foot into the carpet mid-bend and it's the tail that lets go in an almost cartoonish fog of tyre smoke and noise. In a Ferrari or a McLaren, you concentrate when you are driving quickly. In the Gallardo, you can't. You're too busy laughing.

Oh, and there's one more important point. Ferrari recently started to offer a seven-year warranty, which suggests that it has great faith in the quality. But in the past few years, since Audi took over the factory, I've never experienced any mechanical malfunction at all in a Lambo.

Go on. Buy one. You may think it's a stupid idea now, but trust me on this. On your deathbed, you'll remember a drive you took in it. And you'll go through the Pearly Gates smiling.

11 September 2011

Oh, grunting frump, you looked so fine on the catwalk

Jaguar XF 2.2 Diesel Premium Luxury

Back in 1995, Ford toured the world, showing off an exciting new concept car – a small two-seat roadster that was made from carbon fibre. Yum, yum, we all thought. We shall be very interested in buying that should the bigwigs in Detroit decide to put it into production.

Sadly, though, by the time it reached the showrooms, it had sprouted a roof, a hatchback and acres of pleblon upholstery. Furthermore, it was made from steel instead of carbon fibre and it looked like a teapot. It was called the Ka.

Ten years later Ford did it again, showing us a fantastic-looking concept called Iosis. It said at the time that the next version of the Mondeo would look very similar. And it did, except for every single detail.

This is the trouble with concept cars. They do not have to adhere to pesky EU rules about how high the headlamps must be from the ground and how much of the tyres' width must be covered by bodywork. They don't have to be crash-tested, and neither does every single piece have to pass through the company's accounts department. They are freestyle cars. Flights of fancy.

They don't even have to work. Many years ago Peugeot turned up at the British motor show with a concept car that looked like a cross between an America's Cup catamaran, the glider Pierce Brosnan used in the remake of *The Thomas Crown Affair* and a sex toy. However, on the downside, it didn't have an engine. It didn't even have a space where an engine could go.

In recent times concept cars have started to look a bit more

like the cars you and I do buy. But even so, all of the little details –
the fat tyres and the funky lighting and the weird door
handles – are still rejected by the bean counters for being too
expensive, or by the production line manager for being too complex to fit. This means the car that finally makes it to the
showroom never looks quite as good as the car that appeared
under a sea of girly flesh at a motor show. Concept cars, then,
are the font of disappointment.

By far the worst offender in this is Jaguar. Almost without
exception, every one of its new cars in recent times has been a
shoulder-sagging visual let-down because, just before it was
unveiled, the company had produced a concept to show how
brilliant it could have looked if only there were no rules. In
short, Jag's designers have spent the past twenty years writing
cheques that the rest of the company cannot cash.

However, a couple of weeks ago Jaguar unveiled a concept
car called the C-X16, and if you examine it very carefully you
will see that there are no details that are obviously impossible to
mass-produce. Maybe the sideways-opening rear window will
have to go because of some obscure bit of legislation from
Brussels, but other than that, it looks real. It looks possible. And,
more than that, it looks absolutely sensational.

It is quite similar in appearance to both the Jaguar XK and the
Aston Martin V8, which is perhaps unsurprising since all three
were styled by the same man. But it's smaller than both of those,
and cheaper, too. They're talking about a price tag in the region
of £55,000. For that, you would get a supercharged V6 engine,
which would then be boosted further by a Formula One-style
KERS, or kinetic energy recovery system. Engage this by pushing a little button on the steering wheel and the 375 horsepower
coming at you from the petrol engine would be increased
momentarily by 94 more from an electric motor. Will that be a
showroom feature? Who knows? Price Waterhouse Coopers,
probably.

I'll be honest. I'm very excited about this car and especially the convertible version that's bound to follow. There's just one request, and I'm directing this at Jag's chassis people, who have been a bit hardcore of late. While it is very important to keep the oversteer-crazed helmsmen at *Autocar* happy, can I please remind you that most of the people who'll want to buy this car will be middle-aged with bad backs? They will want, therefore, a decent ride. This has to be your priority.

Anyway, that's then, this is now and we have a new Jaguar XF to think about. Recently, when reviewing the new Audi A6, I said the Jag was not as good for a number of reasons. And then, in a shoddy piece of journalism, didn't go on to say what they were. Truth is, I couldn't remember. It's just that the XF is a bit like Cheryl Cole. I recognize that she blows up many frocks, but I don't see what the fuss is about, frankly.

Now, however, there's a revamped model. It has a restyled bonnet and tweaked front end and new gills in the front wings. It looks fine, but outside a red carpet event it doesn't look quite as fine as the BMW 5-series or the Audi A6. Somehow they look more modern and more expensive.

It's the same story on the inside. I like the minimalism Jaguar's designers tried to achieve, but it would have been better if they'd succeeded. I'm loath to say this, but it all looks a bit cheap. The headlamp switch, for instance, is on the indicator stalk. There's only one reason to put it there: to save money. That's why Mercedes and BMW don't. Because they know we know.

Still, the most important new feature in this car is the engine. It's a 2.2-litre four-cylinder diesel, and this is the first time Jag has ever used a four-pot paraffin stove in any of its cars. I was expecting great things because other diesel engines in the Jag and Land Rover range are excellent.

Unfortunately, this is not a Jag engine. In essence, it's the same unit Ford, Citroën and Peugeot use and I'm afraid it's not very good. It's not refined and it's not as economical as the engine BMW fits. What's more, these days the government – idiotically – taxes you according to the composition of gases coming out of your tailpipe. And the fact is that the Jag's engine produces way more CO_2 than BMW's equivalent.

There's more. Every few thousandths of a second, the computer that runs the engine in a modern car takes stock of the prevailing conditions. It checks the temperature, the position of the driver's foot, the gear he's selected and the barometric pressure, and it compares its findings with a programmed map so that it knows precisely how much fuel to squirt into the cylinder to provide the perfect balance between power and economy.

To try to get the emissions down, Jaguar has very obviously fitted a map that demands the absolute barest minimum of fuel to keep the engine alive. As a result, around town it feels constantly on the verge of stalling. You can get round the problem by switching the eight-speed gearbox (why?) to 'sport' mode, but that rather negates the reason you bought a diesel in the first place.

All in all, then, this car is not Jag's finest hour. At the risk of sounding like a stuck record, the BMW 5-series is better in almost every way.

However, it has at least given me an idea. What if car companies started making concept cars that were uglier and less exciting than the actual cars they spawned? That way we'd always view a new car in the showroom with delight, rather than a tinge of disappointment.

18 September 2011

Now we're flying

Mercedes-Benz SLS Roadster

With a deserved reputation for being a bit hopeless with my hands, I approached the job of building a dam very seriously. So I climbed onto my new six-wheel-drive ex-army Alvis Supacat and spent the morning driving around the farm looking for suitable stones.

When my hands and lungs were bleeding from the effort of loading them into the 'boot', I headed off to my newly dredged pond to start work. Using skills I'd learnt from watching documentaries on the building of the Hoover Dam in America, I started by erecting a temporary blockage using bits of old skirting board I'd found in a skip. This didn't work.

So I gave up with the idea of a temporary dam and plunged straight in with the real one. And after about an hour, I realized I wasn't making much progress at all. Even though I had many stones and some of them were quite large, they didn't fit together very well, or they sank into the ooze. Either way, the water carried on flowing, oblivious to my efforts.

This caused me to break out a spade. And what a stupid, terrible, ungainly thing this is. You plunge it into the ooze, and everything you pick up simply falls back into the water again. Pretty soon I was sweating like an Egyptian boilerman and my back muscles felt like they might actually be on fire. And still the water kept on coming.

However, all the while, I'd had my eye on a mini-digger that the pond dredgers had left behind. 'You can use it, if you want,' one of them had said as he left for the weekend. And I'd nodded and said, 'Sure,' not being prepared to admit to another man, especially not a son of the soil, that I had absolutely no idea how it worked.

Men are supposed to understand diggers like we understand testicles. But I don't. The last time I'd used one, I accidentally pulled a seal's head off, and damn nearly killed my son. And that was a tiny little thing. The one sitting by my pond, winking at me, was much larger and had many more levers, all of which were incomprehensible. But I decided to give it a try anyway.

It took a while to realize that the big scoopy thing on the front wouldn't move unless both the big red levers were pushed forwards, and that if I wanted the digger itself to move forwards, I had to pull the levers in front of me backwards. But pretty soon I had it by the water's edge, scooping up silt like my actual name was Seamus O'Gallagher. And ten minutes after that, the dam was built. I was very proud.

I was so brimming with confidence that after I'd celebrated with a crusty sandwich, some pickle and a cold beer, I started up the dredger man's dumper truck and spent a little while using it to knock down various dead trees. Then I cleared a spring that was jammed with flotsam from the woods. Then I made a waterfall. I was in heaven. A man and his machines at one with nature. I was an artist and internal combustion was my brush. Later today I'm going to buy a chainsaw.

It really is hard to think of any machine that provides more enjoyment than the tools of landscape architecture. Diggers. Bulldozers. Dumper trucks. My beloved Supacat. There's an effortlessness to the hydraulics, which means that at tickover there's enough power to shift in a moment what took God himself 450 million years to create. When you are in a JCB, you really are a master of the universe.

You have much the same feeling in the new Mercedes SLS roadster, which comes with a bonnet that seems to cover a slightly larger area than Wyoming. That's one of the many, many things I love about this car. Because you sit right at the back, in the boot almost, you have a sense that you could have a huge accident and simply not know about it for a week or two. Imperious: that's how you feel.

As I have said on many occasions, the coupé version of this car is one of my all-time favourites. But it does have a drawback: its gull-wing doors. Yes, they look very interesting at a motor show, and yes, they hark back to the original Mercedes 300, but when you are in Wolverhampton, and people are looking, you feel like a complete knob every time you get into and out of your car. The only reason doors like this are fitted is for showing off. And in this country, show-offs are held in dim regard.

There's another problem, too. If you roll an SLS and end up on the roof, with petrol sloshing about, life is very tricky because, of course, you can't open the doors to get out. To get round this issue, the hinges are fitted with explosive bolts that fire when the car is upside down. That means a) you are driving around with a bomb right next to your head and b) the car is heavier than necessary.

Happily, the roadster – for obvious reasons – has conventional doors. This means no embarrassment on the high street, no bombs and no excess weight. Yes, the car is heavier than the coupé, thanks to specially strengthened sills. But because the doors are lighter, the overall increase is negligible. Which means the performance is not affected in any way you'd notice.

It'd be a good test of Wayne Rooney's new barnet, the roadster, because my God it's fast. And with the roof down it feels almost ridiculous. You put your foot down, the double-clutch gearbox responds in an instant and you are catapulted towards the horizon in what feels like a category 70 hurricane.

Then there's the AMG soundtrack. This is a car that assaults all of your senses. It batters you. Think of it this way. The coupé is a Gulfstream biz jet. Sleek. Fast. Comfortable (ish). The roadster is a Gulfstream biz jet, too. Only with no roof. Imagine that and you have an idea of what it's like to drive this car down the A361.

But it's not all about straight-line speed. By mounting the gearbox at the back, and using a carbon-fibre prop shaft that weighs about the same as a mouse, the engineers have made sure the weight is distributed perfectly. You feel that when you hustle it and it surprises you: because you can't really understand how something with a bonnet that vast can feel so small and agile. And you especially can't understand when your ears are bleeding and your hair is in the slipstream three counties back.

It's not a Ferrari. It won't grip like that. It isn't supposed to. It doesn't grip at all, if I'm honest. Turn the traction control off and it will spend most of its time sideways. But it all feels easy. It's not a sports car, and it's not a GT car. The ride's too hard for that. It's a muscle car, really.

Bad things? Well, there is a hint of scuttle shake – you can feel it through the wheel – the lumbar support controls look as if they were bought from a motorist discount store in the 1970s and while the gearbox may work well at speed, it's dim-witted and a bit jerky around town. Minor things, really.

Good things? Everything else. The roof is fast, you get all the usual Mercedes trimmings and the price is about £30,000 less than I was expecting.

I would quite understand why you'd buy a Ferrari 458 instead. It's a better driver's car. And I'd be stumped if someone asked why the SLS costs twice as much as the ostensibly similar Jaguar XKR. Really, it's a question of each to his own. Some people like to play golf. Some like to build dams. Some prefer the delicacy of an Italian supercar. Some, the charm of a Jag. But if you 'get' the SLS, nothing else will do.

25 September 2011

The topless tease luring men to ridicule

VW Golf Cabriolet GT

You may dream of driving a convertible car through the mountains of southern France on a beautiful summer's day. But, having done this sort of thing on many occasions, I'm able to tell you that you will arrive at your destination with a comically red nose and a shirt that appears to have spent the past few months at the bottom of a stagnant pond.

As a general rule, you should not drive a convertible with the roof down when it is more than 70 degrees. But here's the problem. If you try to drive a convertible when it's less than 70 degrees, pretty soon you will notice that your fingers have frostnip, that your nose has fallen off and that you are in the advanced stages of acute hypothermia.

It's not just the temperature that causes problems, either. It's also the wind. Wind is the most debilitating of all climatic phenomena. I have experienced extreme cold at the North Pole, and, while it's extremely unpleasant, the human frame can cope. It's the same story with extreme heat. I once interviewed a man, in December, in the Australian outback, and at no point did I feel I was suffering from the onset of madness. I've also been drenched without problem. But wind is different. In one hour at our holiday cottage on a peninsula on the Isle of Man in March, I became a drooling vegetable, incapable of rational thought.

Wind is like a barking dog, or an early-morning Italian strimmer. It's inconsistent. There's no regularity or predictability. The noise goes away but never for very long, and you never know when it will be back. I don't know why the CIA uses waterboarding to extract information from suspects; just put them on a boat in the Irish Sea for ten minutes and they'll tell you anything you want to know.

Wind also messes up a girl's hair. This is a fact. Every girl I know loves the idea of driving with the roof down, but after just a few moments every single one wants to put it up again.

Even if you are impervious to heat, cold and the constant pounding from Uncle Hurricane, you still have the problem of embarrassment. This is very real. You may think, as you cruise about in your convertible, that you look good. But unless you are Angelina Jolie or Pierce Brosnan, which you are not, I can assure you that actually you look a tool.

What message are you giving out? That you are carefree? That you are young at heart? That you are available? But you aren't. You're middle aged and a bit pathetic. Driving with the roof down when you are paunchy and balding is like having a sign on your desk that says, 'You don't have to be mad to work here . . .' There is a rule we devised on *Top Gear*. Once you are past the age of twenty-seven, you can drive alfresco only when it is safe to drive naked. In other words, when no one is looking. Because if people are looking, they will laugh at you.

To sum up, then, driving a convertible is uncomfortable and will cause other road users to think that you are a prat with manhood issues. And yet, despite this, we buy more convertibles in Britain than any other country in Europe. And we are certainly the only country where you will find people going to work dressed as Scott of the Antarctic simply so they can get the roof down. We are all mad.

Perhaps this is why Volkswagen is now offering two versions of the Golf convertible. There's the Eos, which is a Golf with a folding metal roof, and the Golf, which is a Golf with a folding canvas roof. I've always rather liked the Eos, but I will concede that it is quite expensive. The new Golf is not: £25,000 for the GT version I drove is really not bad at all. And the diesels cost considerably less.

The roof on the new car is a triumph. Even though it's the

size of a family tent from Millets, it can be raised or lowered in less than ten seconds. And, what's more, even if the lights turn green before the manoeuvre is completed, you needn't worry. Because it all still works provided you keep your speed below a precise and Germanic 18 mph.

When it's up, the refinement is genuinely astonishing.

You can drive at motorway speeds and there is absolutely no sense at all that you're separated from the slipstream by only a pac-a-mac. And things are similarly amazing when you fold it away. Gone is the hysterical, hair-whipping madness you endured in the original Golf cabrio. Provided you use a sturdy hairspray, you have no worries at all.

Another thing that's gone from the original is the rather awkward-looking roll-over hoop. That car was built by Karmann, but you would have been forgiven for thinking that actually it had been constructed by Silver Cross. The new car, built by VW itself, also has a roll-over hoop, but it's hidden away and will burst forth only if it thinks you're about to turn the car over.

Sadly, however, there's something else that has gone missing: any sense of joie de vivre. Those early Golf drop-tops were joyously raucous and wonderful, with their peppy 1588 cc injected engines. The car I drove may have been badged as a GT and it may have developed 158 horsepower – nearly 50 more than the original – but it's not what you'd call peppy.

It's an interesting engine. Although it's small – just 1.4 litres – it is fitted with a supercharger, which you don't notice, and a turbocharger, which you do. On a motorway, if you caress the throttle pedal, you feel it working, girding its loins, preparing extra boost for when you stop the foreplay and dive right on in.

However, when you do dive in, it pulls up its knickers and won't deliver. I suppose the fact is that the Golf cabrio is a heavy car and a 1.4-litre engine simply can't deliver enough oomph to make you nod appreciatively.

That said, it's refined and economical, and the moderate

power sort of suits the comfy ride. This is a car for cruising rather than blasting about.

The interior? Well, although the boot's quite small, the car's fairly spacious. However, like all Golfs, it's a bit like the inside of Eeyore's head. Gloomy and dark. You even have to pay £130 extra for what VW calls the 'luxury pack'. Although, as far as I can see, all this does is give you fancy door mirrors.

There is one stupid thing. The radio tells you what song you are listening to at any given moment, which is a nice touch, but it relays the message by scrolling huge flashing letters across the screen. It's as if you're sitting in Times Square.

It's actually quite easy to sum this car up, though. In the olden days the Volkswagen Golf was a fun, lively little thing, and the original cabrio was an extension of that. Today the Golf is a byword for common sense and dour practicality. And the new soft-top is a reflection of that.

I can't see the point, frankly. If you want a convertible, you want something that's a bit daft. And this new car . . . just isn't.

1 October 2011

I'm sold, Mrs Beckham – I want your baby

Range Rover Evoque Prestige SD4 auto

Douglas Adams said the answer was forty-two. He was wrong, though. It doesn't matter what question you are posing; the answer is always a diesel-powered Range Rover Vogue SE.

What's the best car for taking the children to school? What's the best car for a day's shooting? What's the best car for a drive to Scotland? What's the best car for a quiet drive home after work? What's the best car for crossing Africa? What looks best in a field? Or in Knightsbridge? Range Rover. Range Rover. Range Rover.

This week I drove the new and completely insane Mercedes C 63 Black Series. It is a car designed and built specifically to eat its own tyres. One set lasted just twenty-five minutes. I absolutely loved the madness of the thing. It's a hoot. But for life in the real world? No. I'd rather have a Range Rover. I'm not alone, either. Just recently I was at the home of a leading light of what the *Daily Mail* calls the Chipping Norton set. Fourteen couples were present, and every single one of them had turned up in a Range Rover.

We have a little secret on *Top Gear*. Well, Hammond and I do. We know that, no matter what car we review, it's not as good as a diesel Range Rover. We daren't ever say this out loud, though, because it would render the whole show pointless.

Hammond has about 700 cars, several thousand motorcycles and a helicopter. And you may think he spends many hours in the day wondering what to drive next. He doesn't, though. He always uses his Range Rover. I always use mine. Because whatever we're doing, it's the answer.

That said, Land Rover has been trying to spoil it with a chintzy Wilmslowfication programme of adding completely unnecessary bling. The company has it in its mind that a Range Rover does not need to look good everywhere. Only behind the electrically operated golden gates of a brick and plastic many-pillared mansion on the Prestbury Road. And there's an issue with the battery on new models, too. They go flat for no reason.

There have been other mistakes as well, chief among which is the Range Rover Sport. It's no such thing. Underneath, it's a Land Rover Discovery, which means it weighs more than Dorset. So it's not a Range Rover. And it's not sporty, either. And it doesn't have a split, folding tailgate, which means there's nowhere to sit at a point-to-point. It's a silly car.

But not half as silly as the Evoque seemed to be when I first heard of it. It would be a Ford Mondeo chassis on stilts, with a four-cylinder engine and an interior designed by Victoria Beckham. It sounded as if Land Rover had taken leave of its senses. This tanning salon with windscreen wipers would ruin the whole brand.

I was wrong, though, because the Evoque is brilliant. It's one of those cars in which I had to spend hours trying to find something – anything – about which I could complain. And all I could come up with is the dip switch, which requires a bit more effort to operate than is strictly necessary.

Some say that the plastics you can't see are a bit flimsy. But who cares about that? The plastics you can see look great and are mostly covered in nicely stitched leather. The interior is fabulous and, in the five-door model I tried, spacious, too.

Oh, and some have been saying it's too expensive. But if that were true, Land Rover wouldn't have already taken 30,000 orders. You mark my words on this. Soon you will not be able to move for Evoques. It will overtake hydrogen to become the most abundant element in the universe.

The car I tested had a 2.2-litre diesel engine, which was

smooth, returned 44.1 mpg and provided enough oomph to get from 0 to 60 faster than an original Golf GTI.

The handling was good, too; and the ride. But only in Normal mode. If you engage the Sport setting, the whole car starts immediately to pogo. This, however, is the mode you should select, because when you push the button the dials change from a silvery blue to a vivid scarlet.

You can also change the cabin lighting from a cool vodka-bar blue to a burlesque red. And this is just the tip of a techno iceberg, which must mean there is more wiring behind the scenes in this car than in an Airbus A380. Take the television for example. It sits in the middle of the dash and it's capable of showing two things at the same time.

This means my wife was able to watch the French beat the Welsh while I looked at the satnav. How's that possible? Even James May is stumped.

Or I could choose to look at the live feed coming from one of the five cameras mounted on the outside of the car. Or the trip computer. The central command unit in an Evoque is the best in the world. You spend so much time playing with it that journeys pass in a flash. And, because you are rarely looking where you're going, a bang and a wallop as well.

Of course, Land Rover wasn't content to put this excellent car on sale and revel in the plaudits and the profit. So a big cheese said at its launch that off-road capability wasn't really important any more. It's a silly thing to say when you are running Land Rover. And doubly silly because it's so obviously not true.

That's why the Evoque rides farther from the ground than even the Freelander, with which it shares many components. It's why the angles of attack are so good, allowing you to climb and depart from steep inclines without biffing the front and the rear. And the Evoque is fitted with the same off-road electronics program as you find in the big Range Rover.

What we have here, then, is a proper Range Rover that is also an Audi TT, a hot hatch, an off-roader and a branch of Dixons all rolled into one tiny, easy-to-park package. If I had a job sell-ing BMW X3s or Ford Kugas or any other high-riding semi-off-road car, I'd be on the lavatory, whimpering. Because anyone who wants such a car and doesn't choose the Evoque is so mad, they will have had their driving licence taken away.

Actually, it's like an iPad. The truth is that if you have a smart-phone and a laptop – which you do – you don't need one. But I bet that didn't stop you splashing out, did it?

I have the same problem with the Evoque. I have a seven-seat Volvo and a big Range Rover and a fast Mercedes. I have abso-lutely no need in my life for an Evoque, but I want one. And you will, too. Especially when I tell you that there's talk of a hot ver-sion with the engine from the Ford Focus RS.

I may have to invent a new star rating for that. Because this morning's plain Jane diesel – despite the wonky dip switch – is an easy five-star car. It may even be more than that. It may be the new forty-two.

30 October 2011

I say, chaps, who needs a fourth wheel?

Morgan Three Wheeler

Almost no one wakes up in the morning and thinks, I know. Today, I shall start a car company. And those who do make this curious lifestyle choice never decide to make a small hatchback or a solar-powered trike that could be used in the emerging world. No. They always, always, always think, I shall make a supercar.

Usually, this is foolish. Oh, you may have a mate who is a dab hand with glass fibre and you may have a considerate bank manager who did a bit of racing in his day and likes the idea of your quad-turbo, multi-supercharged 300 mph road rocket. But what you are actually starting is a corner shop. And I'm sorry but Ferrari and Lamborghini are the supermarkets. And, as a result, their carrots are going to be more orange and cheaper than yours. Which means that pretty soon you will get a letter from your previously supportive bank manager that begins thus, 'I am disappointed to note . . .'

I look at the efforts from Noble and Koenigsegg and Zenvo and Spyker and Saleen and I'm afraid I can't help thinking that these cars, while interesting and commendable, are ultimately a shoreline on which some poor blighter's hopes will one day be dashed.

You go to the Geneva motor show and every year there's some poor chap in a bad suit, sitting in the unlit lowlands of the hall, desperately hoping that someone will notice the terrible car into which he's ploughed his life savings. And you always think, Why?

The Ferrari 458 is a stunning, bewildering, brilliant, intoxicating blend of power, finesse, poise, technology, styling, rage, speed and g. It was created by some of the most extraordinary minds in the automotive world in one of the most advanced factories. And forgive me, but you aren't going to be able to make something better in a shed at the bottom of your garden.

Which brings us neatly on to Morgan. Unlike any other small car company, it does not try to beat the big boys. It simply makes stuff that you can't get anywhere else. Sound business, if you ask me.

What Morgan makes is a range of cars for people who still believe it's 1938. People who use the word 'bally'. Enthusiasts of the side parting. Fans of sheepdog trials who like to get under the 'old girl' at weekends to do a bit of burnishing. Not me, in other words. In recent years there have been attempts to bring the company to a point where the second world war has actually begun, with cars such as its Aero. But this is dangerous because when you lose that traditional Morgan 'look', you're going to alienate your customer base. 'Pah. The old girl looks like a bally Nissan,' is what they'd say.

Plainly, the people at Morgan thought the same thing, which is why they've now decided to go back to their roots, to a time when someone had invented the wheel . . . but not four of them. Morgan began life making three-wheelers and the company is at it again with what is surely the most preposterous car on the market today.

Imaginatively called the Three Wheeler, it started out as an American engineer's homage to Morgan's Neolithic approach to car design and manufacture. He built a bike-engined three-wheeler and the powers that be at Ye Olde Workshoppe in Malvern thought, 'Golly. That bally Yank may be on to something here.' They went over there and bought him out for a reputed sum of twenty guineas. And some beads.

First, Morgan's engineers ditched his Harley-Davidson engine and replaced it with something called the X-Wedge. It's a 2-litre air-cooled V2 with a solid forged crank and three belt-driven camshafts. But the layout is nothing compared with where it is. In short, it's not in the car. It's slung out in front, where it sits like a big, complicated bumper. There is, so far as I can see, absolutely no reason for this.

Enthusiasts say that because the engine is air-cooled it's better that it sits exposed, but I don't buy this. The engine in a Volkswagen Beetle is air-cooled and that sat inside the car, not overheating, just fine. I suspect it's not in the car so that people can look at it and get all adenoidal and nostalgic about how life was better in black and white.

Of course, putting a two-cylinder engine in front of the car is nothing compared with what they've done at the back, which is to fit just one wheel.

I should imagine that when Morgan enthusiasts see this, many will quickly develop a noticeable bulge in their Rohans. Whereas I stood there thinking, 'Have these people never seen a three-legged dog? It doesn't work. And neither will that.'

Amazingly, though, it does. I know better than most that a Reliant Robin falls over whenever it is presented with any sort of curve and any sort of forward momentum. That's because Reliant chose to fit a single wheel at the front. Morgan, however, has turned everything around and fitted a single wheel at the back. The stability is remarkable. It takes a while to get the confidence to push, but push you can until, eventually, you discover that it will get round Donington's Old Hairpin at 80 mph. At almost exactly three-quarters of the speed that would be possible if it were an actual car.

Other things worthy of note? Well, the vibrations are bad, and if you are more than, say, 3 foot tall, you may have to take a leaf out of the car's book and leave a limb at home. Also, at £30,000, it is expensive.

However, I'm afraid to admit I rather liked it. I like the way Morgan painted it to look like a second world war fighter plane – something most Morgan owners think has only just been invented – but most of all I like the way that it feels so completely and absolutely different from anything else that is allowed on the road.

One of the big differences is that it's very difficult to reach the brake pedal. Another is that your head's in the slipstream and your right arm is like the engine, sitting outside the bodywork.

Even the engine feels weird. Because there are only two cylinders, the torque comes in staccato bursts. One second you have enough to fell a tree; the next you're becalmed. Morgan even had to fit a cushioning device to the running gear so that the Mazda MX-5 gearbox could cope.

And yet, you can do a doughnut in it. And you can leave the lights in a cloud of smoke as that single rear tyre does its best and fails to put the power on the road. I bet if you really wanted, you could make it buzz the bally tower. After five minutes behind the wheel, I began to think I might be Kenneth More.

Is it fast? No. Is it safe? Perhaps not. Is it practical? No. Is it comfortable? Yes . . . compared with being stabbed. But did I enjoy myself in it?

Absobloodylutely. Let me put it to you this way. You have a choice of going to Paris this afternoon on a once-in-a-lifetime trip. Would you prefer to make the journey in a comfortable Airbus A320, or a draughty, noisy Spitfire? My case rests.

6 November 2011

Beach beauties love my bucking bronto

Lamborghini Aventador LP 700-4

As a general rule, American cities are all exactly the same. There's a pointy bit in the middle, which is ringed by large shops selling tasteless food in vast quantities. The hotels are all the same, too, and you can forget about finding a charming, family-run restaurant in the back streets. Because it's not there.

That's why Miami always comes as a pleasant surprise. It is different. The strip of land known as Miami Beach is home to hundreds of art-deco hotels and apartment blocks, which you will find nowhere else, and if you squint – which you will because it's impossibly sunny – you can imagine that at any minute you will see Gus Grissom and Alan Shepard prowling past in their Corvettes.

Elsewhere in the world the late 1950s were smoky and awful and full of misery, but in America they were a time of hope and adventure and brave young men drinking and driving and drinking and balling and drinking and dreaming of going into space. It was a time of Cocoa Beach and people with shiny smiles partying. And you still get that flavour in Miami Beach today. I like it there.

Unfortunately, there is a problem. You can't just turn up with your dowdy English hair and your flabby breasts and your pot belly, because you will look foolish. In Miami you need to make an effort.

So. It's no good just having a speedboat. It must have three big engines in the back and an enormous pouncing tiger painted down the side. Likewise, you can't just have a motorcycle. It must be as customized as your girlfriend's face, with

9 foot-long forks, a saddle made from the foreskin of a whale and exhausts that do absolutely nothing to mute the sound of the 7-litre V8 engine around which you simply cannot get your legs.

You might imagine that all of this would come to a shuddering halt on the golf course; that it would be impossible to stand out in the excess-all-areas environment of a Florida fairway. But you'd be wrong, because in Miami you can buy a customised golf buggy with 20-inch chromed rims and a painting of a snake on the bonnet. Rolls-Royce grille? Certainly, sir.

I went out for dinner at a restaurant called Prime One Twelve. As this is regarded as the hottest place in town, getting in wasn't easy. 'Do you have a reservation?' said the impossibly beautiful, stick-thin girl at the reception desk. Having established that I didn't, she looked me up and down, saw that I was fat and that my teeth were the colour of a pub ceiling and decided that, contrary to all the evidence, the place was full.

Well, of course it wasn't, so a few minutes later the waiter was running me through the menu. It was all about the size of the cut and the amount you got on the plate. The tomatoes were bigger than Richard Hammond's head. My side order of spinach was delivered in a bathtub. And the steak? Holy mother of God. It was as though everything had been sourced in one of those Hollywood B-movie valleys where the ants are the size of men and the grapefruit are bigger than airliners.

However, I didn't really notice the dead brontosaurus on my plate because I was way too captivated by the spectacle that was unfolding outside.

Prime One Twelve attracts the crème de la crème of show-offs. The cars in which they were arriving were mad. Jacked-up Camaro convertibles with spinners. Bentleys on 24-inch rims. One man arrived in a neon insect. Another in a lowered Rolls-Royce. I can only begin to imagine how terrible these cars must have been to drive – cars always are when you fit wheels that could roll a cricket pitch – but that doesn't matter. In Miami cars are not for driving. They are for arriving.

I'm not making this next bit up. Couples were appearing in the lobby of the apartment block across the street from the restaurant. They would then wait five or 10 minutes for the porter to fetch their car from the underground car park. And then they'd drive it 50 yards to the valet at the restaurant.

So who cares that the ride of the Roller is ruined? Who cares that you need a step ladder to get out of the Camaro? And who cares that your Porsche's modified exhaust system could make you deaf after five miles? You will never go that far.

Elsewhere in the world people buy cars for all sorts of reasons. Value. Economy. Speed. Space. Comfort. But in Miami people buy, or rent, cars for showing off; for demonstrating that back home in Philadelphia their shower curtain ring factory is doing pretty well.

The new Lamborghini Aventador would fit into the mix jolly well. Even before you add silly wheels and a custom paint job it's £250,000, and people will certainly see it coming. It's 2 inches wider than a Range Rover – or about the same width as a London bus. Plus it has a 6.5-litre V12. I doubt we will ever see a new engine such as this again. Today, thanks to Euro emissions regulations, turbocharging is the only realistic answer.

It's a shame, because the immediacy of the thrust is intoxicating. The acceleration is as vivid as Miami's sea front, and the top speed is about 212 mph greater than it would ever go there.

But what about here? In the civilized world of old money and taste?

Lamborghini is keen to stress that the engine is its first all-new V12 since that of the Miura, that the four-wheel-drive system is as advanced as technology allows, that it has carbon ceramic brakes and that Formula One-style pushrod suspension is also part of the recipe. Now the only reason you would fit this is to reduce unsprung weight. But I can't see how that matters in a car that weighs about the same as the Empire State Building.

I suspect the real reason it's fitted is so that owners can feel that behind the vastness and underneath the flamboyant, mad, brilliant show-off styling, there's some Ferrari-style cred.

Honestly? There isn't. This is a brute of a car. You don't drive it. You wrestle with it. It's more refined than Lambos of old and less inclined to want to kill you. But when you open the taps, you are no longer driving. You are hanging on.

I drove it very extensively across Italy, and although it's quiet and surprisingly comfortable, and although it has an Audi sat nav and Audi controls, you are never allowed to forget that it's a raging monster. And for that I absolutely loved it.

I loved the speed. I loved the styling – it's probably the best-looking car ever made – and I loved the sheer stupidity and silliness of its dash and its face and its insane rear end.

I don't want one. I'd rather have herpes. That said, I do want to live in a world where I can sit outside a restaurant in Miami and watch some poor girl trying to get out of the passenger seat without flashing her knickers.

20 November 2011

Hop in, Charles, it's a Luddite's dream

Mercedes C 63 AMG coupé Black Series

It seems that a few years ago Prince Charles asked scientists to research the concept of alternative medicine to discover why acupuncture, holistic healing, leaves and sitting cross-legged on the floor can be used to cure hay fever, eczema and bowel disorders.

Well, a British science writer did just that and, after exhaustive research, he has published a book saying that alternative medicine mostly doesn't work and that when it does, the results are 'dismal'.

Naturally, you would assume that Prince Charles would read the carefully researched points and, with upturned palms, explain that, contrary to what he's been saying for many years, the best cure for a headache is a couple of Nurofen, rather than a balm made from soil and bits of armpit hair.

However, he appears not to have done this. Despite all the scientific evidence, it seems he continues to believe in the power of humming and peace oils. We see this a lot. People use science when it agrees with their opinion but dismiss it as nonsense when it does not.

This is hardly surprising really since what we know as science fact today is almost always science fiction tomorrow. The earth was flat. And then it wasn't. Matter cannot travel faster than light. And then it can. Well, maybe. Man is causing global warming . . . Watch this space.

I'm glad that there are scientists. I'm glad they go to work in their laboratories with their Bunsen burners and their tweezers. It's important that mankind strives to understand where he came from and where he's going. But scientists should not try to explain their findings to Prince Charles. Or me. Or you. Because we're thick and we won't understand what they are on about.

However, because we don't know we are thick, we will listen to a bit of what they have to say, take away a nugget and, if we are a politician, or a prince, try to do something about it.

Not that long ago everyone was worried that particles and gases coming out of exhaust pipes were making people who lived near motorways very stupid. This was a big concern for the residents of Gravelly Hill, north of Birmingham. Ford's scientists decided the answer was lean burn technology, engines that ran mostly on air, and their view was shared by Mrs Thatcher, then prime minister, who had a chemistry degree.

However, lean burn was a few years away from reality, and people who were thick demanded results immediately. So all car makers had to fit off-the-shelf catalytic converters instead.

Sadly, what catalytic converters do is turn all the gunk that was making people in Birmingham stupid into carbon dioxide, which science now says is turning polar bears into man-eating were-wolves. So, to get round that, cars are now being fitted with electric power steering and flappy-paddle gearboxes – anything that cuts the amount of fuel used and therefore the amount of CO_2 coming out of the tailpipe.

But it's all hopeless because catalytic converters are made from platinum, which is fast running out. So then we will be back in 1985, in jerky cars with crappy steering, with people in Birmingham looking like jacket potatoes and Ford saying, 'I told you so.'

There's a lot of science in the new BMW M5. Before setting off, you can tweak the engine, the gearbox, the suspension and the seats. And it's all very impressive. But I much prefer the approach taken by the skunkworks deep inside Merc's special forces division, AMG. Which is: no science at all.

Recently AMG's cars have started to become a bit soft. The raw, visceral engines have been muted with turbochargers to keep the emissions down, and the standard automatic gearboxes have been replaced with flappy-paddle-operated manuals that work just fine on the open road but with an epileptic jerkiness in town.

AMG says that the new C 63 Black Series is softer, too. It says it's listened to criticism – from me mostly – that the old CLK Black was too hard for the real world and that the new model is kinder to your spine.

Well, you could have fooled me. It's like being inside a fridge-freezer that's tumbling end over end down the side of a rocky escarpment. Is it softer than the old car? Perhaps, but not in a way that the human bottom could ever detect.

It even looks more racy. Like the old car it has flared wheelarches to accommodate the wide track, but in addition there are now nostrils in the bonnet and, if you tick the right option boxes, all sorts of carbon-fibre winglets designed to show other traffic that you are in agony.

Of course, you may think that all of this is necessary and important at the track. You can even specify the new car with timers that measure your performance not only against previous laps, but also against professional racers. However, it's all a complete waste of time because on a track this car is not an M5. It is not designed to post the best possible lap time. It is designed purely and simply to eat its own tyres. I got through one set in just fifteen minutes.

This hunger for rubber is caused by the engine. It hasn't been softened, which means you get 6.2 litres of basic, unturbocharged power and torque. The diff may come from the sort of engineering that's used to open lock gates and the chassis from a book called *How Victorians Did Things*, but none of it can cope with that V8 firepower.

Drive with the traction control on and you won't go anywhere at all. Drive with it off and you will go everywhere sideways. This, then, is my sort of car. It's almost childlike in its honesty.

And here's the best bit. Despite the madness, it is not a

stripped-out racing car. Unlike Porsche, which removes even the satnav and the radio from its track cars and replaces the back seats with scaffolding, Mercedes has left pretty much everything you could reasonably expect in place. It will even, if you ask nicely and cross its palms with silver, put the rear seats back.

If you learn, as I have with my Black, to steer round manhole covers and potholes, you can use the car on a day-to-day basis. It has iPod connectivity. It has air-conditioning. It has a boot. You just have to remember that at no point, on the road, can you ever use full throttle. Unless you've just found your wife in bed with your mother, you've lost your job and your kids have been arrested.

This car is unique. It's something that's not made any more and will probably – because of the world's poor grasp of science – never be made again. On the face of it, it's a German DTM racer, a road rocket from the country that invented such things. But, deep down, it's on Telegraph Road, eight miles north of Detroit, on a Saturday night, with its cap on back to front and a bottle of Coors in its big, working-class paw. Deep down, it's a muscle car.

And Prince Charles will like this. Sometimes it can also be a muscle relaxant.

27 November 2011

It's no cruiser but it can doggy-paddle

Jeep Grand Cherokee 3.0 CRD V6 Overland

Last weekend AA Gill wrote about New York in such glowing terms that I wanted to drop what I was doing and go there immediately. He admitted, of course, that in summer the humidity makes the sky feel as thick as wallpaper paste and in winter it turns slate grey and comes right down to the ground, blotting out the city's raison d'être. But he talked of an autumn day when there's a crimson tide in Central Park and the sky appears to have been bleached, perhaps sucked of its richness by the vibrancy of the people below.

I would agree with Adrian that New York is one of the world's greatest cities. Along with, in no particular order, San Francisco, Rome, Tokyo and Cape Town. But last weekend, after much quiet contemplation, I decided that none of them can hold a candle to what is unquestionably the best of them all: London.

I came to this conclusion because I was working at the ExCeL exhibition centre, which is located about 6,000 miles to the east of the City, near where they are holding the running and jumping competitions next year.

Boris Johnson will tell you that the best way of getting there is by the Sustainable Light Railway. But he's wrong. The best way of getting there is via the Thames, on a Fairline Targa 47 luxury yacht.

I should make it plain from the outset that it is extremely difficult to drive a boat through the middle of London. The civil engineer Joseph Bazalgette made sure of that, separating what at the time was an open sewer from the City with the Thames Embankment. And today the authorities keep his spirit alive with an astonishingly well thought-out plan to make absolutely certain that no private individual can park their boat even for a moment, at any point, anywhere at all. You can come, but you must go away again straight away.

Two years ago the Port of London Authority introduced a 12-knot speed limit on parts of the river and I'm not surprised, really, because out there in the chop, among the pleasure boats and the barges, which are operated by men who live by rules laid down in the 19th century, you feel completely at sea. Taking a pleasure boat into this alien world full of people who make a living there is like riding a tricycle along the Central Line. It's going to end up with a crash. Despite all this, you must try it. It is No 1 in the things you have to do before you die. Wait for a crisp autumn day when there isn't a cloud in the sky and go through the middle of London on a boat. And then you will realize why nowhere else gets close.

It's the variety, really. New York is a monoculture. It does one thing. So is San Francisco. And so is Rome. London, though, is different. You have the Tower, which is almost a thousand years old, right next to the brand new glass-and-steel cathedrals to capitalism. You peep past converted 19th-century warehouses to catch a glimpse of a 17th-century cathedral. And then it's eyes left for the Palace of Westminster and eyes right for the London Eye. Pretty soon your head is rotating like that little girl's in *The Exorcist* and you're wishing you had compound vision like a fly.

One minute you are gliding past *HMS Belfast*, a second world war battle cruiser, and no sooner have you taken stock of its rear end than it's time to swivel round and gawp at London's signature dish – Tower Bridge. It's not a river cruise, really; it's time travel.

And I'm willing to bet that there's no other city in the world where anyone would point excitedly at a power station. But that's what you do in Battersea.

However, it's the return journey, at night, that leaves you breathless. Many cities tart themselves up for one-off events.

London looks like it's New Year's Eve all the time. It is lit absolutely beautifully. From the laser marking the meridian at Greenwich, up past Canary Wharf and under the blue-white lighting at Tower Bridge, you arrive once more at Westminster, where the clock tower is a warm blend of vivid green and sodium orange, and, opposite, the London Eye appears to be a portal into space. There was even a gorgeous new moon to top the scene off.

Then there's Albert Bridge. It's lit so vividly, and the reflection is so strong, you feel as if you're steaming along in a vast river of custard. And to the left and right are the new blocks of executive flats, each with its floor-to-ceiling windows. No point closing the curtains. Who's looking? We were. And that was great, too.

There is no one-hour trip in the world in which you see so much, and so much of what you see is so wildly different. Every day I'd spot something else. Another frilly Victorian detail. Another elegant way of converting the warehouses that served the industry that built them into flats for the industry that fuels Britain today. A hundred years ago they were full of cotton and tobacco. Today they house bankers who've come over from New York for six months and can be seen on their balconies, gawping and not quite believing their eyes. Fifth Avenue is a modern-day wonder. But the canyon carved by Old Father Thames – that's on a different plane altogether.

Sadly, because the boat was so sublime and the view so moving, I used it every day to commute to the ExCeL. Which meant the new version of the Jeep Grand Cherokee I am supposed to be writing about this morning spent most of its time sitting in a lock-up garage.

It had a difficult birth, this car. Conceived when its parent firm Chrysler was married to Daimler, which owns Mercedes-Benz, it shares many of its underpinnings with the Mercedes ML. But before it was born, its parents divorced and its mum ran off with a man from Fiat. As a result of her new liaison it has a 3-litre Italian diesel engine. No other option is available.

The figures look quite good, and off road there is enough engineering – both electrical and mechanical – to ensure progress will be maintained long after most of today's high-riding urban-crossover sports-utility offerings have sunk without trace into the mud. But one brief drive was enough to confirm that while the Jeep may be fine in the big outdoors, it's a very old-fashioned Hector everywhere else. The automatic gearbox operates on geological time and the engine is about as refined as a scaffolding company's tearoom.

And while the interior is festooned with all manner of luxury items and electrically operated toys, there is a very real sense – as is the way with all American cars – that a million pounds has been spent at Woolworths.

The exterior is different. It looks quite good, and, unlike Land Rover, Jeep offers a tasteful range of colours. Mostly, though, this car is big and it's cheap and it's simple. Which, of course, makes it absolutely ideal for someone who for some reason doesn't want a Toyota Land Cruiser.

4 December 2011

Uh-oh, some fool's hit the panic button

Chevrolet Orlando 1.8 LTZ

We're told that between Christmas and the new year, 8 billion British people have defied the troublesome economy and, between them, spent £70 trillion on mildly discounted products in the sales. This sounds like good news. But if you examine the pictures of those rampaging around Oxford Street you will notice quite quickly that every single one of them was Chinese.

The Chinese love a bargain. But what they love even more than that is a brand name. I was in Beijing last month and was told, time and time again, that local produce had no appeal at all to the country's new rich. They want Tommy Hilfiger and Prada and Ray-Ban. And if they can get these badges at the lowest possible prices – well, that's got to be worth the cost of a return ticket to Heathrow.

That's what you need these days to survive out there on the high street: a name that's known. Because a name that's known is a name that can be trusted. Fairy Journeyman HoneyWasp perfume may be excellent and good value but it cuts no ice with a bottle bearing the Chanel legend.

It's not just the Chinese, either. I have a friend who dresses in quite the most hideous clothes you have ever seen. They look like they have been made either as a joke or by someone who was being deliberately stupid. But when I explain this to him, he always points to the label and says, with a hurt tremor in his voice, 'But it's Dolce & Gabbana.'

I guess I'm just as bad really. I only buy Ray-Ban sunglasses and Sony televisions. Not because I know they're the best but because the names have a Ready Brek aura of comfort-blanket warmth about them. And don't claim you're immune. Because I bet you'd rather do business with a man called Victor than a man called Vince.

In the world of cars, a brand name is everything. While in China I drove a car called the Trumpchi. Want one? Of course you don't, because who's behind it? What's it made from? And where? I could tell you that it's hewn from a gold bar, costs six pence and runs on water and you'd say, 'How intriguing,' as you wrote out a cheque for a Volkswagen.

You know where you are with a Volkswagen and you're right. Every day, thousands of engineers work to the best of their abilities to make sure that every single car they make upholds the company's reputation for durability and safety. Protecting the brand name: it's everything.

Unless you are running General Motors. Protecting all the brands it controls is nowhere near as important as making any damn thing to keep the bankruptcy wolf from eating the company's front door. Which explains the Chevrolet Orlando LTZ in which I endured a mercifully brief drive recently.

Louis Chevrolet was born in Switzerland and after a brief spell in Canada arrived in America where he drove racing cars for Fiat and made road cars for himself. Fairly soon, though, he sold his car company to GM and went off to have fun. In 1929 he lost every cent he had made in the stock market crash and ended up in Detroit working on the Chevrolet production line. That was sad, but worse was to come . . .

Because today, he is six feet under the ground in Indianapolis, spinning wildly at the Orlando people carrier that bears both his name and the stylized Swiss flag badge that he designed. The company he founded has always had one eye on budget performance. This was its guiding principle as far back as the 1930s, when it offered the cheapest six-cylinder car in the world.

In the 1950s it came up with the plastic Corvette and in the 1960s it was among the first companies in the world to fit a production car with a turbocharger. But all the while, it was plugging away with its small block V8. The mainstay of blue-collar speed. The heart that pumped the Camaro and the Nova SS into the world's consciousness.

This is what we think about when we think of Chevrolet. Men with tattoos and their hats on back to front, whooping wildly as their thunderous and wondrous machines skittle off the line in a shuddering roar of smoke, axle tramp and more smoke. Not sophisticated. But nice.

But, we are told, there is no place for this sort of thing in a country full of Al Gore, windmills, rising oil prices and Mexican pool cleaners whose houses are now worthless.

And to make matters worse, Chevrolet has always been run by the Flat Earth Society. The company may have been founded by a Swiss but nobody who followed in his footsteps ever had an atlas. To them, the world started at Los Angeles and ended at Boston. They didn't sell cars outside America because, as far as they were concerned, it was the fifty states . . . and then some jungle.

Well, they've had a wake-up call now and what they've done is panic. Instead of sitting down and thinking, Right, we must protect the brand with a range of fun, fast, quintessentially American cars that we must sell in places such as Englandland and the People's Republic of Japanland, they've run around like headless chickens, being chased from pillar to post by clueless government wallahs who want a return on the bailout cash now. 'Now!!! D'you hear?'

The result is the Orlando. Built in South Korea from the same platform that props up the Vauxhall Astra, it is a 15½ foot, seven-seater people carrier of monumental awfulness.

We will start with the seats. Yes, there are seven but there is no one alive today that could fit in any of the five in the back. And there is no boot at all, unless you fold the two rearmost chairs into the floor. It's hopeless.

But it's not as bad as the engine. For the first mile, I was absolutely sure it was a diesel, but then I noticed that the rev counter read to 6,000. Dear God in heaven, I thought. This ailing cement mixer is running on petrol. It's a 1.8-litre four-cylinder unit that does nothing well. Even movement is a struggle. I was staggered to notice the car was fitted with traction control. Why? That's like fitting traction control to a chest freezer.

On top of the lack of power, it's also thirsty, unrefined and sounds like a wounded whale. And none of that should surprise you. Because asking a Chevy engineer to design a four-cylinder engine is like asking a man in a burger van to poach a halibut. It's still cooking, but it's not the sort of cooking he's used to.

I should say at this point that the prices are quite low. The LTZ model is just £18,310, which doesn't sound too bad. But if you want any colour other than white, you must pay an extra £410, and if you want satnav, then that's another £765. What are they thinking of? Why fit traction control, which is unnecessary, and make us pay extra for a road map, which is?

Handling? That's terrible. The ride? Terrible. Seat comfort? Terrible. And to top it all off, it was plainly styled by a man who gets tumescent at the thought of house bricks, and finished off on the inside with a range of plastics that feel like Cellophane.

Some people may buy this car so they can tell their friends they have a Chevrolet. They won't buy another.

1 January 2012

Simply no use for taking the kids to see Granny

Audi R8 GT

We like to imagine these days that we live in a global village and that everything is the same wherever we go. But this isn't actually the case. The Coca-Cola you drink in Russia tastes completely different from the Coca-Cola you drink in Ross-on-Wye. The Big Mac you eat in Cape Town tastes nothing like the Big Mac you eat in Cape Canaveral. The girl to whom you make love in Greece will be different from the girl to whom you make love in Grimsby.

You would imagine also that when it comes to cars the world is one harmonious, homogenized lake of similar goals, similar machinery and similar driving styles.

Not so. In the past eight weeks I've been to India, Italy, America, China and Australia. And although all these places are huddled together on one tiny blue pinprick in the vastness of space, they might as well have been in different galaxies.

In Australia, for instance, people drive much the same sorts of car as we do, on the correct side of the road and in a similar fashion.

However, every single Bruce is infected with a partisan attitude to motoring that you will find nowhere else. In Scotland there is Celtic and Rangers. In America there is north and south. In Australia there is Ford and Holden.

An Aussie friend of mine – a barrister – tried to argue recently that this was a working-class issue, but within moments he was busy proving himself wrong. 'I mean, I drive an Audi, so I couldn't care less about the Ford and Holden war,' he said, before going on: 'I mean, deep down, I'm a Ford man, like my father before me. I'm from a Ford family and, given the choice, would have a Falcon. I actually hate Holdens. Hate them, d'you hear?'

By this stage he was banging the table. 'I would bleed blue blood for Ford and I would strike down with furious anger those in Holdens who would attempt to poison my brothers. Holdens are the worst things on God's green earth. The worst!'

There is no equivalent of this in China, although it seems there are two very different types of Chinese motorist. You have the rich ones, who buy expensive European cars that they drive as fast as they'll go, and the not-so-rich ones, who buy Trumpchis and Roewes and dither about at junctions, terrified that they are about to be mown down by a twelve-year-old in a 140 mph Range Rover Sport.

In India the car is not a device for moving you and your family from A to B. It is a device designed exclusively to take you and your family into the next life. You buy a car. You set off. You have an accident and that's that. Someone is killed on the roads there every three seconds.

Then you have America, where, generally speaking, the car is a horse. And you don't wear the poor creature out by galloping everywhere. You plod at about 35 mph, pausing only to shoot at someone coming the other way. Against that background there is no call for finely balanced handling and good looks. It's why there has never been an American sports car. Not one. Ever.

Whereas in Europe, with our twisting roads and active minds, every single car ever made – with the possible exception of the Austin Allegro – has the spirit of the Mille Miglia at its heart. Here, a car that won't handle properly is as wrong as a horse with no legs would be to the Yanks.

Let me put it this way. I can think of only two mid-engined cars from America's mainstream manufacturers and only two from Japan. Australia has never made one – or not a serious one, at least. Nor has India. Nor has China. Whereas here there have been mid-engined cars from Ferrari, Lamborghini, Maserati, De

Tomaso, Pagani, MG, Porsche, Rover, Renault, Peugeot, Audi, BMW, Jaguar and, if you stretch it a bit, Volkswagen, Skoda and even Hillman.

With my Chinese or Indian hat on, I can tell you that mid-engined cars are a damn nuisance. Because there will come a day when two people will want a lift. Or when you realize that your shopping won't fit in the boot. They also scrape their noses on even the tiniest of speed bumps and are very uncomfortable.

But we Europeans like them, nevertheless, because the weight is at the centre, which means they're bound to be better balanced than cars with their powerplants at the front. We also like the styling that results.

And the sheer frivolity of a machine that has no practical purpose. Mid-engined cars make us priapic. Which is a good thing because they also tell the world that we haven't had kids yet.

Occasionally, though, a car maker will forget all this and try to make a mid-engined supercar that is more than just a toy. And that brings me neatly on to Audi and the R8. It's very good. With a V10 it's spectacular to drive, but because it's all so sensible it doesn't ignite the small boy that lives in us all. It feels like a big TT.

Well, now Audi has obviously realized the futility of venturing further down this road – an everyday supercar is as silly as an everyday ball gown – which is why it has produced the limited-edition R8 GT.

GT stands for grand tourer but it isn't that at all. It's a harder, lighter, louder, more powerful version of the original and it comes in two specs. You can have a road-going version, which I recommend, or the more focused track-day version that came to my house.

This has a switch in the ashtray that does nothing at all, a fire extinguisher, some scaffolding in the back and idiotic four-point racing harnesses that take six hours to adjust and three hours to fasten up and mean that when you're driving along you can't reach half the controls or the glove box.

There was much to hate about this car, then. But if you peel away the track-day frippery and nonsense, there was actually a lot to love. They may have only skimmed this and chiselled away at that to save a miserable 100 kg – most of which was then off-set by the silly roll cage in my car – but the GT feels light. Much lighter than the standard car. And as a result of that, it feels more awake, more eager.

This is something Europeans like. American testers found that on a track, if you pushed it, there was a lot of understeer, but I fail to see how this is possible. Yes, the R8 GT is four-wheel drive, and that is usually a recipe for the nose to run wide. But so little power is sent up front, and the chassis is so good, I found that, if anything, it was the other way around. It was the back that stepped out of line. We Europeans like that, too.

Then there's the engine. This is a 5.2-litre Lamborghini-based V10 that has been fettled so that it now produces 552 horse-power. And it's power that all comes in a rush near the top of the rev band. That encourages you to hang on to each gear as long as possible, and that in turn unleashes a soundtrack that's totally at odds with the featherweight feel of the car. You expect it to howl, or shriek. But it bellows – a deep, low, frightening sound. It's fantastic.

So's the ride. So's the driving position. So's the way Audi has sensibly left most of the luxury equipment on board rather than follow Porsche's lead by taking it all out. And then charging you to put it back.

My only real gripe with this car is that the single-clutch flappy-paddle gearbox is a bit dim-witted. And maybe £152,000 is a bit much when for a little bit more you could have a Ferrari 458. Which is a tiny bit better.

8 January 2012

Amazing where bottle tops and string will get you

Hyundai i40 1.7 CRDi 136PS Style

Back in the mid-1960s a South Korean construction company called Hyundai decided it would like to start making cars. With the financial muscle of Manchester City, it could have employed anyone to lead this new venture, but, for reasons that are not clear, the man it eventually selected was an Englander called George Turnbull.

George had made a name for himself as managing director of Austin Morris, where he had steered the Morris Marina from what must have been a drunken doodle on the back of a beer mat into what passed in the Midlands back then for 'production'.

Rather than work with Koreans, who obviously wouldn't know a shock absorber from a stick of rhubarb, George brought with him a team of other Englanders, including the man who'd designed the chassis of Princess Anne's car of choice, the Reliant Scimitar.

The car they created was called the Pony – rhyming slang, probably – and apparently it was exported to Britain, although I don't remember hearing about it, seeing one or having met anyone who had decided that their life was incomplete without an oriental Marina on the drive. Certainly I never had a picture of it on my bedroom wall.

No matter. Thanks to healthy sales in motoring meccas such as Egypt and Ecuador, it was a success, and even though Mr Turnbull had gone off by this stage to make Hillman Hunters in Iran, among other things, Hyundai launched a replacement model. Through a lack of imagination, this too was called the

Pony and was described in the brochure as having 'rectangular halogen headlamps' and 'easy-to-read gauges'.

The company was reduced to mentioning these rather trifling things because there really wasn't much else of note. I drove this car way back, and still, even to this day, I can recall the horror. Everything from the vinyl upholstery to the ungainly exterior styling was hideous, and, worse, none of the components seemed to be joined up.

There was a lever that sprouted out of the floor, but using it to engage a gear was so hard that by the time you found something that felt promising, the car had coasted to a halt and you faced the prospect of trying to find first again.

I believe that the gearlever was connected to the box with string. And that the steering column ended up in a big box of yogurt. If you twirled it fast enough, the yogurt would spin and centrifugal force would eventually cause the front wheels to change direction. Often one would go left and one right, but you usually didn't notice, partly because you were too exhausted by twirling the wheel and partly because, by this stage, you'd have run over a small piece of grit and your back would have broken.

Then there was the engine. Oh dear. I distinctly remember thinking after just a few miles, by which time I'd reached 6 or 7 mph, that oil was a much better lubricant than the garden furniture Hyundai had plainly decided to use instead. The only good news was that you could never reach a high-enough speed to be worried by the fact that the brakes were made from old milk bottle tops and didn't work.

What more could we have expected, though? Asking Korea to make a car at this time was like asking the residents of a sink estate in Algeria to make a space shuttle today. It's not going to end well, especially if you employ the father of the Morris Marina to help out. And yet somehow it has ended well, because here we are, several decades down the line, and while many European and American car companies have foundered, Hyundai has become the fifth-biggest car maker in the world.

Its latest offering is called the i40 and it's not bad at all. It's similar in size to a Vauxhall Insignia and Ford Mondeo, although it is better-looking and, if you go for the eco-models, just as economical. It's also a tiny bit cheaper, but gone are the days when a 'made in Korea' tag meant huge savings. You pay European prices these days because the cars are just as good.

The diesel engine in my test car was not enormously refined but it managed to pump out plenty of horsepower while not suffering from a prodigious thirst. The steering's a bit odd, but at least when you turn the steering wheel, both front wheels turn in the same direction. And when you press the brake pedal, you slow down. There are no surprises, nasty or otherwise. The interior feels as if it were made by Honda. The suspension is conventional. The gearbox is unremarkable.

Hyundai said that when it was designing the i40, it took as its benchmark the Volkswagen Passat and Toyota Avensis. And that's what it has ended up with: an inoffensive, odourless blend of the two. It is just 'some car'. And as a tool it's absolutely fine. Which is why, of course, it's not fine at all.

In the olden days, when Korea was a war, not a country, everyone had a four-door saloon. We viewed anything with a hatchback as being rather suspicious, and perhaps even a bit French. Nowadays, though, the market for conventional saloons has all but gone. We like our cars to be tall or big or roofless or made to look like something from the 1950s. We like them to be interesting, so conventional front-drive turbodiesel saloons just don't cut it. They're like going out in M&S trousers.

It takes a special kind of dullard to look at the wealth of possibilities in the land of internal combustion and say, 'Yes. I want a four-door saloon.' And it takes an even more unusual soul to say, 'And the four-door saloon I want is a Hyundai i40.'

The biggest problem is that all the great car companies were founded by a single man with a vision. Colin Chapman, Enzo Ferrari, Sir William Lyons and so on. Hyundai, on the other hand . . . Well, it was started by a man called Chung Ju-yung, who was born in what's now North Korea, into extreme poverty. It was the Four Yorkshiremen sketch, and then some. On one occasion he managed to raise enough for a train ticket out of his village only by nicking a cow from his dad and selling it.

Later, in the big city, he drifted from job to job – one minute a docker, the next a handyman. Eventually he started a car-repair workshop, but during the war the Japanese colonial government merged his fledgling business with a steel company. And that was that. Mr Chung ended up back where he'd started. In a rural village in the north, with a dad who had no cow.

When the war ended, he started a construction firm he felt would be able to help out with the rebuilding needed. He was right, and today Hyundai is a multinational conglomerate. And its car business makes Mercedes-Benz look like Plymouth Argyle. One factory alone is capable of making 1.6 million vehicles a year.

It is a remarkable rags-to-riches story and I should like to have met the man who started it all. But do I want to buy one of his cars? No. For the same reason that I'd rather buy cheese in a delicatessen than in Walmart. The ingredients may be the same. The taste may be pretty similar too. But you would rather buy from someone who wants to make cheese than from someone who makes cheese mainly to make money.

22 January 2012

Bong! I won't let you go until you love me

BMW M5

Once, I drove on a highway near Atlanta, in Georgia, that runs through a wood for about 16,000 miles. It was very boring. But after a quick trip to Wales last week I've decided that the M4 is even worse.

Some say that the most boring motorway in the world is the M1, but actually it's not dull at all. It has a history and even a hint of romance. People have written songs about it and you pass many exciting places such as Gulliver's Kingdom and the Billing Aquadrome. You're tempted at every junction to get off and have a snout around. Except, perhaps, junction 22. Coalville's not that appealing.

Plus the M1 makes your blood boil, especially at the moment, because almost all of it is coned off and subjected to a rigorously enforced 50-mph speed limit. Driving on this section is like reading the *Guardian*. It gets the adrenaline going and it makes your teeth itch with impotent rage.

The M4, on the other hand, is like a book with nothing written on any of the pages. You pass Bracknell and Reading and Swindon, and you would rather die, screaming, than go to any of these places. Eventually there's a signpost to Theale, which doesn't even exist. Have you ever met anyone from Theale? Been there? Read about it? I think it may have been hit seventy years ago by the friendly German bombs meant for Slough.

Then you have the 'works' exit, which looks as though it might lead to the sort of place where they keep grit and snow-ploughs for the winter. It looks harmless, municipal. But try to find it on Google Earth. Go on. Try. Because it's not there. It's a slip road the authorities want us to believe goes nowhere. Spooky, eh?

After this non-event you are plunged into a nothingness that goes on for a light year until eventually you are asked to pay £6 to cross a bridge. Why is that? How can it cost £6 to use a fairly crummy bridge when for £10 you are given access to the whole of London?

I hate the M4, and what makes it even worse is that there's no car made that makes it even remotely interesting. If you drive down it in something powerful, you will be caught by the speed cameras. If you opt for something comfortable, you will fall asleep and crash. If you go for something economical and sensible, you will become tired of the engine moaning out its one long song and deliberately run into the crash barrier to end it all.

There is, however, one car with just the right combination of features to keep you awake – to keep you interested in being alive – for just two hours more. The BMW M5.

I have always been a fan. The original, a 282-bhp version of the boxy 1980s 5-series, came out of nowhere and redefined what we thought might be possible from a saloon. And since then every single version has pulled off the same trick.

The latest is even more of a star. Unlike the equivalent AMG Mercedes, which looks like a street brawler, the BMW is a bit like a bouncer at a 'nite' spot that wants to be seen as posh. It's wearing a tux, and you have to look hard to notice the neck like a birthday cake, the chest like a butter churn, the thighs like tugboats. It's a big bruiser, this car, but there are a few clues. Just the blue brake callipers. And some *Wallpaper* magazine-style LEDs in the door handles.

Under the bonnet the old V10 is gone. In its place is a turbocharged V8, which is good if you are a polar bear but bad, in theory, if you are a petrolhead. However, in practice, you don't notice the turbo lag. There must be a gap between you pressing the throttle and the warfare beginning but you never, ever, feel it. It's not a great engine. But it's very good.

My friends, who are not very interested in cars, said that the ride was 'bumpy' and that I ought to let some air out of the tyres. They were wrong. The ride is actually quite good. Another friend, who writes about restaurants, said it sounded like a diesel. He was wrong, too. It sounds like a tool for scaring dogs.

Out of town and away from my metrosexual mates, the new M5 continues to amaze. It feels heavy and the front is maybe a bit over-tyred, but it's just so fast and so composed and so balanced and so wonderful that you even find yourself grinning a deeply contented grin when you are on the M4 and it's raining and the burglar in the toll booth is still a hundred miles away.

There are so many toys to play with, so many things to do. You can set the engine up in sport mode and the gearbox in race mode and the suspension in comfort. And then, when you've found a setup you really like, you can store it in a single-button memory.

And then you can decide that actually, on some days, you prefer everything to be slightly different, so you can store this as well. Then you can choose precisely what information you would like on the head-up display. And then you can dive into all the submenus on the iDrive system and change everything up to the shape of the car itself.

So what we have here is a genuine four-seat limo. A car from which you would be pleased to emerge at a film premiere. But then this same car is also a tail-out, smoke-and-wail drift machine. And a finely balanced road racer, and a gadget. It's everything. And it's only £73,040. We're talking five stars and then some.

However. There are a few problems. There are so many gadgets that some of the features are not very easy to use. Such as: last night I arrived at the *Top Gear* edit, put the gearbox in neutral and started to get out. 'Bong,' said the warning buzzer and 'flash' went the dash display. 'You have not put the vehicle in Park. It may roll away. You may not lock the doors until you have put it in Park. Bong.'

'Bong.'

'Bong.'

'Bong.'

I got desperate. There was no Park position on the flappy-paddle gearbox. 'Bong.' I applied the electronic handbrake so the car couldn't roll away. 'Bong. Yes it can,' insisted the machine. 'Bong.'

Eventually I went into the edit suite and said, quite crossly, 'Does anyone here know how to put an M5 in Park?' They were all very amused, but an hour later we were still none the wiser. We were so desperate that we even resorted to the handbook. But still there were no answers.

It turns out that you must turn the engine off when it is in gear. Then Park is applied automatically. If you turn off the motor when the box is in neutral, you are bonged at until the end of time. I would very much like to meet the man who designed this system. So that I can jab some cocktail sticks into his eyes.

I suppose that eventually you would become used to this. But there is something else that would always be a nag. The problem is twofold. In the olden days the M5's price tag was justifiable because the car was much better than the standard 5-series. That simply isn't true any more. The 530d M Sport is very possibly the best, most complete car in the world right now and it's hard to see why the M5 costs £31,000 more.

It gets worse. Today the M5 is a cruiser. A bruiser. A heavyweight. A very different animal from the original. It's a wonderful thing, be in no doubt about that, but if you hanker after the olden days, you can have an M3, which is still lithe and sharp and crisp. And it's almost £20,000 less.

In short, then, the M5 is still a great car. But these days BMW makes other great cars that are considerably less expensive.

The M version of the 1-series is another example. Happily there is no M4.

29 January 2012

A heart transplant sexes up Wayne's pet moose

Bentley Continental GT V8

I'm in northern Sweden – really northern Sweden – slap bang in the middle of Europe's last great wilderness. For 300 miles in any direction there is nothing but fir trees and snow. Very late in the morning the sun heaves itself above the horizon, hovers for a moment and then slumps out of sight again. It's pale, the colour of custard, and entirely devoid of warmth. It's cold here. Really cold. And I absolutely love it.

I love the crisp whip crack of a Scandinavian winter morning; no matter how many beers you've drunk the night before, even the most savage, lightning-bolt hangover is swatted into history by the dry, needle-cold crispness. I love the endlessness of the sky. I love the jumpers. I love the way everyone is so tall. But most of all I love the way that people up here drive their cars.

I was picked up yesterday at Luleå airport by a man in a van. Ahead lay a four-hour journey into the middle of nowhere. 'We'll do it in three,' he said, slewing out of the short-term car park and pointing the nose of his Sprinter north. He was wrong. By travelling at a steady 80 mph, we were at the hotel in just two and half hours.

What makes this remarkable is that while the snowploughs had cleared the roads of drifts, they were still covered in a 4-inch veneer of sheet ice. In Britain such conditions would have had Sally Traffic urging everyone to stay at home and not venture out unless the journey was 'absolutely necessary'.

But in Sweden my man in a van drove as though it were mid-summer in Corsica. Occasionally he'd have to put his mobile phone down for a moment to correct a small tailslide, but you could see in his eyes the road surface was of absolutely no concern at all. Small wonder the world of motor sports is littered with people from these parts.

Of course, you've heard all this before. How the Swedes are born with an ability to deal with an oversteer. That they can drift before they can walk. But I'll let you into a little secret. The Swedes crash. A lot. Last night I saw two trucks in ditches and a car that had hit a moose – messy. Especially for the car.

This morning I watched the driver of a BMW X3 struggle for twenty minutes to retrieve his hapless little box from a drift before a kindly soul in a passing Mercedes offered to give him a tow. Eventually the Beemer popped out of its icy trap like the cork from a well-shaken bottle of Moët and rocketed backwards straight into the side of the Good Samaritan's Merc. I tried not to laugh but it was impossible. In fact, I laughed so hard the membranes around my brain came out of my nose.

I admire the Swedes for the stoic way they continue to go about their business. How they dismiss prangs as part of the price you pay for living among the Arctic foxes and the wolves. But let's dispel any myth that they have superhuman gifts behind the wheel. Because they don't.

Oh, and while we're at it, can we stop pointing the finger at local authorities in Britain for never being ready when the snow comes? That's the same as pointing the finger at the highway people in northern Sweden for not being ready to deal with a swarm of locusts.

And can we also pause for a moment to think about the real benefits of four-wheel drive. If it were necessary in inclement weather, the Swedes would have embraced it warmly as a tool for masking their perfectly ordinary driving skills. But they haven't. You rarely see an off-roader up here. Mostly they drive what people drive in Leamington Spa.

Which brings me rather clumsily to a new version of the all-wheel-drive Bentley Continental, a car I have never liked very much.

The first car I drove was my grandfather's Bentley, and it was a joyful experience, sitting on the old man's knee, steering that vast bonnet around what was then the Socialist Republic of South Yorkshire. It should therefore be a brand that makes me feel all warm and fuzzy.

It doesn't, though, because in recent times the Bentley has been hijacked by Paris Hilton and Wayne Rooney and others of a vajazzle persuasion. When I heard that the Manchester City player with the Brazilian on his head had recently had eggs thrown at his car during a night out, I knew immediately what sort of car it would be. And I was right. Mario Balotelli drives a Continental. Of course he does.

There's more. This is not a car that claims to be especially fast. Nor is it designed to be especially sporty. So why does it need a huge, twin-turbocharged 6-litre W12 engine? I'll tell you why. So that the people who buy it can boast about it to their friends down at the Dog and Footballer: 'You've only got a V. I've gorra bloody W.'

Well, in the new model it's gone, and in its stead there's a lighter, nimbler, sportier twin-turbo V8. It's loosely related to the engine you'll find in the new Audi S8, only in the Bentley it sounds dirty. As though it smokes sixty a day. It is a glorious sound. And many times I cracked the double-glazed side windows to revel in its earthiness.

It hasn't only transformed the way the car sounds. It's also transformed the way it feels. I'm not going to call it chuckable, but it's way, way better than Mr Rooney's house brick. And way, way more economical, too. Wayne will get 17 mpg if he's lucky. The new car will do 27 mpg.

Of course, you may imagine that with 2,000 more cubic centimetres and four more cylinders, Wayne and Paris are going to leave you far behind at the lights. Not so. From rest to 60 mph the new car is hardly any slower than the W12 version, and at the top end it'll still nudge 190 mph. What's more, it makes a better noise, it's more fun and it's more economical. And £12,000 cheaper.

Inside, it's much as you'd expect. Quilted leather, heater knobs that appear to have been mounted in honey, a dash made from turned aluminium and electronics from that endless well of common sense, Volkswagen.

It's strange. All the things I hated in the big-engined car, I found myself liking in the new one. Four-wheel drive? Used to be pointless. Now it's just right and exactly what's needed. The overly sensitive parking distance control? Used to drive me mad. Not any more. The weight? Ridiculous. Now? It's excellent – it makes the car feel solid and planted.

I suppose it's a bit like, say, Derby. On a grey, wet, chilly day, it's the worst place on earth. But change the sky to blue and all of a sudden it's lovely. That's what happened with the Continental. One detail has changed and it's made everything seem better.

Except the styling. I'm sorry, but to look at, this car has all the appeal of a gangrenous wound. I'd sooner put the headless body of a moose on my drive.

Bentley needs to address this soon, and while it's on the subject it also needs to do something about its brand ambassadors. In short, the next time a Manchester-based footballer steps into one of its showrooms, the company should give him a cheque for £200,000 and tell him to buy something else.

12 February 2012

The arms race is over and Vera Lynn has won

Aston Martin DBS Carbon Edition

For the past couple of weeks, everyone has been running around, waving their arms in the air and wondering what in God's name possessed the Indian government to express a preference for a bunch of Frog warplanes rather than the faster, more manoeuvrable and far deadlier Eurofighter Typhoon.

Some have said this is India sticking two fingers up at its former colonial masters. Some have said that France has offered to do some kind of behind-the-scenes nuclear deal as a sweetener. Some blame David Cameron, and others will undoubtedly point the finger at the *Top Gear* Christmas special.

But I suspect the real reason the Indians look set to go with the French is this: their Rafale plane is around £20 million cheaper than a Eurofighter. And it's all very well saying that the British-backed plane has an operational ceiling of 55,000 feet – 5,000 higher than the French jet – because so what?

The Eurofighter was designed to hold back an invasion of western Europe. It's designed to reach the dark blue edge of space so that it can shoot down spy planes. It's designed to reach Mach 2 so that it can intercept Soviet bombers coming in low over the North Sea. It's designed to twist and turn so that it can enter a dogfight with a fearsome MiG 29.

All of which is of no use to the Indians because the Tamil Tigers do not have any spy planes and the Bangladeshis do not have any low-level nuclear bombers. Yes, Pakistan can muster quite a few American-made F-16s, but these have a notoriously short range before they have to be refuelled. India, then, doesn't need technology. It just needs a plane with a gun on it. And frankly, the Rafale will do.

Today, with the arms race over, I have no idea why the jet fighter is being developed at all. The Americans announced recently that their new F-35 is bedevilled with many technical issues – such as it disintegrating every time it lands. And I can't help thinking, What are you doing?

The US navy has its F-14 Tomcats and F/A-18 Hornets, and it's hard to see why these need replacing. Because it's not like any other country in the world has anything even half as capable.

I must confess I sometimes think along the same lines when I'm musing about cars. Because where we are now is plainly good enough, and recent attempts to make cars better have made them, actually, a little bit worse.

Take the horsepower race as an example. Audi, BMW, Mercedes and to a certain extent Jaguar have been trying to outdo one another for years, so that each new model had to be more powerful than anything else on the market. Sounds fine. But we have reached a point where a road car simply cannot handle the potency of the engine under the bonnet. Try driving an AMG Mercedes up a gravel drive to see what I mean. It just digs a hole.

And then try doing a full-bore standing start in a Nissan GT-R. It's so vicious, it hurts your head. Why would you want that? Why would you want a car that gives you whiplash and light bruising every time it starts moving?

The fact is that 350 horsepower is probably enough. And with only that much under the bonnet, there'd be no need for extremely expensive brakes, tyres and suspension components. Furthermore, less power from under the bonnet would mean better emissions, which would in turn remove the need for electronic power steering and flappy-paddle gearboxes, and engines that shut down at the lights and all the other green-sop trinketry that's only necessary because of the sheer amount of fuel needed to produce a shedload of power that can never be used.

Maybe, then, in order to go forwards, we ought to go backwards a little bit. And that brings me neatly to the door of the Aston Martin DBS Carbon Edition. There are mutterings in the rectory that Aston Martin needs to make a new car fairly soon, that its current range has been around for too long. And received wisdom would suggest these mutterings are quite correct. Aston may be able to say that it has launched both the DBS and the Virage but both of these are DB9s with fancy bits added on. And we're not fooled.

Now comes the DBS Carbon Edition, which sounds intriguing, but actually it's just a standard car that has special paint. I do not know why the paint is special but there you are. So as I pulled out of my gates, I was ready to be a bit cynical, to suggest that the time has come for Aston's Kuwaiti backers to cough up some development readies. But then I started to think, Right. And if they did, how would they improve on what we have here?

Would the new car be better looking? That's highly unlikely. The DBS is about as close to perfect as a car can be. Would it be more powerful? The marketing department would insist on it, and would that be a good thing? Probably not. Because more power would mean more torque and that would mean the traction control would have to work overtime just to keep you pointing in vaguely the right direction.

No. The fact is that when the DBS first saw the light of day, I fell head over heels in love with it, and while it has been overshadowed in recent months by various new Ferraris and the Mercedes SLS, it remains an absolute joy.

The big V12 engine snarls and growls when you poke it with a stick but it settles down to a distant hum when you are just loping down the motorway. It's much the same story with the suspension. Put it in sport mode and this is a proper racing car but leave it in 'normal' and it cruises. I know of no other car that pulls off this trick quite so well. Some are quite good at being Dr Jekyll and Mr Hyde. But the DBS? It's Dr Jekyll and Vera Lynn.

Better still, though, is the reaction this car prompts from other road users. Normally, when I'm driving something very expensive, people in vans are quick with a belittling observation about price, practicality, insurance or economy. But when they see the DBS, to a man and woman, they all say the same thing, 'That is gorgeous.'

It's not just here, either. It doesn't matter where in the world we take our *Top Gear Live* show, one thing is constant: people always like the Aston best.

There are some issues, chief among which is the satnav system. It's far better than the Volvo unit that was fitted in earlier models but it's still way too complicated. Why, for instance, is the rocker switch that adjusts the scale of the map hidden behind the steering wheel? And why, just after you've found it, does it decide that, actually, it's the switch for changing radio stations.

Then there's the price. Mainly because we sort of know the DBS is just a DB9, we can't understand why it costs far more. And nor do we fully understand why this Carbon Edition adds yet more. Mostly, though, the problem is that it feels like you are driving around in an old car. There's no sense that you're riding the technological wave.

You may feel that, as a result, your friends will not wish to speak with you any more. However, the fact is that sometimes, old school is better than new school. For proof of this, ring the Royal Navy and ask if it wishes it had kept its Harriers.

19 February 2012

Good doggy – let's give the bark plugs a workout

Suzuki Swift Sport 1.6

Over the years I have watched several hundred games of car football. It's a staple ingredient of the *Top Gear Live* show and is much like normal football, except in every single detail. There are, for instance, three players on either side, and each of them is in a car. The ball is 4 foot across. There are no goalkeepers and the game lasts until James May's team has lost.

We have used many types of car, and, as a result, I'm in the unique position of being able to say what's best if you want to go down to the park this morning for a drive-about with your son and his mates. I recognize this is a fairly limited sliver of the market, but I try to cater for everyone, so here goes. The Reliant Robin is hopeless. While the engine and in particular the gearbox are strong, it has a habit of falling over every time you try to cut inside an opposing player. Or track back. Or shoot. Or defend. Or do anything, really.

The Smart Fortwo is even worse. This also has a habit of falling over and, to complicate matters, the gearbox doesn't allow you to switch from first to reverse quickly enough. Sometimes all six players can be stationary, as they wait for the car's brain to allow the transmission to shift. This makes for a boring spectacle.

The Toyota Aygo is much better, as is the Austin Landcrab, but the best of the lot is the old Suzuki Swift. It's small, nimble and tough. The only real problem is the windscreen washer bottle, which is located just behind the front bumper, where it is easily damaged.

And it's not just car football in which that little car scored well. It was also extremely good at ice hockey. I know this because, when we filmed the *Top Gear Winter Olympics* show some years ago, we used Swifts in the rink and they were epic, whizzing hither and thither in a blizzard of snow, with cheeky exhaust noises and panache.

The Swift is a rare thing – one of only a trio of cars that all three of us on *Top Gear* like. The others, in case you're interested, are the Ford Mondeo and the Subaru Legacy Outback. I liked the Swift so much that when I reviewed it on these pages back in 2006 I toyed seriously with giving it five stars. It was as good as a Mini Cooper, I reckoned, but it cost a lot less.

Well, now there's a new model. It's noticeably bigger than before, which means it'll be less manoeuvrable on the football pitch and more exposed as well. It also means it'll be harder to park in a town. That would be a price worth paying if the extra length and width translated into more cabin space. It doesn't. The rear seats are very cramped and the boot is almost amusingly small.

Happily, however, the extra size doesn't seem to have affected the weight very badly. This is an unbelievably light car, which is good for fuel economy and good for speed. Very good, actually. I've done some checking and it has a similar power-to-weight ratio to that of the old Peugeot 205 GTI. And that's a vital comparison . . .

I miss the old Peugeot and all of the other hot hatchbacks from the Eighties. There was a time when 12 per cent of all Ford Escorts sold in Britain were hotted-up XR3s and you couldn't park in Fulham because every street was rammed with Lhasa green Volkswagen Golf GTIs. Every car maker made a hot hatchback then because we all wanted one.

Today everyone seems to have forgotten the recipe. The current Golf GTI is a lovely thing to drive, but all the telltale styling details that set the original apart from the standard models are gone. Take away the badge and the car now looks almost identical to the diesel.

Ford has gone the other way. Its hot Focuses look as though they've crashed into a motorists' discount shop and every single thing in there has become attached. Spoilers, vivid brake callipers, scoops, vents. They're vajazzles with wheels.

Then you have Renault. It does an excellent range of hot hatches but they're all stripped out – racy, knowing. They seem to be saying that if you buy one, you are interested only in high-g cornering, and that was never the point of a hot hatch.

They were supposed to do everything. They were fast when you wanted them to be and restrained when you didn't. They were supposed to stand out from the crowd but only if you were concentrating. And they were supposed to be cheap. Which counts out the Citroën DS3. And the Mini. And the forthcoming Audi A1 quattro.

The car makers would probably argue that there's no point making a crash-'em-and-bash-'em hot hatch these days because people are more interested in space and style and they all want a Range Rover Evoque. But the only reason we all want an Evoque is that you aren't making a thrill-a-minute, bung-it-into-a-parking-space and put-the-dog-in-the-back hot hatches any more.

Which brings me back to the Swift Sport. The twin-cam 1.6-litre engine is a little gem. It has a variable intake system and variable valve timing, so although it's small, it's nicely techie. Starting it – with a button these days rather than a key – is like poking a small dog with a stick. Immediately, it's keen to be off, jumping up and down and making excited whimpering noises.

To keep that feeling alive, the gearing is incredibly short. Think west highland terrier rather than alsatian, or even spaniel. This means the motor is always on the boil, always at peak revs and therefore always ready to go. From standstill to 62 mph takes 8.7 seconds. Pretty much exactly how long it took a 1.6-litre 205 GTI, incidentally.

But there is a bit of a drawback here. In sixth, at 85 mph, it's doing 3000 rpm, which makes it noisy. Very noisy. I'm tempted to say too noisy. It gave me a headache.

The upside comes, though, when you get off the motorway, because, wahey, this is a car that takes you back to the mid-Eighties. It feels eager and crisp, so it turns even the most dreary journey to buy milk into a fun-filled extravaganza of puppy-dog enthusiasm, squealy tyres and grinning.

It's not that fast, but this, remember, is a £13,499 car. To my mind, Suzuki has got the blend between cheap and cheerful exactly right.

The styling recipe is bang-on as well. The alloy wheels are just the right design and size, and it has two exhausts – on either side of the car.

Inside, the seats are buckety without being stupid, and you get exactly the right level of equipment – cruise control, air-conditioning, USB connectivity and seven airbags. There's also a good radio.

The new car has had rave reviews from all the motoring magazines and specialist writers, and it's got one from me, too. It's noisy and the boot is microscopic. But other than these things, it's brilliant and almost certainly the best small car on the market today.

26 February 2012

Look what oi got, Farmer Giles: diamanté wellies

Jeep Wrangler 2.8 CRD Sahara Auto 4-door

Recently Bristol crown court sentenced a bus driver to seventeen months in prison after he was caught on CCTV deliberately ramming his vehicle into a cyclist. I've watched the footage and it's extraordinary. The bus driver comes alongside the bicyclist and then veers sharply to the left, flinging the poor man 10 feet through the air, breaking his leg and crushing his precious bike. It's plainly a moment of madness. A temper tantrum. A spot of road rage that got out of hand.

But if you watch the footage again, carefully, you see that, to begin with, the cyclist is pedalling along quite slowly in front of the bus. And when the bus moves right to overtake, the cyclist appears to go right as well. Was it an act of provocation on the cyclist's part?

For many years bus drivers have been told by the authorities and those who read the *Guardian* that they are knights of the road, eco-warriors on a mission from God. They were given their own lanes, and car drivers were ordered by Her Majesty's government to get out of their way.

This went to their heads. So as soon as the last passenger was seated, they would simply pull out, even if a car was alongside. On many occasions I've been forced to swerve into the path of oncoming traffic by a bus that's set off without warning. And, of course, if there had been a trial or an inquest, its driver would have been given a tree or some tofu for taking the good fight to those whose cars were making life so unpleasant for the world's polar bears.

But then someone of a *Guardian* disposition decided that,

actually, bicycles were an even better way of going to work than the bus. So cyclists were suddenly given their own lanes, and their own special spaces at junctions. And there was talk that in any impact between a motorized vehicle and a bike, the driver would – no matter what the circumstances – be blamed.

So, all of a sudden the roads are filled with two groups of people who believe they have right on their side. It's the Judean People's Front and the People's Front of Judea. It's all the animals being equal but some wishing to be more equal than others. And the consequence is inevitable. One man is in prison. The other suffered a broken leg.

The only solution is to take away their special lanes and their priorities. It's to make them understand that they may use the roads but only if they're jolly careful, because roads have always been for 'the people'. And the vast majority of people have cars.

Mind you, on Kensington High Street last week it didn't feel that way. I thought at one point I'd become involved in a west London étape of the Tour de France. It was 7 p.m. and there were hundreds and hundreds of people with wizened bottoms and beards and idiotic hats and luminous clothing, cycling through red lights at way beyond the speed limit.

As they passed, many shouted abuse at me for daring to be there, stationary at a red light or cruising along at a mere 25 mph. Some banged on my doors. Some bared their teeth. It was awful and I considered carefully the idea of running one of them down. Maybe two.

Mind you, some of their rage may have had something to do with the fact that I was driving a Jeep Wrangler, which is a big American off-roader. They obviously hated it very much and that's why I decided to leave them alone. Because I did too.

As we know, Jeep began by making rugged military 4x4s in 1941. General Dwight D. Eisenhower said the second world war could not have been won without the company's first effort. It was a simple thing, too, which is why many claim Jeep stands for 'just enough essential parts'.

Over the years, that original morphed gradually into the Wrangler. This was often converted by its fanbase into a high-riding, doorless, roofless monster with a V8 under the bonnet. It was usually to be found cruising around in Key West in Florida with a giant purple eagle on the bonnet. But, crucially, it still worked well off road. Indeed, one of the most enjoyable drives I've ever had in any car was in a Wrangler, going up the craggy Rubicon trail over the Sierra Nevada mountains in California. Its heart may have been in San Francisco, but its soul was still in the back country.

Unfortunately Jeep decided to start selling its Wrangler in other countries – countries in which people do not talk loudly around the swimming pool and giant purple eagles are considered poor form. In Britain, for example, we have the Land Rover. And Germany has the Mercedes G-wagen. So Jeep has decided its Wrangler should become more restrained. More practical. More European. And it hasn't worked at all. First, it is extremely ugly. And, second, you can't see out of it. The blind spots are so big, bicycles are invisible. So are buses. So is the Albert Hall.

There's more. The only way Jeep has been able to fit in rear doors and seats is by shunting the front seat so far forwards that you can – and must, in fact – operate all the controls with your face. To make matters worse, the satnav screen is so bright that once the sun has set, it's as if you're driving into the beam of an alien spaceship's searchlight.

Of course, there are many levers and switches that mean it still works off road, but I'm afraid that on the road it's not good at all. The 2.8-litre diesel engine feels as though it was designed by people who had no real concept of how such a thing might work. We have a name for these people: Americans.

Then there's the suspension. It is very soft. So soft, in fact, that you can drive over even the most alarming hump at whatever speed takes your fancy and you won't notice it at all. On the downside, it feels as if you may fall out on every bend. And the steering is woeful. And, strangely, the ride on the motorway is unbelievably fidgety.

Then again, a Land Rover Defender is pretty hopeless on the road as well. But that doesn't pretend to be a luxury tool, whereas the Wrangler I tested does. It has roof panels that lift out – if you have a PhD in engineering – and cruise control and lots of gizmos. It's like a diamanté wellington. A gold-plated cowpat. A village idiot at the Savoy.

There's a sign picked out on the dash that says, 'Since 1941'. What it should say is 'Mechanically unchanged since 1941'.

It's a shame. I used to like the old Wrangler. I know it was a bit, ahem, Venice Beach, and that if it were a man, it would shave its scrotum and enjoy going to the gym. But it was a good and interesting take on the 4x4 theme.

The new version has lost all that. It's trying to be a Land Rover for those who also want a few creature comforts. And for £28,000 it's not as if you're short of alternatives. All of which are considerably better. Except cycling, of course, or using the bus.

4 March 2012

Powered by beetroot, the hand-me-down that keeps Russia rolling

Lada Riva

We see many Russians in the big, wide world these days and they appear to be extremely well off. They always have enormous watches, huge cars and embroidered jeans. Many also have football clubs.

So you would imagine that if you were running an airline, you would try to impress these newly moneyed people by lavishing your Moscow service with your latest, newest, shiniest aircraft. Weirdly, British Airways has chosen to do the exact opposite.

In my experience, BA puts its best aeroplanes on the transatlantic routes, and then, when the fittings and fixtures are a bit tired, the aircraft are relieved of their JFK duties and are used to ferry holidaymakers to the Caribbean. When they are too decrepit even for that, they take people to and from Uganda, and I thought that afterwards they were scrapped – or sold to Angola. But no.

It seems they are then given to the fire service, which uses them to train crews on the art of passenger evacuation, and then to the SAS, who run about in the burnt carcasses, shooting at imaginary terrorists. After that, they are used on the Moscow route.

I recently flew with BA to Russia, and to give you an idea of how old the plane was, I will tell you the on-board entertainment system used VHS tapes. And to make the quality even less impressive, my television screen was less than 2 inches across. And it was located on the bulkhead, several feet in front of my face. Also, it was broken. So was the lavatory seat.

I was going to write a letter to the chairman of BA, explaining that he's got it all the wrong way round. Using your best aircraft to compete with America's airlines – which are exclusively staffed by fat, bossy women and serve rubbish food – is like Chelsea fielding their best team to take on Doncaster Rovers.

I was going to point out to him – because plainly he doesn't know – that the Berlin Wall has gone, and that the Russians are no longer queuing for six years to buy a beetroot and then being shot for saying it's a bit warty.

I was also going to invite him to take a look around Gum, the department store in Red Square. There was a time when people would come from thousands of miles away because it had just taken delivery of some pencils. Now it makes the Westfield shopping centres in London look like an Ethiopian's larder. The smallest watch on display is bigger than the TV screen I hadn't been able to watch on my flight, and the underwear costs more than the ticket.

It's not just materialism, either. In Russia people are now free to say absolutely anything that comes into their heads. Talk as Russians do in Britain and you'd be hauled over the coals for racism and branded a bigot. You want to suggest the legal age of consent should be lowered to twelve? Go right ahead. People won't call you a paedo; they'll be interested to hear why you think that way. They went seventy years without being able to discuss anything. Now they want to discuss everything.

Of course, you are not able to write too disparagingly about Vladimir Putin, unless you want some radioactivity with your bacon and egg, but you can sure as hell say what you like – to whoever you're with. I found it fantastically liberating.

There are other things, too. In Britain if Sir Philip Green or Lord Sir Sugar were to spend an evening at the Wolseley restaurant in the middle of London playing tonsil hockey with a phalanx of 6-foot hookers, tongues would wag. In Russia that sort of thing appears to be quite normal.

A friend texted while I was there to say, 'Be careful. Moscow is bad for your soul.' He's wrong. It's not bad for your soul, but

I bet it could be very bad for your marriage, your bank balance and your gentleman's area.

Moscow buzzes and hums. You should try the bone marrow in Cafe Pushkin and spend a few minutes at the side of the road seeing if you can spot a car that would cost less than £50,000 if you'd bought it in Britain. Then check out the pavements and see if you can find one single girl who's fat or less than 6 foot tall or not wearing a beautifully cut pair of jeans. I have no idea what Hugh Hefner's wet dreams are like. But I bet they'd be along these lines.

I went to the Kremlin at one point to discover it's all been done up and refurbished. Not so it resembles how it might have looked in the past but so that foreign diplomats are blown into the middle of next week by the grandeur. Every room is like being inside the mind of a gold-obsessed four-year-old princess.

And then, just as I'd decided that Russia looked like the love child of Monte Carlo and Kuwait, with a little help from Onyx on Thames, someone leant over and told me the Lada Riva was still in production. And I'm sorry, but that's like being told that the king of Saudi Arabia does his own washing, with a tub and a mangle.

Why would Lada still be making the Riva? What could anyone I'd seen in my whole visit want with a car as nasty as that? Or has it been improved radically since it was the staple wheeled diet of Mr Arthur Scargill's disciples? I had to find out. So I did. And it hasn't. In fact, I think it's become worse.

The Riva began its life in 1966 in Turin, where it was known as the Fiat 124. Fiat did a deal with the Communists, helping to build a factory in Russia in which the company's old design would continue to be produced. This became the Lada Riva.

Fans will tell you that much changed over the years, but I can report that actually nothing changed at all. Except that now the Riva is also made in the great car-producing nations of Ukraine and Egypt.

I don't know where the car I drove was made. Or who made

it. But I suspect he was very angry about something because it was horrific. The steering column appeared to have been welded to the dashboard so that it wouldn't turn. The brakes caused the car to speed up a bit and turn left, violently, at the same time.

The buttons on the dash appeared to have been put in place by Janet Ellis from *Blue Peter*, and the engine had plainly been lifted from a cement mixer that had spent the past thirty years chewing up rebel soldiers in southern Sudan.

It would get from 0 to 60 mph. But only when it was built by Fiat. Since it became a Lada, it hasn't really been able to move at all. And, boy, is it badly made.

When I eventually ran it over with a monster truck, it folded in half. And to put that in perspective, let me explain at this point that when the very same monster truck ran over an Indian-made CityRover recently, the car was pretty much OK afterwards.

Why, then, is the Riva still being manufactured? Why are there people in Russia still buying it? Could it be, perhaps, that behind the white-toothed, gold-capped Moscow smile, the rest of the country is – how can I put this – a bit poor?

Maybe, in other words, the chairman of BA knows something I didn't realize: that those who can afford to fly in Russia have their own planes.

And those who can't are stuck out in the middle of nowhere boiling swede in the hope that one day they'll be able to afford a car that's forty-five years old before it's left the bloody showroom.

11 March 2012

The yummiest of ingredients but the soufflé's gone flat

Porsche 911 Carrera

When I was growing up in the 1960s and 1970s, labour- and time-saving devices were all the rage. The Ronco Buttoneer, for instance, made putting on a button a quick and easy job. Which was just as well because the button you'd just attached often came adrift again in a matter of moments.

The top-loading washing machine had replaced the front step, and then came the remote-control box for the television, which meant we no longer had to sit through *Nationwide* because we couldn't be bothered to get off our backsides. We also waved goodbye to the punka wallah with the invention of the Pifco fan. Life was very good.

But at some point in recent years someone decided to put the complication back. So now, instead of adding boiling water to a spoonful of instant coffee, we have machines that require constant attention. Every single morning mine wants more water, or more beans. Then it wants me to empty its trays and clean its pipes and decalcify its innards. Making a simple cup of coffee has become a thirty-minute palaver.

It's much the same story with my mobile phone. Because it turns out that even when you are not using an application, it's still open, in the background, chewing the battery. And shutting it down is a complex procedure that usually ends up with you taking a photograph of your own nose.

Televisions are massively complicated now. And gone are the days when you simply loaded a VHS tape and watched a movie. Now, with Blu-ray, the machinery takes ten minutes to warm up and you have to sit through hours and hours of waivers, copyright threats and trailers.

My dishwasher is more complex than Apollo 11, my juicer has a 200-page instruction book and have you tried to use a pay-by-phone parking meter? Of course not, or you'd still be out there, in the street, asking yourself what on earth was wrong with putting a pound coin in a little slot.

Naturally, cars are now very complicated as well. It's almost certainly true to say that the ignition key for your modern car is more complex than the whole of an Austin A35. Which means, of course, it rarely works. I've lost count of the number of times I've been in a car that keeps flashing up a message saying, 'No key detected,' when I'm sitting there waving the damn thing in front of its dash, whimpering slightly and wondering out loud what was wrong with the old system.

Then there's the BMW M5, which can get from 0 to 62 mph in about thirteen minutes. You spend twelve minutes and 55.7 seconds telling the on-board computer what sort of setting you'd like from the gearbox, the chassis and the engine, and then 4.3 seconds going from 0 to 62.

You might imagine that the new Porsche 911 had been spared all this nonsense – 911s, after all, are meant to be pure, clean, unfettered sports cars. And there is no place for complexity in such things.

Well, dream on, because the new 911 is a geek's fantasy. Every component can be tuned while you're on the move to deliver something different, and there are now two readouts on the dash telling you what gear you're in. Which seems a bit odd in a manual. I know I'm in third. I just moved the lever.

The thing is, though, this being a Porsche, it's all very instinctive and commonsensical. Amazingly, since there are no buttons on the steering wheel itself, you don't have to go into submenus or hold knobs down for two seconds to make stuff happen. I hate to admit it, but I thought it was brilliant. But that's probably because I never bought the whole 911 sports-car thing in the first place.

There was a lot more I liked as well. The styling may be ludicrously similar to that of the previous model. And the one before that. And the one before that as well. But the little things that have changed have given the new model some nice new curves. You could even call it good-looking.

The big debate about this car is its electric steering. Because of European Union rules on emissions, manufacturers are under pressure to introduce systems that use less energy, whether or not they are better at the job. So the conventional hydraulic power-steering setup has been ditched in favour of one that works off the battery.

In the same way as Neil Young keeps banging on about the awfulness of digital sound compared with vinyl, various 911 purists say that the classic 'feel' of a 911 is now gone. And I'd agree with that. But since I'm not a 911 purist, I must say I think the new system is better. For sure, you are getting an artificial sense of how the tyres are interacting with the road and, yes, on a track you can spot this. But for everyday driving, the electric system is meaty and tremendous.

Emissions regulations have had other effects as well. The engine now shuts down at the lights and Porsche has had to fit a seven-speed gearbox. In theory this is fine. You lope up the motorway at tickover, sipping fuel like a vicar sips sherry. But when you're in seventh, doing 60 mph, you don't get the twitching and fizzing you expect from a car of this type. It feels a bit puddingy.

Of course, when you get off the motorway and realize you're running late and you need to make up some time, it's not puddingy at all. It's just delightful. That said, I would opt for the bigger-engined S model. The standard car I drove, while lovely, sometimes didn't feel as fast as I'd been expecting.

Now normally when I've reviewed 911s in the past, I'd get to this point and say that while the car is jolly clever, it's not for me. The rear-engined Porsche is like Greece and marzipan and Piers Morgan. Simply not my cup of tea.

But this one is different. Over the years, the engine has crept forward in the chassis so that it's no longer slung behind the rear axle waiting to become a giant pendulum. It's water-cooled, too, these days, which means the Volkswagen air-cooled clatter is gone.

Inside, the silly buttons that looked like half-sucked boiled sweets and felt about as cheap as an Albanian's suit have been replaced with good, high-quality items. The driving position is better, the seats are wonderful and though the car is now bigger than ever, it's still small compared with all its rivals. That's a good thing.

Drawbacks? Two, as I see it. The boot's at the front, which means you get dirty fingers every time you open it; and Porsche has never shaken off the City boy braces-and-Bollinger image it earned in the Eighties. Which means you are never, ever, let out of side turnings.

OK. Two and a half. The engine isn't quite gutsy enough. But go for the S and that's resolved. In spades. Just avoid the convertibles. Unless you enjoy looking a plonker.

I'm sure there is much that will disappoint the diehard 911 fan in the new effort. But there is so much to delight those of us who have never liked 911s. I could even see myself buying one. It's a fab car. Really, really fab. And, all things considered, good value as well.

PS: Since finishing this piece, I've realized the Porsche actually gets no stars at all because it's useless. Last Sunday the tyre went flat. There is no spare. And no depot carried anything that would fit.

Recently a friend of mind had a flat tyre in his 911 and it took

Porsche two weeks to find a replacement. Unless the manufacturer can address this, there is simply no point buying its cars. Because one day you will need, say, to take your mum to hospital and you will have to phone and cancel.

18 March 2012

I ran into an EU busybody and didn't feel a thing

BMW 640d (with M Sport package)

After the recent and very sad deaths of six British soldiers in Afghanistan, questions were immediately asked about the worthiness of the Warrior armoured vehicle in which they were travelling when the bomb went off.

And equally immediately they were answered. The Warrior fleet in Afghanistan was upgraded last June at a cost of more than half a million quid a pop, with armour better able to deal with an explosion and improved seating to protect those inside from the shockwave.

The trouble is, of course, that the men who go to war in beach footwear and skirts know full well that this has happened and are now using bigger bombs. This means the Warriors will have to be upgraded again, which will mean more explosives are needed to blow them up. It's a problem that's faced military commanders since the dawn of time. And it's a problem that will never end.

Each time there's a tragedy, coroners can point the finger of blame. They can accuse defence chiefs of penny-pinching and the engineers who design these vehicles of incompetence. But the reality is very simple. If a bomb is big enough, it will tear through anything. And there's nothing that can be done to change that.

Or is there? Because the truth is that man is constantly faced with seemingly insurmountable problems, and we have a habit of working out a solution. We devised ways of getting iron to float and to fly. We developed antibiotics to combat disease. We are clever. And nowhere is this truism more evident than in the car industry.

Every year, the European Union erects a set of ecological fuel-saving goalposts through which it demands car makers must pass if they want to continue doing business. And every year the motor manufacturers squeal and whimper and claim it can't be done. Then they do it. And then the EU responds by moving the goalposts further away.

The most recent move was a big one, and it's having a profound effect on how cars feel. You may wonder, for instance, why the easy-to-use automatic gearbox is now being ditched so eagerly in favour of a robotized manual system. These gearboxes invariably make town driving jerkier and I hate them with a passion. But an engine sending its power to the wheels through this system uses less fuel than an engine sending its power to the wheels through a torque converter.

And the car makers that are sticking with the traditional auto are now offering a setup with eight speeds. This means the car is constantly changing up or down. It's very annoying. But with more cogs, the engine has to work less hard. And that means more mpg, which means the oil supplies will last a little longer.

It gets worse. Even Porsche has now started to use electric power steering, which means you only get a digital interpretation of what's happening up front rather than the real deal.

And then we get to the starter motor. That now has to be the strongest component in many cars because their engines shut down when you stop at a set of lights. Then start again when you want to set off. I don't know why this irritates me so much, especially as usually you can turn the system off, but it does.

Not half as much, though, as the trend towards dashboard read-outs telling me what gear I should be in. You're behind a Peugeot on an A-road and are waiting for an opportunity to overtake. This means you are in third. 'You should be in fifth,' it says. It's wrong. It doesn't know what I'm going to do next. It doesn't know I'm being delayed by an old man. It even has an opinion on changing down for a corner. 'Nope,' it says as you slide it into fourth for a long left. And this is a system you can't turn off. Unless you have a hammer to hand.

Another effect of the legislation is the trend for engine designers to replace cubic capacity with turbocharging. A turbo engine uses waste exhaust gases to spin a fan, which is then used to force air and fuel into the engine under pressure. Sounds great. But a turbo engine cannot have the immediacy of a free-breather. There has to be a delay between putting your foot down and actually going, as you wait for the exhaust gases to gather enough force to spin that fan.

These, then, are just some of the tricks being used by car makers to shoot their cars through the EU's goalposts. And every single one of them makes a car a little bit worse.

Which means it's now the job of car makers to mask the problems with the Elastoplast of ingenuity. And that brings me on to the diesel engine under the bonnet of the BMW 640d.

Yup. It's a diesel in a BMW sports coupé, and if that isn't a sign of the times, I don't know what is. But it's a BMW diesel, which means it hums rather than clatters. And it's fitted with two turbochargers. A small one that gets going almost immediately, and then a bigger one to give you some oomph on the open road. Clever, eh?

Less clever is the name. Why is this called the 640d when the 535d has exactly the same engine? Not very logical for a company famed for its obsessive-compulsive nomenclature.

Naturally the engine is mated to an eight-speed box but to make sure you don't notice the constant cog-swapping, the changes are so smooth, you don't feel anything at all. It's like being on a never-ending helter-skelter of torque.

This car is like an old house. It's riddled with cracks but because of some extremely skilful plasterwork, you can't spot them. BMW has addressed the problems presented by the EU and, in less than a year, has masked the efforts it made to overcome them with the silky smile of German efficiency.

But what of the car itself? Well, there are plenty of other

coupés on the market, but none has so much space in the back and none is anything like as squidgy Slumberdown-soft.

There's a misplaced line of thinking in the car industry that anyone who buys a good-looking two-door coupé must therefore want a bone-shaking sporty ride. Some do, for sure, but plenty don't. And that's where the 6-series is brilliant. It is good-looking. It is a two-door coupé. It is stylish. And yet it rides like a hovercraft. Fit the optional £1,485 Comfort seats and it's like driving around in a cloud.

There is a button that allows you to firm everything up, and even a sub-menu in the computer that lets you choose which bits of the package you want to be sporty and which you do not. And I recommend that on day one, you glue the switch in Comfort mode and leave it there. The Sport setting just makes you uncomfortable for no real gain in terms of handling.

What we have here, then, is a car that doesn't just get through the EU's goalposts but also goalposts that no other car maker has spotted. Goalposts for people who want good looks, comfort and economy.

25 March 2012

Blimey, you've got this mouse to roar, Fritz

Volkswagen High Up!

You don't need an iPad. You can watch films on your laptop, you can store data on your phone and for taking pictures it'd be easier to set up an easel and break out the oils. An iPad is stupid. A complete waste of money, especially if you already have an iPhone, which does the same job. And can be used for speaking to other people, too. So why did I trot quite vigorously to the shop and buy one? Simple: iPads look nice. That's it. The end.

We see exactly the same thing going on with Fiat 500s. Why do people buy them? Because they are spacious? Because they are fast? Because they are economical? No. They are surprisingly uneconomical, in fact. The only reason the little Italian cutester is to be found clogging up every chic street this side of the Urals is that it looks nice. That's it. The end.

I could sit here now and tell you that, mechanically, the 500 is identical to various cheaper or more practical cars such as the Ford Ka and the Fiat Panda. I could tell you too that there is very little space in the back, that some drivers have found even the two-cylinder TwinAir eco-version struggles to do 36 mpg and that the 500 is made in Poland by people who just want to go home and watch telly.

But it'll make no difference. The Fiat 500 is like a useless little mongrel at the dog's home. The one with the wonky ears and the sad eyes. You know that it'll be a bad buy. You know it will leak. But you soooo want it. It's soooo sweeeet. And that's the end of that.

Except it isn't, because wading into the fray is the new

Volkswagen Up!, which unlike the Fiat 500 comes as standard with its very own exclamation mark.

There are plenty of variants from which to choose. There's a Take Up!, a Move Up! and a High Up!, and then you have the colour-based special editions, which were going to be called the White Up! and the Black Up!. Until a VW bigwig realized a company that made Hitler's favourite runaround shouldn't really be selling a car called the Black Up!. So now it's called the Up! Black. Clear? Good.

So let's move on to the engines. There's one. It's a naturally aspirated three-cylinder 1-litre unit that is available in two states of tune. These are: not powerful enough, and nowhere near powerful enough. Prices start at less than £8,000, although the car I tested – the 74-horsepower High Up! – was £10,390, plus an extra £35 for carpets, £350 for cruise control and £225 for a laser to stop it crashing into things in towns. And what I'm going to do now is waste your time and several hundred of my words explaining why you should buy the Up! rather than the little Fiat.

First of all, the Up! is a lot more spacious inside. It has the longest wheelbase of any city car, and that means there really is space in the back for two children. Yes. Of course you can also get two children in the back of a Fiat, but only if you kill them and chop them up first.

Moving further forwards, I will agree the Fiat has a funky and attractive dash. But the VW's is cleverer because in the High Up! it comes with a detachable 'maps and more' touchscreen that can be used outside the car like a TomTom satnav and then, when it's clipped in place, as a phone interface, a satnav, an entertainment system or to show driving statistics. It works brilliantly.

I like the way the dash appears to be a big slab of painted metal, too, and even though this little car is made in Bratislava, it still feels Germanically, Speerishly well put together. I'd love to say the same of the Fiat but I can't. So I won't.

Further forwards still, we get to the engine, and here we stumble over the VW's first black mark. Even though I was driving the most powerful Up!, I couldn't even think about pulling into the outside lane of a motorway unless there had been a terrible crash and the whole carriageway was blocked.

It doesn't matter how far back the faster traffic may be – you will soon be in its way. Really, VW should have called it the Hold Up!.

It's fairly pedestrian from 0 to 62 mph, but it's the time it takes to get from 62 to 70 mph that really alarms. We're talking hours and hours. The problem was that I had only 70 torques, whereas the little Fiat, which has one cylinder fewer, has 107.

Naturally Volkswagen will argue that the Up! is a city car and that this lack of oomph is of no consequence. But that's rubbish. It's OK to have a pair of city shoes and a city suit, but when you are spending £10,000 on a car, you expect it to be able to deal with cities and the countryside equally well.

Still, if you are happy to mix it with the trucks and the Peugeots in what Michael McIntyre calls the 'loser lane', it hums along and sips fuel like a mouse drinking sherry through the end of a hypodermic needle. In the real-world economy stakes, it's a full-on Alcoholics Anonymous co-ordinator and the TwinAir Fiat is Oliver Reed.

To drive, it's even Stevens, really. Both handle nicely. And both make fabulous noises. The Fiat sounds like a lollipop stick in a set of bicycle spokes whereas the VW sounds like Androcles's friendly lion. It's the pipsqueak that roared.

Apparently VW didn't feel the need to fit its inherently unbalanced three-cylinder engine with balancer shafts because, it said, it was so small it wasn't really necessary. And I'm glad because the end result is just so characterful. It put me in mind of the old three-cylinder Daihatsu Charade GTti, the first production car to offer 100 bhp per litre.

That, too, sounded fantastic, although, if memory serves, I was enjoying the noise so much, I crashed it into a wall. Something that's not possible in the VW, thanks to its special laser option.

I could ramble on in this vein pointlessly for hours, likening the Up! to other rivals from Toyota, Kia and Citroën, but you're not interested, are you? This is all just blah, blah, blah. Because while the VW is a demonstrably better car than the little Fiat, apart from the speed issues, you're only really interested in how it looks.

I think the Up! looks fab. It was styled by Walter de Silva, who used to be in charge of design at Alfa Romeo and knows what he's doing. I think the front manages to be cute, conventional and futuristic all at the same time, and I think the rest of it is a remarkable achievement – it's a box but it doesn't look that way.

The problem is, of course, the Fiat looks better. And while the Up! comes as standard with an exclamation mark, this is no match for Fiat's vast range of scorpion stickers and Italian racing stripes and snazzy wheels. In short, the VW is a bloody good little car. But the Fiat's quite a lot more than that.

That's why you're going to say Up! yours to the Volkswagen and buy the 500 instead.

1 April 2012

Styled for mercenaries. Driven by mummy

Ford Kuga 2.0 TDCi Titanium X PowerShift

Pretty much every week we are told the best place in the world to live, work and raise a family is not St Tropez or Tuscany or California but . . . wait for it . . . drum roll . . . Denmark. The reasoning behind this is always the same. There's no crime, no unemployment, no obesity, no vandalism, no jealousy, no angst and no dog dirt on the pavement.

However, the trouble with places where nothing bad ever happens is that nothing good ever happens, either. Living in Denmark, I've always thought, would be like living in a ping-pong ball. It's a missionary position country, sitting in the kaleidoscope of nation states like a low-tar, semi-skimmed splodge of beige.

What does it have that we are all supposed to covet? Extremely high taxes. Iffy weather. Lots of Lego to tread on in the middle of the night. And quite the worst pop music it is possible to imagine. The only Danish tune ever deemed good enough to be played on the international stage was 'Barbie Girl'.

And it's all very well telling us we should live there but it has an immigration policy that prevents anyone from doing any such thing.

I suspect even the Danes are baffled about why they keep being picked out as a shining example of humanity at its best. Just last week a newspaper in Copenhagen suggested it must be because, while cycling from place to place, visitors enjoy looking at all the pretty Danish girls' bottoms.

Well, after a quick visit last weekend, I can confirm it is right. It is that. But there are other things, too. Lots and lots of other things. The best restaurant in the world is in Copenhagen, everyone looks like Helena Christensen, in smaller bars you are allowed to smoke, the bacon is superb and, despite the cold, there's a thriving cafe society. Every modern building looks as if it were designed by Bang & Olufsen. The water in the harbour is as clear as gin and the whole city even smells clean. It's like breathing lemon juice.

In fact, I've decided that the world's five best cities are, in order: San Francisco, London, Damascus, Rome and Copenhagen. It's fan-bleeding-tastic. And best of all: there are no bloody cars cluttering the place up. Almost everyone goes almost everywhere on a bicycle.

Now I know that sounds like the ninth circle of hell, but that's because you live in Britain, where cars and bikes share the road space. This cannot and does not work. It's like putting a dog and a cat in a cage and expecting them to get along. They won't, and as a result London is currently hosting an undeclared war. I am constantly irritated by cyclists and I'm sure they're constantly irritated by me.

City fathers have to choose. Cars or bicycles. And in Copenhagen they've gone for the bike. There are many reasons for this. In Denmark cars are taxed to the point where they are fundamentally unaffordable to everyone except Georg Jensen, Ron Bang and Colin Lurpak. A base-model Volkswagen Golf is more than £25,000. A Ferrari 458 Italia is about £500,000. A Range Rover is essentially a Fabergé egg.

Then there's Denmark's obsession with wind turbines. There are so many, it's actually quite difficult to land an aeroplane. In fact, the Danes have run out of land on which to build them so they are hurling them up in the sea. This means that if you're out sailing, you don't tack because of the wind. You tack because you are on an obstacle course. This has given the Danes a sense that they are on point in the war against carbon dioxide, even though not a single conventional power station over there has been turned off yet.

I should also point out that, geologically, Denmark makes Holland look like Nepal. The whole country is on its tiptoes, just managing to peep out of the sea. Which means cycling is always easy. They have a liberal attitude to bikes as well. So when yours is stolen, you simply help yourself to someone else's. You can actually be fined £55 if you ride home at night without stealing someone else's lights before setting off.

Best of all, though, in Britain cycling is a political statement. You have a camera on your helmet so that motorists who carve you up can be pilloried on YouTube. You have shorts. You have a beard and an attitude. You wear a uniform. Cycling has become the outdoorsy wing of the NUM and CND.

In Copenhagen it's just a pleasant way of getting about. Nobody wears a helmet. Nobody wears high-visibility clothing. You just wear what you need to be wearing at your destination. For girls that appears to be very short skirts. And nobody rides their bike as if they're in the Tour de France. This would make them sweaty and unattractive, so they travel just fast enough to maintain their balance.

The upshot is a city that works. It's pleasing to look at. It's astonishingly quiet. It's safe. And no one wastes half their life looking for a parking space. I'd live there in a heartbeat.

But I don't live there. I live in a country with Ben Nevis in it. And as a result we have to have cars. Which brings me on to the impressive-sounding Ford Kuga Titanium X, with a Duratorq engine and a PowerShift gearbox. This new addition to the Kuga line-up sounds like a tool for Clint Thrust, CIA agent, assassin and all-round soldier of fortune.

It looks like one, too, with power bulges on the bonnet, chunky alloys and an aggressive, lean-forward, 'Do you want a fight?' stance. It's a car that in white would not look out of place on Martin Landau's Moonbase Alpha in the Space 1999 television series. But, actually, it's a school-run mummymobile.

Well, that's what it's supposed to be. But there's no getting around the fact that while it appears to be a big, tall car on the outside, it's titchy on the inside. This is a problem because any dog big enough to climb into the boot isn't going to fit when it's there.

There's another problem, too. The car I tested was priced, with options, at £31,840. That is a colossal amount of money, but despite this, it didn't even have a satnav system.

As a tool, it's not bad at all. There's a bit of wind noise and the tyres make a racket, but it's comfortable and apparently well put together and what equipment you do get was plainly designed by someone who recognizes the fact that not everyone in the world understands bits and submenus. It's all very easy to use.

But I kept asking myself: What's the point? Because all you are buying here is a Ford Focus on stilts. It doesn't drive as well as a Focus, it uses more fuel, it costs a shedload more and there doesn't seem to be an upside. Yes, some Kuga models come with four-wheel drive, but others don't, so all you are doing is saying to Mr Ford, 'Yes, I want a Focus, but can I spend a lot more on buying and running it, for no reason?'

In many ways, then, this car is like a Bang & Olufsen stereo. Humdrum innards, made appealing only by a bit of cunning styling. With the Kuga, however, the styling's not quite cunning enough.

8 April 2012

Simply the best, but so bashful buying one is *verboten*

BMW 328i Modern

Almost every country has a unique detail that sets it apart from anywhere else. In France, for instance, you can't walk for more than 100 yards without treading on a dog turd. Australia has too many dangerous animals. Germany has too much armpit hair. India needs a spring clean. And then we get to Sweden, where I spent a recent weekend. The little detail here is odd: there aren't enough chairs.

I stayed at a boutique hotel and on the first day met colleagues in the dining room. After a while we were asked to move because the table had been reserved by someone else for dinner. This was fine, except the only other available seating in the whole building was two ornamental sofas on the second-floor landing.

They didn't appeal, so we moved next door to the Grand. This is a big old-fashioned hotel and quickly we found a table with enough seating for all of us. However, each time one of us went to the bar or the lavatory, the waiter would take his or her chair away and give it to someone else.

Later that night we arrived at a lovely restaurant where we had booked a table for twelve. And it did indeed have a table around which twelve people could sit. But there were only ten chairs. The same thing happened the next night, and the next.

In our green room, backstage at the city's ice hockey arena, we had a sofa and a chair. About half what we needed. A call went out for more seating, and two hours before we left to come home, a man arrived with a moth-eaten leatherette beanbag.

Someone suggested that Sweden used to have enough chairs for everyone but Ikea had exported all of them to Britain. I think, however, the real reason is that, in a socially democratic utopia such as this, it would be considered bourgeois if everyone could sit down at the same time.

In Sweden everyone's car was either light grey or dark grey. The sky was grey, too. And the sea. No one appeared to be rich and no one looked poor. The girls were pretty but not too pretty. And the buildings around the harbour were lovely in an unmemorable way.

They've built a museum to house a ship that sank 380 years ago and were expecting 200,000 visitors a year. In fact they're getting that many every two months and you sense they are actually quite embarrassed about the success.

We see this sort of thing with the Swedish boat I used to take a tour of the archipelago. It was a 40-foot carbon-fibre twin-engined cruiser with a price tag of £400,000. That's a huge amount of money for a boat of this size, so to make sure it appealed to the locals it had been styled to look like a Somalian's fridge and fitted with an interior that put me in mind of a budget French hotel. Naturally, it was licensed to carry eight but there were only six seats.

This is what you have to remember about Sweden. You can have money but you'd better not let it show. They would find a Sunseeker 'revolting'. They would think *My Big Fat Gypsy Wedding* was science fiction. They've even named themselves after a nondescript vegetable.

Which brings me on to the ideal Swedish car. It's the new five-seat BMW 328i, which in Stockholm is probably sold to school-run mums as an eight-seat MPV.

The BMW 3-series is now one of only two traditional four-door saloons to feature in Britain's top 10 bestseller list, which means it must strike a chord here, too. And it's not hard to see why: it's the modern-day Ford Cortina. A no-nonsense design, done well.

Well, when I say no-nonsense . . . In the olden days a BMW 328i would have had a 2.8-litre engine and it would have had six cylinders. But, to keep the European Union green counters happy, the latest 328i has a 2-litre four-cylinder turbo. Despite the smallness, you get 241 bhp, a bit more than in the equivalent Audi, but the BMW produces only 147 carbon dioxides. For a car with this much oomph, that is deeply impressive.

There are many reasons for this. One is the cleverness of the turbocharger design, which not only keeps the polar bears happy but also eliminates lag. It must be there – the chasm between putting your foot down and picking up speed – but you really can't feel it.

Then there's weight. Even though the new car is bigger than the old one, it weighs about 40 kg less. That's good for the ecos, and as a bonus it makes the whole package feel livelier. And it really does feel very lively indeed.

It doesn't tear your face off, and it doesn't make much of a noise, but this car can make serious progress, blurring its way though the eight cogs in the optional automatic gearbox and humming a happy little tune to itself as you scythe past other traffic and arc through corners as if you were a world championship water-ski-ist. This car is more like a scientific instrument than a means of transport. It's delightful.

The gear lever is a bit annoying. It always bongs at you when you try to move it about, but the Sport/Normal selector is a joy. You simply press a button and then choose which bit of the car you'd like to be what. The best solution? Lots of speed and a nice comfy ride. Then it's even better than delightful.

However, there are one or two issues that need to be addressed. First of all, it looks pinched. In the past, all BMWs looked as if their body had been stretched to fit over the wheels. It's what made them look purposeful. There was a sense the shell could barely contain the power that lay within.

But the new car looks pinched – like an elephant on a unicycle. And it takes a very keen eye to tell the fast 328i from the cement salesman's diesel. I'm all in favour of quiet restraint and hiding your light under a bushel when you are out and about. But BMW has gone too far with this new car. It's a bit too Swedish.

The interior is beautifully organized and well made, but the 328i I tested was fitted with a steering wheel that felt as if it was covered in sandpaper. Cheap doesn't really begin to describe the pound-shop nature of this item. And it gets worse because my car was equipped with optional wood trim of such monumental terribleness, I longed for every journey to end so I could get out and not look at it any more.

It looks exactly like the 'wood' used to make a Disneyland log canoe. In other words, it doesn't look like wood at all. It looks like Fred Flintstone's club. Like a giant Cadbury Flake. The sort of thing that no one, not even Wayne Rooney, would find appealing, attractive, interesting, tasteful, desirable, nice or real.

Then there's the problem with buying a 3-series. Go on, try it. Engage your internet, go to BMW's website and try to make sense of what's there. You can't. Not till you've found your reading glasses, and then gone to Boots to buy a pair that is even more powerful. And even when you are able to read the micro-dot typeface, your computer won't have the plug-in necessary to enjoy any of the site's features. Not that you will understand what's on offer anyway, because it's either flowery rubbish or techno gobbledygook.

Soon you will give up with the complexity and buy something else. Well, I would, and that's a shame because whatever you buy will be worse.

15 April 2012

Click away, paparazzi, I've got nice clean Y-fronts

Audi A8 3.0 TFSI

Until quite recently it was pretty easy to run the public relations department of a car company. You organized foreign jollies for journalists, you got one of them to translate the vehicle's publicity pack into something close to English and then you ran a fleet of press demonstrators.

And your boss was happy if the journalist you flew out to St Tropez, and furnished with a fully fuelled car for the week, gave it a friendly notice in his paper. Even if the paper in question was the *Welsh Pig Breeders' Gazette*.

But then Audi employed a man called Jon Zammett as its head of PR, and he decided he wasn't really that bothered about small puff pieces in provincial farming magazines. What Zammett wanted was to see Audi in *Hello!*.

So on the quiet he began to furnish various celebrities with Audis. He has been so successful that now pretty well every star we put in *Top Gear*'s Reasonably Priced Car tells us that he has an Audi and that he's very pleased with it. And it's not just celebrities, either.

Why do you think Zammett was invited to last year's royal wedding? Why does he now appear on red-carpet party guest lists more than Jordan and Victoria Beckham combined? Simple. You have a face? You want wheels? He is a one-stop shop in a suit.

It was a brilliant wheeze, a fairly low-cost plan that took Audi out of the oily rags and into the diamond-encrusted, pap-spattered glitter ball of celebrity. Frankly, the man's a genius.

Providing stars with cars was only part of his headline-grabbing antics. Because in the past celebrities were expected to make their own way from their sumptuous homes to the glittering gala do. This meant they would turn up in front of the flashguns in whatever their local chauffeur company happened to be running at the time – an S-class Mercedes, usually.

Zammett realized this was a lost opportunity, and so at a secret location – in Warwickshire – he keeps a vast flotilla of Audi A8s and the contact details of a hundred or so former coppers who can be called upon at a moment's notice to fire up the fleet and descend on the Empire in Leicester Square.

Just go and check all those old copies of *Hello!* that you keep by the lavatory. Notice how the car from which a knees-together star with a Daz-white smile is climbing is always an Audi. Zammett did that.

It's had a marked effect on sales. After the collapse of Lehman Brothers, when every car firm had its back to the wall, Audi actually shifted more metal than ever before. One company chief said, 'We note that there is a recession in full swing at the moment. But we have decided not to take part.'

Last year in Britain alone Audi sold 113,797 cars. That's almost 32,000 more than Mercedes and a staggering 73,000 more than it sold back in 1999.

That's the result of today's strange obsession with celebrity. Or is it? Could it be that the new Audi A8 is simply better than its mighty rival the S-class?

In terms of looks, no. If you take away the Audi's grille, which looks like George Michael's beard, it could be a Toyota or a Honda. That's fine if you want to maintain a low profile, but if you want to cut a dash, you'd be better off with the Merc. That thing's got serious presence.

Value? It's hard to say, really, because there are countless models and each is available with a vast array of options. The car I tested was a four-wheel-drive petrol-powered 3.0 TFSI, and that's just shy of £60,000 – a tiny bit less than an entry-level S-class.

So what about space? Well, I was recently chauffeured in an A8 to the ballet and I fell asleep in the back. So it's fine. It's also fine in the front. But then it would be. It's a really, really big car.

So now we must consider what it's like to drive, and this is where Audis in recent years have come a cropper. The company's engineers have never understood that road-worker Johnnys in Britain are not quite as thorough as their opposite numbers in Germany. Which means that big Audis in the past have always been way too firmly sprung. Or, to put it another way, uncomfortable.

The new model is different because the driver is allowed to choose just how soft and gooey he wants the ride home to be. And we're not talking here simply about the suspension. Oh no.

You've various settings for that, including Comfort, Automatic and I-Want-to-Go-Around-the-Nürburgring. You have a similar variety of choices for the engine and gearbox, the steering, the differential, the lights and even the seatbelts. Why? This is a large car, designed for large people who just want to get home after a large lunch. If they'd wanted a bone-hard ride with electric performance, they'd have bought a BMW M3.

In a bit of a huff, I put everything in Comfort mode and set off up the M40. It was utterly delightful. As relaxing as a happy ending. Smooth, quiet, soft – exactly how a big car should feel.

But then I turned off the motorway, and oh dear. All of a sudden the suspension and the steering seemed to lose control of the bulk. It was like trying to drive home on a slightly decomposed hippo. So I dived back into the menu and chose the Dynamic setting, and suddenly everything was worse.

Eventually I realized that it's best to let the computer choose a setting to suit the conditions. But even here there's an issue. Because the steering system constantly flicks from Dynamic to Comfort, you are never sure how much effort you should use to turn the wheel. Sometimes you think just a bit will be required, and then just as you spot a bus coming the other way, you realize it should have been a lot.

There are other small irritations, too. The gear selector is too fiddly, the steering-wheel-mounted buttons feel cheap, the dash is made from wood (very 1986) and when you select reverse, the radio turns itself down. Is this so you can hear when your dog's head bursts? Surely it's too late then.

Another point I should make at this stage. Don't bother with the 3-litre petrol I drove. It's quiet and refined, but in all honesty the diesel provides all the get-up-and-go, with less thirst. And a better resale value.

It sounds here as though I have a downer on the new A8, but that's not strictly accurate. Because when it's bad, it's not really very bad at all. And when it's good, it's fantastic. It is so quiet and so comfortable on the motorway, you can set the cruise control, sit back and use the on-board wi-fi to get on with some emails. Just remember that if you've selected the auto steering, it doesn't actually mean it will steer automatically.

I also loved the quality of the stereo and the DAB radio system that let me listen to Christian FM. This is much better than normal radio because you are not warned about traffic jams ahead. Only the fact that you will soon be engulfed by God's fiery love.

Truth be told, though, you get Christian radio in a Mercedes S-class as well. And with that car you will always have upmarket mini-cabbers queuing around the clock when the time comes to sell. It's a more sensible buy.

The trouble is that in a Merc you look like a fat man on his way to a meeting. In the Audi, thanks to the efforts of Mr Zammett, you look like Jude Law.

29 April 2012

Get a grip – it's only a Roller

Rolls-Royce Phantom II

You can't really relax when it's your daughter's eighteenth birthday party and your house is rammed to bursting point with a cocktail of rampaging testosterone and vodka. Certainly you can't just go to bed, partly because of the worry that everyone is going to get pregnant, but mostly because of the noise.

So I didn't. I stayed up all night, totally forgetting that at eleven o'clock the next morning I was due at the Emirates stadium in the nuclear-free, vegan outreaches of north London. Happily, I had booked a driver. Unhappily, he turned up in the brand-new, second-generation Rolls-Royce Phantom.

At first, all was well. Buoyed by a drink-fuelled content-ment that nobody had cut their head off or given birth, I slumped into the vast rear seat in a Ready Brek glow of warm fuzziness.

However, about twenty minutes later this had begun to wear off. And as we reached London, I started to worry that I might die. Ten minutes after that, I was worried I might not.

There was a rolling tide of nausea in my head that manifested itself in waves of great pain and an all-over veneer of perspir-ation. I desperately wanted to go to sleep but the driver was unfamiliar with Islington – there isn't much call for Rolls-Royce test drives there – so I needed to help him find the best route. 'Can you drive as fast as possible,' I asked, 'into a lamppost?'

Eventually we arrived and I discovered something interesting. When you step out of a Rolls-Royce into a mooching herd of football fans, they become united in a certain knowledge that you are an onanist. They voice this opinion loudly and often, and since you are going in the same direction as them, it doesn't stop.

I arrived at my host's box in a blizzard of sweat, sickness and abuse, only to discover that one of the other guests was a motoring writer who once told his readers that my opinion was worthless because I was a multi-millionaire tax exile who lived on the Isle of Man.

I've wanted for some time to hear him explain why the opinion of a 'multi-millionaire' is somehow less relevant than the opinion of, say, a schoolteacher, and how he got it into his head that I was a tax exile. But, sadly, when the moment arrived, I was otherwise engaged, trying to stop myself fainting.

The match was dismal. There were no goals. And then I was faced with the problem of getting through the crowds to the waiting Rolls. And here's a funny thing. As we all walked along, everyone was jolly friendly. There was some good-natured joshing about my support for Chelsea and a few questions about Richard Hammond's teeth, and all was well . . .

Until I stepped into the Rolls, whereupon I suddenly became an onanist again. So there I was, feeling like a skin bag full of sick, in the back of a Phantom that was going nowhere because of a vast horde of Islingtonites who were making hand gestures and chanting. Oh, and one thing you might like to know: if you push the button that draws a curtain over the back window, you make everything ten times worse.

I'm not quite sure why, but today you can be a bank robber or a pugilist or a benefits cheat, and that's fine. You can be a drug addict or a Peeping Tom. But woe betide anyone who is rich.

Every day the *Daily Mail* finds someone on a high salary and mocks them mercilessly. David Cameron's ability to lead is questioned simply because he's perceived as being wealthy. *Autocar* reckons that because my DVDs have been big sellers, I'm no longer capable of rational thought.

A far-left candidate in France's presidential elections proposed a 100 per cent tax on all earnings above €360,000 (£300,000) a year, and I bet if such a scheme were introduced here, it would receive almost unanimous support. There's a sense, and it's completely wrong-headed, of course, that in these difficult economic times anyone who has a bob or two must have stolen it from a charity box or a nurse.

And naturally there is no statement of wealth that even gets close to a Rolls-Royce Phantom. Which is why Arsenal's whisper-quiet peace-and-love brigade turned into an army that would have warmed the heart of even Stalin. Top tip, then: if you're going to buy this latest version of the Phantom, for God's sake stay away from the mob.

At first glance the new car seems to be pretty much identical to the old one. At the front the headlamps are slightly different and at the back there's a chrome strip on the bumper. There are some new wheels as well but, really, it's just a slight change of wardrobe rather than a full liposuction, boob enhancement and tummy tuck.

It's much the same story on the inside, too. A raft of tiny little cosmetic alterations that caused me to think, Oh no, I'm going to be sick. You, on the other hand, will sit there and wonder, I know the last Phantom was pretty bloody good but surely there was scope for a bit more improvement than this. Well, there has been improvement. It's just that you can't see it.

Nine years ago, when the Phantom first slithered out of the factory in Goodwood, West Sussex, Rolls-Royce was at great pains to point out that, although the company was owned by BMW, the car shared only 15 per cent of its components with a 7-series. Never mind that one of the components was 'the engine'; the manufacturer made a good point: the Phantom didn't feel, look or drive anything like a Beemer.

However, since the Phantom's launch, BMW has developed a raft of electronic improvements that are now available on an £18,000 1-series. But not its £350,000 Roller.

So. What to do? Go to all the trouble and expense of designing new electronics for the Rolls? Or simply use BMW items?

That's what the company has done. The swivelling headlamps. The 3-D satnav. The USB port. It's a forest of BMW technology in there, and you know what? It's sacrilege and it's wrong – and I don't actually care.

Because even though there is now a rather worrying Dynamic option for those who wish to take their Rolls-Royce around the Nürburgring, the Phantom still feels, drives and looks like nothing else. It is a sublime experience, like getting into a warm bubble bath and then getting out and finding yourself somewhere else.

The quality is unmatched. The eighteen cows, for instance, that donate their skin to make the seats in a single car are kept far away from barbed wire fences and anything else that might make them uneasy. And Rolls has developed a new colouring process in which the dye permeates the entire hide, ensuring it will never crack. You don't get that attention to detail in even a palace.

The carpets are thicker than anything you have at home, the wood veneer is peerless, the art deco light fittings are wondrous to behold, the V12 engine makes no noise at all, the ride comfort is straight from the pages of *Aladdin*, and while there are many gizmos, they're all hidden away. We see this with the gear lever, which may have eight speeds at its disposal but offers you a choice of only forwards, backwards or neither.

The new Phantom, then, is an intelligent and discreet step forward for what was – and still is – the only car in the world that completely detaches you from reality. Just remember, though, that if you go to the wrong place in it, it will detach otherwise normal people from their sanity.

6 May 2012

I know about your frilly knickers, Butch

Mercedes SLK 55 AMG

Ever since it minced into the marketplace sixteen years ago, Mercedes' little SLK has been the world's only transgender car. Even though it was born with an Adam's apple, dressed in shorts and trained to use the urinals, it has always been as girlie as a pink bedroom full of soft toys.

If I'd been running Mercedes-Benz, I'd have been quite pleased about this. I'd have accepted that the car was a ladyboy and changed its name immediately to the Fluffy Rabbit or E. L. James. I'd have offered it in a range of pastel colours and employed Stella McCartney to design a range of interior fabrics.

But no. Mercedes could not accept that its child was a bit light in its loafers. So as it grew, the company fitted it with a massive V8 engine and changed its exhaust note from Barbra Streisand to Ted Nugent. This was unwise and unfair – like forcing Freddie Mercury to get a job as a scaffolder.

Undaunted, Mercedes called the new car the SLK 55 AMG and sent it out into the world with a simple message. 'Now look. We've given you an enormous penis. Go and use it.'

It certainly wooed me, because I bought one. Of course, my colleagues thought I'd taken leave of my senses and laughed openly in my face. So did all the nation's van drivers, and in every petrol station people would point and suggest loudly that my salon must be doing well. The codpiece front and the baritone rear fooled nobody.

But I didn't care because I like small cars. I like convertibles. And I like big engines. And the SLK was the only car on the market that met all three of those requirements – and a few more besides. It had an automatic gearbox and, though it was fast and hard and brutal, it came with all the usual Mercedes refinements including a DVD player, a TV, electric seats, cruise control and so on. It was a doddle in town, brilliant on a sunny day, easy to park, as fast as a comet, good-looking, exciting, noisy and enormous fun. Who cared that it enjoyed musicals and went to bed at night in its sister's knickers? I even ordered mine in black.

Mercedes, though, was still not satisfied. It knew that when it wasn't looking, its scaffolder was endlessly watching the shower scene from *Top Gun*, so with its replacement the company has gone mental. The car has the haunches of a hyena, the snout of a racer, flaps, ducts and claws. It's a low-profile, full-fat, high-octane he-man. I'm surprised the advertising slogan isn't 'Are you a woman? Well, you can eff off.'

Let's start with the engine. In essence, it's the same 5.5-litre unit that you get in other, bigger AMG cars, only without the turbocharging. Do not think, however, that the lack of forced induction means you will be bouncing up and down in your seat when leaving the lights to try to conjure up some extra wallop.

Because of new air-intake ducting, new cylinder heads and a modified valve drive, you are presented with 416 brake horsepower. That is about 60 more than you were given in the old SLK 55, and in a car this size it means the performance is very nearly insane.

However, because the new engine is fitted with a feature that shuts down either two or four of the cylinders when they're not needed, it produces only 195 carbon dioxides and should be good for more than 33 mpg. The Lord giveth and then the Lord giveth even more.

Handling? Well, now, let's be clear on this: if you want finesse and delicacy, buy a BMW. In a straight line, an AMG car is an easy match for anything made by BMW's M – or motor sport – division. But through the corners the Mercedes will be left far behind. This is not a criticism. Because although the Merc may not be able to tame the laws of physics quite as well as an M car, it will put a much bigger smile on your face. BMWs reward your skill. Fast Mercs make you laugh.

And so it goes with the SLK. Mercedes may have fiddled with the camber and perforated the brake discs. The little convertible may have all the racing paraphernalia, but it's still a car you have to wrestle if you want to get the most from it. It's a car that's happiest when it's a little bit sideways.

Inside, however, there's no evidence of this at all. The gearbox is a proper auto. The radio is digital. The headrests are fitted with ducts that feed warm air to your neck. The car I drove was even equipped with a device that suggested when I might like a cup of coffee.

However, while there's one improvement over the old model – the can-of-pop-holders are no longer located in front of the heater vents – there is one step backwards. If you push the seat all the way back, the leather rubs against the rear bulkhead and squeaks every time you go over a bump. It's very annoying.

It sounds, then, as if this new car is much the same as the old one, albeit a bit faster and quite a lot more economical. But I'm afraid that's not strictly accurate. Because where's the noise? The old car crackled when you started it, roared when it was moving and ticked when it wasn't. And without this soundtrack the excitement has gone. It means you never feel inclined to put your foot down. I spent my week just pottering about. At one point I found myself doing 60 mph on the motorway. On the Burford road in Oxfordshire the other night I was overtaken by a Fiat 500.

Really, it should come with a cattle prod and a device that tells the driver to pull over and get some bloody Red Bull down his neck. Sometimes I'd try to go a bit faster, but there seemed to be little reward, and as soon as I stopped concentrating I went back into Peugeot mode.

So we're left with a big question. At £54,965 the new SLK 55 costs less than I was expecting. But why pay this much for a car that doesn't raise the hairs on the back of your neck? If you want a pottering-about, top-down cruiser, why not buy one of the much cheaper, smaller-engined versions?

Because they're for girls? OK, then, why not buy a BMW Z4? This is a much underrated car. At less than £40,000 for a twin-turbo 3-litre, it has the same hard folding roof as the Mercedes but is better looking and much less of a handful. Oh, and there's one more thing. It's the only car in the world that was designed by women. I like it very much.

13 May 2012

Fritz calls it a soft-roader. I call him soft in the head

Audi Q3 2.0 TDI quattro SE S tronic

It is extraordinary how often a room full of well-qualified adults can discuss a subject in their chosen field and arrive at a conclusion that's completely muddle-headed and stupid.

We see this a lot in politics. Only recently an MP called Keith Vaz went on television to say the immigration desks at Heathrow needed to be 'personed up'. I actually went back and watched the moment again. But there was no mistaking it. This man – an MP with a first-class degree from Cambridge – had obviously been to a meeting where other sentient beings had convinced him to use words that no one else understands.

Then there was the war in Iraq. Wise, clear-thinking people had access to all the information that the satellites could provide. And yet still they made a decision that was idiotic and wrong.

A few years ago Coca-Cola did the same thing, albeit with less important ramifications, when it decided to make Coke taste like a used swab. BA did it with its tailfins. Gerald Ratner described a product he sold as 'total crap', Paul McCartney recorded 'Ebony and Ivory'. Philips pioneered the laserdisc. Clive Sinclair decided to put his all into an electric slipper. John Prescott invented the M4 bus lane. The *San Francisco Chronicle* turned down the syndication of the Watergate story, saying it would only interest people on the east coast. And *Top Gear* made a film about an art gallery in Middlesbrough.

I'm to blame. I brought it up in a meeting and instead of getting insects to lay eggs in my hair, the production team nodded sagely. We'd take over an art gallery, fill it with automotive-based art and prove that cars bring in bigger crowds than unmade beds and pickled sharks. Somehow, though, it didn't occur to any of us that this would be a very long and boring film until after it appeared on the show. 'That was very long and boring,' we all said afterwards.

Of course, the motoring world is rammed with more mistakes than almost any other industry. Someone at Pontiac looked at the design for the Aztek and said, 'Mmm. Yes. Excellent.' And there were similar noises in the Ford boardroom when the stylists mistakenly unveiled their joke plans for what became the Scorpio.

Daimler really thought it could compete with the Rolls-Royce Phantom by putting some cherry wood in a Mercedes S-class and calling it a Maybach. Toyota launched a car called the MR2 without noticing that when said in French – 'MR deux' – it translates as 'shit', and Audi decided that it could improve on the airbag by developing a system called 'procon-ten', which used a fantastically complicated network of cables to pull the steering wheel forward in any frontal impact.

I could go on, so I will. Austin made a car that was more aerodynamic going backwards than forwards, Ford made a car that blew up if a leaf landed on it and Lancia made a car from Russian steel that was as long-lasting as fruit. And only recently Volkswagen was going to call its new car the Black Up!.

It's almost as though every single meeting in the car industry is specifically designed to exclude rational thought, which brings me on to a gathering of fine minds that must have happened a few years ago in Audi's boardroom.

They obviously decided that it would be a good wheeze to create a new type of medium-sized hatchback that looked like it might be able to go off road but couldn't. 'Yes,' someone must have said. 'That's a brilliant idea. No one else will have thought of making such a thing.'

And they were right. There are no other so-called 'soft-roaders' on the market at all. Apart from the Land Rover Freelander, the

Range Rover Evoque, the Honda CR-V, the Toyota RAV4, the BMW X1, the Nissan Kumquat, the Nissan X-Trail, the Mitsubishi Outlander, the Volkswagen Tiguan, the Citroën Cross-Dresser, the Subaru Forester, the Hyundai Santa Fe, the Volvo XC60, the Kia Sportage, the Vauxhall Antara, the Ford Kuga, the Mazda CX-7, the Kia Sorento and the Jeep Compass.

It's possible they may have known all along that there are many options in this part of the marketplace. But it's unlikely. Because if they had, they would have made damn sure their new car was better than all the others. And it isn't.

Let's start in the boot, which is very small. And the reason it's very small is that under the boot floor, apparently, there is a large subwoofer. What kind of hallucinogenic drug were they taking at the meeting where everyone agreed that this was a good idea? Who stood up and said, 'A good bass sound is more important than an ability to carry dogs, shopping or a spare wheel'?

Further forward, we find the rear seat, which is wide enough for three people but only if they have casters rather than legs. And then up front, at the business end, we find nothing at all, apart from some heater controls that have been designed to be annoying.

My £28,965 quattro SE test car was supplied in loser spec with cruise control as a £225 optional extra and, er, that's it. Every time I selected some tasty-sounding feature from the onboard computer, I was given a message saying, 'You couldn't afford this' or 'You really should have worked harder at school'. It didn't even have satnav.

To drive? Well, it's hard to say because the wheels weren't balanced properly, and trying to be rational when viewing the world through wobble vision is like trying to concentrate on the finer points of someone who's constantly hitting you over the head with an axe.

All I can say is that the engine is rather good. I had the more powerful diesel option that had lots of oomph and the thirst of a bee. It sounded nice, too, in a gravelly, smoky, bluesy kind of way.

However, in the morning, when it had been asleep all night, it did take a second or two to remember what it was and what it's purpose in life might be. You turn the key . . . and nothing happens. And then, shortly before it remembers that it's an engine, you give up and turn the ignition off again. This causes a bit of swearing.

Mind you, for cluelessness, the gearbox is worse. In Sport mode it wouldn't change gear at all, and in Normal mode did nothing else but. Every few seconds. For no discernible reason.

Then there's the Efficiency mode facility that disengages the clutch every time you lift off the throttle. In theory, this fuel-saving measure sounds like a good idea. In practice, it means you simply cannot maintain smooth progress on the motorway.

The Q3, then. Not practical. Not nice to drive. And technologically, not thought out well, either. So what's to be done if you want a car that looks like it could go off road but won't? Especially if you specify the sports suspension that lowers the ride height to that of a centipede.

Well, the obvious answer is the Range Rover Evoque. But if this is too expensive for your taste and not spacious enough, don't worry, because there is a better alternative. It's called a saloon car.

20 May 2012

Cheer up – Napoleon got shorty shrift too

Mini Cooper S roadster

Tall people never really think about how far they are from the ground unless they are presented with an economy-class seat or a row of off-the-peg trousers. With small people, things are different. They think about their height all the time. They think that people like me are tall deliberately, that we do it on purpose just to annoy them. This gives them what doctors call SMS – small man syndrome – and what we call a bad temper.

At parties they feel excluded from conversations as they scuttle about banging their heads on coffee tables. On crowded Tube trains they feel bullied. With girls they feel left out. And when shopping for clothes they quickly become fed up with being directed to Mothercare. This is why most bar-room brawlers and emperors are vertically challenged.

It is quite correct to say that in evolutionary terms they are closer to the amoeba and that tall people sit at the prow of civilization. But these thoughts don't occupy my mind all the time. I don't feel superior to a small person just because my head is nearer to incoming weather systems. But they definitely feel inferior. Which is why they are engaged in a constant and deeply irritating battle to prove themselves worthy.

We see the same problem with dogs. My west highland terrier is in a permanent state of rage. Because she can't climb into the back of a Range Rover by herself or leap over fences, she bites the postman, the paperwoman and people who come to mend the computer. She bites my other dogs, too, and since we haven't seen the milkman for months, we can only assume he's been eaten. She barks a lot as well, making up for the shortness of her legs with volume. If she were a human she'd have been sent to Elba and the world would have been a safer place.

Strangely, at this point, I need to talk about Richard Hammond. He told me the other day that when driving his Fiat 500 he is constantly bullied by other motorists. That he's always being undertaken and tailgated and made to wait longer than is necessary at junctions. And I sighed the sigh of a tall man and thought, It's not the car, sunshine. It's your inferiority complex.

But, having spent a week with the new Mini roadster, I think he may have a point. Small cars do get bullied. Especially when they are pretending to be something that they are not.

My youngest daughter, who is extremely tall, pointed at the little car that had come to our drive and said, 'That is not a Mini.'

Her views were echoed on the road. 'That is ridiculous,' was the most commonly expressed view.

And the roadster is ridiculous because it is about as far from the concept of a Mini as it is possible to get. The genius of the 1950s original was packaging – fitting an engine and four people into a car that was just 2 inches long. The rallying and *The Italian Job* came later.

Well, the new roadster is about as long as the Norwegian coastline but has only two seats. In terms of sensible packaging, it's right up there with the underground bunkers De Beers uses to store a few diamonds. Or those massive boxes that contain nothing but a USB dongle for your laptop.

Of course, the whole point is that it's supposed to be a sports car. But despite the stripes, the lights, the William Wallace war paint and the massive Cuban wheels, it doesn't look anything like, say, a Mazda MX-5. It looks like a Mini. That's been beheaded.

Inside, the news is just as grim. When the new Mini was launched, the wacky interior was interesting. Now it's just annoying. The speedometer, for instance, is the size of Eric Pickles's face, but you have to study it carefully for several minutes to work out how fast you're going.

Then you have the electric-window switches, which look great but are in the wrong place. As is the volume control for the stereo. It's as though one hundred people – all children – have contributed an idea, and they've all been accepted.

Oh, and then we get to the price. The Cooper S version I tested is £20,905. This makes it nearly £500 more expensive than the more practical, less idiotic-looking four-seat convertible. And about £5,000 more than a Mini should be.

Subliminally, other road users know this, too, which is why I spent most of my week in a blizzard of hand gestures and cruelty. I felt like the jack in a game of boules. I felt like Richard Hammond. And the biggest problem with all of this is that the car itself is absolutely fantastic. A genuine gem. A nugget of precious metal in a sea of plastic and Korean facsimiles. I absolutely loved it.

By far the best bit is the engine. It's a turbocharged 1.6-litre that produces 181 bhp. That doesn't sound the most exciting recipe in the world, but after a whisper of lag you barely notice, the torque is immense. It feels as if there's a muscle under the bonnet and you never tire of flexing it.

The only real problem is that on a motorway – and I've noticed this in all Minis – its natural cruising speed is about 110 mph. Because of a combination of where you sit, the angle of the throttle pedal, the gearing and the vibrations, this is how fast you go when you're not concentrating. You need to watch it.

Or get off the motorway. That's a good idea, actually, because although there's a bit of typical big-power-meets-front-wheel-drive torque steer, the chassis is mostly brilliant. It's like an old-fashioned hot hatch: a Volkswagen Golf GTI or Peugeot 205 GTI – the sort of car you can fling into a bend at any speed that takes your fancy.

You would expect the ride to be as awful as the handling is good, especially with all the strengthening needed to make up for the lack of a roof. But no. It's firm, for sure, but it never crashes or shudders in even the worst pothole. It's never uncomfortable. It's a joy. And it's not unduly thirsty, either.

At this point I'd love to tell you all is just as well when you put the roof down. But I can't, I'm afraid. Because in the seven days I spent with this car it never stopped raining even for a moment.

I can tell you, though, the roof is only semi-electric and some of the operation has to be done by hand. That's no biggie. I can also tell you that the boot is much bigger than you might be expecting. But the last thing I have to say is the most surprising of all: this car is worth a serious look.

It's not as well balanced as a Mazda MX-5, but it's faster and it has more soul. In many ways it reminds me of Richard Hammond. It's small and it's annoying and it wears stupid clothes. But when you get to know it, it's a bloody good laugh.

27 May 2012

That funny noise is just Einstein hiding under the bonnet

Ford Focus 1.0 EcoBoost 125PS Titanium

The Volkswagen Golf. The Vauxhall Astra. That medium-sized Toyota that is not called the Corolla any more. What is it called? It's a name that's sadly not on the tip of my tongue. My mind is blank. The Areola? Or is that the ring around your nipple? Whatever, the reason that I rarely test such cars on these pages, or on the television, is simple: what's to say?

For some time car makers have been treading water in a stagnant pool. If they wanted to launch a new car, it was easy. They called a company that made shock absorbers, a company that made pistons and a company that made satnavs. Then got some Poles or Slovaks to Sellotape them together – *et voilà*.

In the mainstream there was no fizz, no drama, no inventiveness and no risk. It got to the point where Ford's engineers made a big noise about the Focus having expensive-to-make independent rear suspension. Yes, this made it lovely to drive at the sort of speed it would never travel but, really, the main reason they were so proud is that they had won a minor internal battle with the bean counters, who doubtless would have wanted them to use a cheaper fixed-axle setup like everyone else. Car making. It had become accountancy.

But a man called Swampy had taken up residence in a tunnel just outside Newbury in Berkshire and started talking about something called 'the environment'. Now there had been lots of anti-state, anti-system Swampies in the past, shouting about workers' rights and peace and communism, but none had gained any traction with the middle classes. So they had remained a noisy but minority interest, like bell-ringing.

Swampy, though, had hit upon an idea that did strike a chord with the nation's jam makers. They liked gardening. They liked peace and quiet. They liked the idea of this young man in his dirty trousers trying to stop the government building a bypass. So suddenly he was joined in his campaign by lots of ladies in camel-hair coats.

It wasn't just in Newbury, either. Environmentalism was taking off all around the world. Leninism had a new face. It was the face of a drowning polar bear. And everyone seemed to like it.

To show they were in tune with the times, politicians started to make green noises, too. Mr Cameron went on a plane to the Arctic to look at a dog and then put a small wind turbine on his house. In America a former presidential candidate called Al Gore made a film called *Workers' Control of Factories*. And global warming became the new terror.

Naturally the motorcar was quickly identified as the main problem. Not only did it allow workers personal freedom but also it produced vast quantities of carbon dioxide from its tailpipe . . . as a direct result of environmentalists insisting in the Eighties that it had to be fitted with a catalytic converter. A device that converts gases that don't warm the planet into CO_2. Which does, apparently.

So every year governments imposed tougher and tougher legislation that forced car makers to wake up. They were being forced by law to make their products chew less fuel. This meant they had to get inventive. And while I don't like the reasoning behind that, I do like the results. Mainstream cars are getting interesting again.

We now have hybrids, and I love the way they obey the letter of the law but completely ignore its spirit. Because how can a car with two power plants possibly be good for the planet? It can't. These cars – they're tools for fools.

More recently we have been seeing some clever variations on the hybrid theme. From Vauxhall there's the Ampera, and from a small firm in America there's the Fisker Karma, which works like a diesel-electric locomotive. Elsewhere people are working on hydrogen fuel cells, and there are pure electric cars, too, such as the Nissan Leaf. But the less we say about those, the better. Because let's be clear. They are interesting to write about, but . . . They. Do. Not. Work.

They are expensive to buy, their ecological benefits are debatable – they run on power that comes from Drax B – and if you want the costly battery pack to last, it takes several hours to charge it up. This means it would take several days to drive from London to Edinburgh.

All of which brings me to the Ford Focus. It's called the EcoBoost and it meets the new green legislation in the cleverest, simplest, bestest way yet. It runs on an engine so small, the cylinder block would sit neatly on a piece of A4 paper. That small.

And before you think that a 999 cc three-cylinder engine could not possibly produce enough power to move a car as large as a Focus, look at the figures. It produces 123 bph – exactly what was delivered by Ford's old 1.6 Focus. But amazingly you get more torque and, of course, greatly reduced fuel consumption.

This best-of-all-worlds solution has been achieved thanks to some extremely clever thinking. The torque comes from a very long piston stroke and a turbocharger that can spin at up to 248,000 rpm. That's sixteen times faster than the blades in a jet engine.

There's more. In most engines the pressure on the top of the piston is around 150 psi (pounds per square inch). In a normal turbo that might get as high as 200 psi. But in the EcoBoost's micro-motor it's more than 350 psi.

Then there's the detailing. The cam belt runs in oil, so it's silent and will last for ever. Ford has even split the cooling system so that the business part of the engine and the people in the car can warm up as quickly as possible on cold mornings. And the exhaust manifold is water-cooled as well. It's probably fair to say that there is more innovation and technology in this engine than you find in a Lamborghini V12.

Which is why I was so cross with the elderly Australian tourist I encountered on London's Kensington High Street recently. 'Why are you driving this piece of shit?' he asked. I explained that I was testing it and that, actually, it was rather interesting. But that didn't calm him down one jot. He was so angry that I should be driving such a thing, he started hitting it with his shopping bags. 'It's shit!' he screamed. 'And you should know better.'

It's not, though. It's great. There's so much torque that you can spin the wheels into second, and it'll easily hold its own with Johnny Van Driver in a traffic-light grand prix. And best of all, because the engine is so light, some of the agility that's gone missing from recent Focuses is back. To drive, it is brilliant, and apart from a gruff but rather endearing three-cylinder engine noise, there's simply no indication at all that you are being pulled along by an engine the size of Richard Hammond's left testicle.

Inside? Well, it's a Focus. It's spacious, and my test car was loaded up with every conceivable extra. The only item I wouldn't bother with is the lane assist. You get a barely detectable wheel wobble and small red light on the dash if you wander out of your lane on the motorway, but, I'm sorry, if you haven't noticed you're about to crash into a bridge parapet, you're unlikely to be brought to your senses by what looks like the standby light on a television.

That's it, though. The only real fault I could find in what's certainly the most important Ford since the Cortina.

And now we get to the clincher. A top-of-the-range Toyota Prius is around £24,910. With the £5,000 government grant, an all-electric Leaf will cost you £25,990 and a Vauxhall Ampera £32,250. Prices for a similarly sized, faster and nicer-to-drive Focus EcoBoost start at £16,445. I could go on. But there seems little point.

Gosh, never thought I'd dump Kate Moss so fast

Citroën DS5 DSport HDi 160 automatic

BMW very kindly agreed to lend me a brand-new V8 X5 while I was in Germany for the recent Champions League final, with the understanding that I would review it if Bayern Munich beat Chelsea. So. Let's have a look, then, at the new Citroën DS5 DSport.

Truth be told, I can't remember much about the BMW. I know it was driven by a polite man called Christian and that it was brown. But engine performance? Seat comfort? Space in the back? Lost, I'm afraid, in the fog of warm, fuzzy satisfaction that England had beaten Germany on penalties. That Chelsea had won the biggest prize in European football. That Didier Drogba – the giant, the colossus – had waved goodbye to his career in a blue shirt by saving the day.

It's strange, isn't it, that twenty-two strangers kicking an inflated sheep's pancreas around a foreign field – that is forever Chelsea, by the way – can elicit such extraordinary emotions in a grown man? It would be like running around in circles because your son beat somebody else's son in a pre-school game of Connect 4. Certainly there's no reason why the win meant I should jump up and down so vigorously that I broke the credit card in my pocket. I may have even hugged a taxi driver, too.

Actually, there is a reason. It's this: back in 1970 some Leeds supporters put dog dirt in my school cap because I had dared to walk through a Yorkshire town sporting a Chelsea scarf. That, then, is what made me so happy in Munich. Because I knew that in a working men's club somewhere, the little gang would be sitting, staring into their stout, feeling terrible. I was happy because they weren't. That's what football is all about.

Anyway, the next day, the elation had been swallowed by a dreary list of appointments and the car that would transport me between them: the aforementioned Citroën.

In the olden days Citroën made its reputation by being different. It would use different engines from everyone else, different suspension, different braking. As a result, it won a strong appeal among oddballs – people who thought whales were intelligent, that vegetables had feelings and that the best way to combat the threat of a Russian attack was to chain yourself to a fence post at Greenham Common.

Of course, when Russia was no longer perceived as a threat and beards had become a joke, the customer base melted away, which meant Citroën had to come up with a new idea. And it did: value for money.

The company would advertise the car for £5,000 and then, with a hysterical television advertisement, explain that you would be entitled to a 100 per cent discount, £1,000 cashback, free financing, no VAT and the opportunity to sleep with any of the sales assistants who took your fancy.

Soon, however, an accountant must have noticed that while many cars were leaving the factories, no actual money was coming back. And anyway, the VFM rug had been snatched away by the likes of Kia and Hyundai, which were offering 2,000 per cent discounts, free holidays in the Far East and £20 million for your old car.

Citroën was forced to come up with a new plan to disguise the fact that underneath, its cars are nothing more than dreary Peugeots. And the plan it came up with was styling. In the company's words, it set about industrializing haute couture.

There's no getting around the fact that the DS5 you see here this morning is extremely striking, and I don't mean 'striking' in the way you'd describe a friend's hopeless attempt at an oil painting. I mean striking as in Kate Moss. This is one good-looking car.

You may, therefore, be interested enough to have a closer look, and when you do, you will not be disappointed. Because inside, if anything, it's even better. You sit behind a styled steering wheel cocooned not only by a high central transmission tunnel but also by a drop-down pod mounted to the roof lining.

And both of these features are festooned with highly stylized buttons. They are arranged in the manner of a Rhodesian ridgeback's neck and look fantastic. But Citroën obviously had a problem.

If you use buttons as a styling feature, you need to give them all a job. Which is why the DS5 DSport that I drove was equipped with every single feature ever fitted to a car, house, spaceship, train, sex toy, fighter jet, submarine, vacuum cleaner, laptop and mobile phone. It's also why there is not a single electrically operated sun blind in the roof. There are three.

This makes for excellent sport in a traffic jam: pushing things to see what happens as a result. I was especially excited to find at one point in a nasty jam on London's Euston Road that I could direct cool air into the central cubbyhole.

Now. As I see it, there are a couple of issues with Citroën doing this. First, this is not a company with the best reputation for electronic reliability, and second, it all adds weight. And more weight means less acceleration, higher fuel bills and the need for firmer suspension.

Couple that need to the fact that this car is sold as a DSport, and the result is an extremely harsh ride. This is a disappointment. Citroën has made fast, even sporty, cars in the past and all have retained the brand's famed reputation for comfort. The DS5 DSport does not. It rides like my AMG Mercedes.

And do not think the upside is a great swathe of agility. Because of the rather vague steering, sharp brakes and all that weight, it's about as much fun to hustle as a hill. Or indeed a Peugeot 308, on which it is based.

The engine's not bad. You can specify a 197 horsepower

hybrid if that's your bag, but to be honest, the 158 horsepower 2-litre turbodiesel in my test car was good enough.

However, even here there's a problem. Because for £29,500 minus the whistles and bells, I'd expect more than a four-pot turbodiesel engine. I'd expect more than 0 to 62 mph in 9.8 seconds.

In fact, the price has got me scratching my chin in many areas. Yes, this is a big car with a big boot and lots of legroom in the rear. But it doesn't look big enough to wear a price tag of nearly £30,000. Especially when we know that Citroëns don't hold their value well.

So what's the alternative? Well, again, I'm stumped because there is no car that looks, drives or feels like this. Maybe the Ford Kuga? But then again, maybe not, because that doesn't have three electrical roof blinds. An Audi Q5? Yes, but that can cost up to £37,310 and doesn't have anything like the Citroën's 'want one' styling.

There's only one way really to sum this car up. It's ideal for those who want a fast-depreciating, possibly unreliable and uncomfortable car that looks fantastic and is unbelievably well equipped and charismatic.

In short, it's undoubtedly a car you want to buy. But I suspect that after a while, it'll be a car you'll want to sell.

10 June 2012

Squeeze in, Queenie, there's space next to Tom Cruise

Kia Cee-d '2' 1.6 GDI

Despite an idiotic decision by some halfwit in the television gallery to run credits over the climactic fireworks display, I think most people were fairly impressed by Gary Barlow's jubilee concert at the top of the Mall.

Certainly, as I watched the elderly gentleman with dyed hair belting out 'Live And Let Die', and Buckingham Palace exploded, and the crowds waved their flags in an orgy of hysteria, the whole extravaganza felt like a giant raised finger to Hank, Pierre, Fritz and Bruce. 'This, people of the world, is how you do it.'

At the time, I was on a hill with some friends from the sleepy town of Chipping Norton, lighting a beacon, setting off fireworks, drinking wine and singing the national anthem. We felt warm. We felt fuzzy. We felt proud to be British.

But we weren't allowed to feel proud or warm or fuzzy for long because by Wednesday the sniping had started. People were writing to the *Guardian* saying that the whole spectacle had been ruined because there were no ethnic minorities in the royal box.

Rubbish. The Queen is German, the Archbishop of Canterbury is Welsh and if Prince Philip had not been in hospital we'd have had a Greek as well. The royal family is, in fact, a shining example of diversity at its absolute best, and the royal box was actually a rainbow nation.

This didn't stop the naysayers. They argued that the whole weekend had been a celebration of class and privilege, that it had failed to elicit much support away from SW1, that it had been too expensive, that the weather had done more damage than anyone would admit and that the BBC's coverage had been lamentable.

Strangely, however, no one seems to have picked up on what, as far as I can tell, was the only mistake of the whole weekend: when the royal family emerged from one of the endless lunches, most were ferried to the next event in a fleet of Volkswagen vans.

I'm sorry, but how did this happen? The courtiers and the advisers left no stone unturned to ensure protocol was followed and dignity maintained. Mrs Queen, for instance, did not clap at the end of that ethnic song in the concert. The Duchess of Cambridge's chapel hat pegs were kept in check. The red carpets were just so, and the show went on even when the archbishop's much talked-about global warming lashed the flotilla with icy winds and torrential rain. And yet when someone suggested the minor royals be ferried about, under the watchful gaze of an admiring world, in a fleet of vans, someone said, 'Yeah. OK.'

It's not OK. You can't put Prince Harry in a van. Not in a country that makes Jaguars and Range Rovers and Aston Martins and Bentleys. Could they not have rustled up a fleet of Rolls-Royces, the second-best-known Anglo-German success story?

Top Gear's live show has been able to borrow a fleet of Morgan three-wheelers, so I feel sure the small company in Malvern, Worcestershire, would have been only too happy to step up to the mark with cars for the jubilee. And Prince Harry would have liked that.

It is extraordinary that people never really think about the wheels on which they will arrive at important events. They think about the frock and the hair and the shoes and the posture. And then they turn up in front of the cameras in a van.

It's not just the royal family who get this wrong, either. Last month officers from Strathclyde police arrived in London to arrest Andy Coulson, having made the journey in a Hyundai people carrier. They reckoned they had enough evidence to charge the former editor of the *News of the World*, and I was thinking, Really? You can't even choose a decent car.

Then we had Jeremy Hunt, the culture secretary, turning up at the Leveson inquiry in a Toyota Prius. What kind of twisted logic was used to make that look a good idea? 'Ha-ha. Mr Hunt will be facing some difficult questions today, but if he arrives in an eco-car, people will feel well disposed towards him.' I didn't. I thought that if he'd really wanted his arrival to take our mind off the issue of the day, he should have rocked up on a white stallion.

We see similar mistakes at film premieres. Recently I went to one for *Prometheus*, Ridley Scott's new blockbuster. The stars had turned out in force. There were several people from *The Voice*, many former *X Factor* semi-finalists and a number of soap stars. And all of them had turned up in gleaming silver Mercedes S-classes. Why? Each of these people is very keen to climb the pole of stardom and therefore each needs to stand out from the crowd. You're not going to do that at a premiere in an S-class. You need to be different. Which brings me neatly on to the subject of this morning's missive: the recently modified Kia Cee'd.

Twelve years ago Kia started to make a hatchback called the Rio, which, along with the three-cylinder Hyundai Accent diesel, was very probably the worst car the world had ever seen. Styled by someone who was either blind or just being stupid, it looked ridiculous and was powered by an engine that belonged in a Russian cement mixer.

The Rio demonstrated to the people of neighbouring North Korea that their leader had a point.

Here it was sold mostly to stupid idiots for whom the attraction of the latest registration digits was far more important than reliability, comfort, economy or speed. People called it 'cheap and cheerful'. But there's no such thing. There is 'expensive and cheerful' or 'cheap and rubbish'. It was the latter.

Today things have changed so drastically that we use a Kia as our Reasonably Priced Car on *Top Gear*. And the new model is the latest incarnation of that, the latest incarnation of a car that has been driven by more stars than almost any other. This, then, is the ideal film premiere arrival car.

But what's it like when you're not pulling up at a red carpet? Well, happily there isn't much space left, because, if I'm honest, there's not much to say.

It costs broadly the same as any of the other thirty-nine mid-sized family hatchbacks from Japan and Europe. It therefore needs to be just as good. And it is.

I explained recently that the car market had come alive in recent months as every manufacturer tried new ideas to meet stringent emissions regulations. But the Kia plays no part in any of this. With the exception of the now ubiquitous feature that stops the engine at the lights and then starts it again when you depress the clutch (huh, call yourself a clutch?), the new Cee'd is about as cutting edge as the bathing platform on a Sunseeker.

It's simply a good-looking collection of what's been learnt over the past 110 or so years. It's plain. If it were a loaf of bread, it would not be sliced. It would not be covered in bits. There'd be no ears of corn in every mouthful. And there's nothing wrong with that.

Except it means I have nothing of any great relevance to say here. There is as much space as you would expect. The back seats fold down. There is a radio. It stops and steers and goes just as well as any other medium-priced hatchback. There's a 1.4-litre petrol engine with 98 brake horsepower and a 1.6 unit that develops 133 bhp. Pretty much the same as every other 1.6 on the market.

There are also two diesel engines on offer – a 1.4 CRDi and a 1.6 CRDi – along with several levels of trim. The layout is good. The radio is not in the roof. The warranty is long. It's safe, so you're unlikely to be killed in action. And that's all we have space for, so let's roll the credits, and move on.

17 June 2012

The wife's away, so come check out my electric extremity

Mercedes-Benz ML 350 BlueTec 4Matic Sport

We refer often in the Clarkson household to people we call 'winners'. It's easy to spot one. He's a man, he has a Montblanc pen and he enjoys playing golf almost as much as he enjoys talking about it, especially to those who aren't interested.

The winner has great hair, thick forearms, a smattering of jewellery and a handshake that could squeeze juice from a tree. He works in marketing, knows what Ebitda means and speaks in a rich, deep voice to demonstrate to everyone within earshot that if his life had taken a different course and he'd ended up in a rock band, he'd have had no need on stage for a length of hosepipe in his trousers. The winner walks with a swagger because not only does he know he has a big package. He knows that we know it, too.

At least once a week the winner leaves what he calls 'the wife' at home so he can meet up with his mates for what they all call 'a few jars'. Mostly this is a competition to see who can order the most idiotic cocktail in the deepest voice, and who's got the most preposterous credit card.

Naturally the winner is very interested in what he wears, what watch he chooses – that'll be a Rolex – and, most important of all, what sort of car he drives. Some would suggest he has an Audi, and it's true: many winners do. But, actually, what he wants most of all is any car with a boot lid that opens and closes electrically.

You and I both know that a boot lid that opens when you press a button on the key fob is monumentally stupid because it means we have to stand in the wind and the rain, waiting for an electric motor to do in half an hour what we could have achieved in about one second.

The winner, though, is not bothered about practicalities. It's why he still uses a fountain pen rather than a biro. So he's perfectly happy to stand around waiting for his boot lid to open. Hell. People can see him. They know they're looking at a man whose life is so complete, he has an electrically opening boot lid. Occasionally, when he catches a girl looking at his tailgate rising, the winner will wink at her. He knows she'll be OK with that. Because he knows that she knows that an electric boot lid is yet another sign his manhood is gigantic.

On that basis the winner will be jolly interested in the new Mercedes M-class, not only because of the electric boot lid but also because of the sheer length of its name. It's the – deep breath – Mercedes-Benz ML 350 BlueTec 4Matic Sport. Get yer chops around that one, love.

There are other things he will enjoy, too. The climate-controlled cupholders, the in-car internet access, the Harman/ Kardon Logic 7 Dolby digital 5.1 surround-sound system and the darkened privacy glass in the back that not only cuts down the glare on the television screens but also makes it hard for prying eyes to see what he's up to back there with the girl who liked his button-operated tailgate.

There is a small problem, however, with some of this stuff. It's quite pricy. The standard car is good value at £48,490, but by the time you've added a selection of 'check it out, chicks' electronics, the bill rockets up to what the winner would call 'north of 60K'.

And this is what makes the ears of the non-winner prick up. A big, well-equipped Mercedes 4x4 for considerably less than a Range Rover. Hmmm . . .

The ML had a fairly poor start in life. It was conceived at a time when Mercedes-Benz had convinced itself that its cars were 'over-engineered' and that it needed to worry less about reliability. Sadly Mercedes addressed this by not worrying about it at all. And that's one of the reasons it decided to build its car at a new factory in Alabama.

Once, while I was driving across this extremely violent state in an early example of the breed, a local asked what it was.

'It's a Mercedes, but it's built here,' I said.

'Oh,' he replied. 'Well, it'll be shit then.'

He was right – it was. A point that was proved just a few miles down the road when the roof fell off.

Quickly Mercedes decided to buck its ideas up, and fairly soon the ML was a lot better. The AMG-powered ML 63 was an absolute gem, in fact. But the car you see here is the new model. And the version I tested was not a tyre-shredding V8 but a 3-litre diesel.

The engine is remarkable. So quiet, even on start-up, that there is simply no evidence at all that the fuel is being ignited by pressure rather than a spark. It's frugal, too. You will get almost 40 mpg, which is incredible for a big four-wheel-drive car that can do 0 to 62 mph in less than eight seconds.

There are more good things, chief among which is the comfort. This particular model may be called Sport but it's no such thing. It's a cruiser, a big, soft old Hector that irons speed bumps into oblivion and soothes its occupants to the point where they need to be reminded with a bit of wheel judder if they nod off and stray out of lane. In this regard it out-Range-Rovers a Range Rover.

I like the simplicity of the controls as well. Particularly good is the column-mounted gear lever, which, as in the Rolls-Royce Phantom, offers you a choice of forwards, backwards or neither. Sport declutch override power zoom? Nope. It doesn't have that.

Mostly the interior is very similar to the interior of any other Merc, and this is an issue. It's like being in an E-class that's on stilts, which would be fine if the ML were a proper off-roader, but, truth be told, it isn't. Not really.

It may have one or two off-road features and it may have just enough ground clearance for a gymkhana car park, but it ain't no

G-wagen. In the rough, it ain't no Range Rover, either. Those of you who shoot? Forget it.

The truth is that you can have just as many seats, the same engine, the same gearbox, the same level of quality, the same ride and even more space for less money if you buy a normal, even-more-economical E-class estate.

The biggest problem, though, is the styling. The original ML was a handsome brute. This looks like a melted Kia. The front and the back are attractive enough, but the sides? Oh dear, no.

So there we are. If you want a big German car, buy an E-class, or better still a BMW 530d. And if you want to shoot a pheasant in the face, I'm afraid you'll have to ignore the ML 350's undoubted strengths and stick with the Range Rover. It's pricier but, as an all-rounder, better.

The new ML, then, is a bit of a loser. A car that's really suitable only for people who are 'winners'.

24 June 2012

If I go back to Africa, will you take it away again?

Porsche 911 Carrera S cabriolet

The Moses Mabhida stadium in Durban, South Africa, is the most beautiful building in the world. Built to host the 2010 World Cup, it's beautiful at night and beautiful in the day. It's beautiful when looked at from far away, or from inside. No other structure I've ever seen gets close. It's a triumph.

Recently it played host to the most ambitious live event *Top Gear* has attempted thus far. Not only would the three of us be performing to a wailing squadron of 15,000 South Africans inside the stadium, but outside there would be a motor show and, on roads closed solely for the event, a 1¼-mile street circuit.

Here there were races between a superbike and Michael Schumacher's Formula One Mercedes, demonstration laps from the Stig in a selection of supercars, full-on inter-nation races and the local hero Jody Scheckter, who never crashed anything even once. Honest.

Richard Hammond, James May and I were very impressed with the line-up but decided that since it was, strictly speaking, our playground, we should be allowed to do some laps ourselves. So we devised a competition: pick any car you like and see who can do the fastest lap time.

May went for the McLaren MP4-12C, Hammond for some kind of Beetle, and me? I went for the love of my life: a Mercedes SLS AMG roadster.

And on my first exploration lap I knew quite quickly that I'd chosen unwisely. Because the track was not only very narrow and tight, but also hemmed in on both sides by concrete barriers. And a tight, concrete-lined street track is not really the main hunting ground for a very large 6.2-litre 563 brake horsepower V8 monster with the growl of a wild animal and the tail of a happy dog. It would be like wrestling with a bear in a phone box.

To make matters worse, the track was a popular attraction for our visitors. Thousands were pressed up to the fences and filling the grandstands. And all of them were thinking the same thing as I roared into view. 'Please. If there's a God in heaven. Make him crash.'

They all had their cameras out, videoing my every move, and not so they could show their friends back home how well I'd done. No. It's so they could put on YouTube the precise moment I hit a wall and my head came off.

Naturally I was extremely unkeen to oblige and decided therefore that the prospect of beating Hammond – May wasn't really a factor – was in no way enough to balance the risk of ending up on the internet in a fireman's bucket. Result: I decided to go slowly.

However, there was a problem. You see, it turns out that when you drive in front of a crowd, and you have testes, you cannot go slowly because you are compelled to show off. This meant that wherever possible I didn't go quickly, or slowly, but sideways, trailing as much smoke as possible from the rear tyres.

That, of course, meant turning off the traction control, which in a car such as this on a track such as that was idiotic. And, worse, I was overcome with an uncontrollable urge to wave all the time. The crowd was waving at me and it seemed rude not to respond, so there I was, waving and power-sliding in a V8 on a track completely unsuited to either of those things.

It's yet another reason I couldn't be an actual racing driver. You're not going to win much if you do a doughnut at every corner and pose for pictures as you go by. And so it turned out to be. Hammond was the fastest.

Which meant that I had to spend four days listening to him

bleating on about how his Beetle is vastly superior to the big Merc. And that's why I wasn't as sad as you might imagine when the *Top Gear* festival ended. We'd had the best time, living like rock stars. But, as with rock stars, the musical differences between Hammond and me had become so enormous, I was beginning to wonder what he'd look like with no skin. I wanted to get home so I didn't have to listen to him gloating any more.

And then, would you Adam and Eve it, when I got home, guess what car was sitting in the drive waiting to be tested. Yup. A bloody Beetle − or, as it would like to be known, a Porsche 911 Carrera S cabriolet. I sank to my knees and wept.

Of course, the good news is that *The Sunday Times* is not available in the swampland where Hammond lives. And even if it were, it's full of big words he wouldn't understand. So since he isn't reading this, I can be honest. The latest 911 is actually a damn good sports car. And the GT3 version is even better than that. It's heavenly.

However, I was being asked to spend a week in the cabriolet, and that's different. Because the truth is, if you remove the roof from a sports car, you are, to a greater or lesser extent, reducing its structural rigidity. And if you attempt to mask that with underfloor strengthening beams, you are adding weight. Which means that you don't end up with a sports car at all.

A Porsche cabriolet, then, is a bit like an Afghan hound that's gone bald. It's still an Afghan hound but the point has been somewhat lost.

Oh sure, on the new cabriolet, Porsche's engineers have devised a lightweight magnesium and aluminium roof frame, along with composite panels, which is said to be 18 per cent more rigid than the one it replaces. But despite everything, the car's 50 kg heavier than its hard-top sister.

Of course, for 99.9 per cent of the time the two versions feel 99.9 per cent identical. But the keen driver will know that for 0.1 per cent of the time, the cabriolet will feel 0.1 per cent worse. And that will be a constant niggle.

If you want a sports car, this is not for you. But if you are the man I met at a golf club in Watford recently, a man who has the old four-wheel-drive convertible, pay attention . . .

The first problem you face with the new car is roof-up visibility. At oblique junctions, you have to rely on your inner Mystic Meg before easing out into the traffic flow. And on motorways you pull over at your peril.

Then there's the gearbox. My car had a seven-speed manual. Now I recognize that a loping seventh is needed to keep fuel use down and the European Union emissions wallahs happy but, ooh, there's a lot of gear-changing. Which would be just about bearable if the clutch weren't both heavy and juddery. This is emphatically not a town car.

Other things? Well, the cupholders are located right in front of the heater vents, which means that when you turn on the air-conditioning, the first thing to get chilled is your mug of tea. And there's a mysterious button that, when pressed, makes the exhaust so loud you can't hear the radio any more. And I didn't like the electric steering. Or the fact the boot is in the front, which means you get dirty fingers every time you need to get something out of it.

Roof down? Don't know, partly because it rained constantly and partly because, as we know, if a grown man drives around with the top down, he looks like the central character in an advertisement for Viagra.

Of course, there are some good things. It's beautifully made. It only ever needs servicing after an ice age. It's not too big. It's not too ostentatious. And it's not that expensive. A standard 3.4-litre Carrera cabriolet is £79,947, whereas the model I have here – the 3.8-litre S – is less than £10,000 on top of that. That may seem a hell of a leap for an extra 400 cc, but having experienced both, I can tell you it's worth every penny. The basic model can feel a bit slow. The S never does.

But when all is said and done, it's only another two-seat

convertible. And if that's all you want, Mercedes and BMW can sell you one just as good for far less.

And now, I'm afraid I must get back to South Africa, because I've just heard that James May is about to finish our three-lap race.

1 July 2012

Oh, Miss Ennis, let's sprint to seventh heaven

Ferrari 458 Spider

When you buy a Nissan Micra or a Volkswagen Golf or a Ford Focus, you expect it to be perfect. The mainstream car makers know this, which is why, before a car is put on sale, it is rigorously tested to make sure the starter motor works when it's below freezing and the air-conditioning can cope even if you drive to the surface of the sun. Every little detail is thought about. Every component tested, and then tested again. And then changed. And then tested again.

However, in the world of very expensive supercars, things have always been rather different. Lamborghini put the Miura on sale knowing full well that if it were driven above 80 mph, it would take off. Then it came up with the Countach, a car with a cockpit so small it could only be driven by either an ant or a foetus. Neither of which would have the strength to move the gear lever, which is seemingly set in concrete, or the steering wheel, which was mostly a piece of decoration. Not that it mattered, anyway, because most days the Countach would not start.

As time went by, the makers of supercars started to think more seriously about longevity and convenience. But even in the 1990s they were still not really there. The Ferrari 355 GTS I once owned was plagued with seatbelts that strummed like guitar strings if you had the roof off, an engine that had to be taken out of the car to be serviced and a two-stage throttle that made only two speeds available: 2 mph and 175 mph.

Later I bought a Lamborghini Gallardo Spyder. This was developed under the watchful gaze of people in sensible shoes from Audi. It would, I figured, be as easy to live with as a Toyota Corolla.

It wasn't. The cupholder was a £600 option and the company simply hadn't thought about the positioning of the pedals in right-hand-drive cars. Which meant that if you bought a manual – as I did – there was nowhere to put your left foot. You had to amputate it.

The problem is this. If you are a small-volume car maker, you simply don't have the funds to design a feature and then redesign it if it doesn't work. So you end up hoping that customers will be so consumed by the speed and the beauty, they won't notice that the door doesn't shut properly and that there is a hippopotamus in the passenger seat.

Happily, today most supercar makers are owned by large-volume manufacturers, which means they have the funds to address little foibles before the machine goes on sale.

Well, that's what you'd think. But after a couple of days with the new Ferrari 458 Spider you realize that things haven't really changed at all.

If the wipers are on full speed they become so hysterical that they bash into the window frame on every sweep. The radio is incapable of finding a signal. When there's no passenger, the unused seatbelt buckle rattles against the back of the seat, and to fill up, you need to hold the nozzle of the pump upside down, or no fuel will be delivered at all.

There's more. If you want to change radio stations – and you will, because whatever you're listening to is mostly hiss – you need to go into the menu, twiddle a knob, push another button, select the station and then repeat the process to get back to the satnav screen.

It gets worse. Because there are no stalks to operate the wipers and indicators, all the main controls – and the starter button, and the horn, and the six-way traction control, and the suspension control, and the radio controls – are on the wheel. My daughter was amazed by this. She said driving the car was like playing Bop It.

We're told by Ferrari that you get used to this after a while, and I don't doubt that's true. In the same way that you can get used to having arthritis.

Make no mistake. The 458 Spider is the most usable and modern of all the supercars. But it's still plagued by the sort of faults that would not be acceptable in a Nissan hatchback.

And it's not cheap. The base car is £198,936, but if you want the steering wheel stitched in cotton the colours of the Italian flag – well, that's an extra £720. That's £720 for some cotton. You want the wheels painted gold? That's £1,238. A premium hi-fi system is £3,411. Titanium wheel bolts are £1,919. Red brake callipers? They're £880. Racing seats? They're £4,961. The end result is that the car I tested would actually cost you £262,266. And that's what an economist would call 'a lot'. But it's worth every single penny. Because this car is simply sublime.

And at this point some of you may accuse me of inconsistency because just recently I said that the new Porsche 911 Carrera S cabriolet does not work as a convertible because the strengthening beams and the structural compromises ruin what was designed to be a pure sports car. Taking the roof off a car such as this is like adding HP Sauce to a quail's egg. It adds to the tang but you lose the delicacy, and with a 911, delicacy is everything.

With a Ferrari, things are different. A 458 is not a purebred sports car. Oh God, it drives like one, but it's also a singer and a model and an athlete. It's a heptathlete with the lungs of Pavarotti and the face of an angel. So you can buy the convertible version because driving this car with the top down adds to the theatre and the pantomime. Who cares that you're going 0.1 mph slower? You're getting a tan.

Made from aluminium, the foldaway top weighs less than the normal roof on the standard car. It even weighs less than it would had it been fashioned from canvas. And it folds away, electrically, in fourteen seconds. You can even drive with the roof up and the back window lowered so you can hear the V8 soundtrack when it's raining. I did that a lot.

Is there a drawback to the new convertible? Well, yes. If you lower the roof at the lights, everyone within 150 yards will tell you that you are a tosser. Plus, in extreme conditions it will be less rewarding than the hard top, and the windscreen is now arched, which looks a bit odd. But what the Lord taketh away at the front, the Lord handeth back at the rear. From behind, it looks like the old Ferrari 250 LM. From behind, it's one of the best-looking cars I've ever seen. It is also extremely comfortable.

As I write now, there are shivers – and I'm not kidding – running up and down my spine as I recall the way it felt on roads near my home. The lightness. The savagery. The noise. The beauty. The trees rushing by, sheltering me from 93 million miles of sky. Then you have the gearbox that changes down not in a few milliseconds but instantly. Bang. Stand on the brakes – bang again. And again. Turn. And POWEEEERRRR. A modern Ferrari feels like no other car on the road. It feels miles better. And this one? Oh, this is the best of the lot.

Sure, you can find rivals that are more technical and even a tiny bit faster. The McLaren MP4-12C (which will soon be available as a convertible, too) is one, and the Bentley Continental Supersports is another. But neither has anything like the lust for life that you find in a 458. They are tools.

I grew out of supercars many years ago. I vowed after the Ford GT that I'd never buy another. And I will stand by that. But if I were to waver, this would be the one. As a car, it would get two stars, for being silly and too expensive. But as a thing. As a celebration of man's ability to be happy. It's in a seven-star class of one.

15 July 2012

Yikes! The plumber's van has put a leak in my wallet

Citroën Berlingo

Eleven years ago I had the most brilliant idea for a new car-related television programme. It would be based in a hangar, it would be presented by four people, including a racing driver who would never speak, and – this was the clincher – it would show only real-world family cars, driving around on the roads of Britain. No silly Ferraris. No expensive foreign travel. No blue skies. I was most insistent on all of that.

And so it was that in the very first feature on the very first show on what would be called *Top Gear*, I drove a Citroën van under leaden skies through Kent. I was very pleased with the results, and so were the commissioning editors of the BBC, because figures showed that more than 3 million people had tuned in.

Later, other people suggested that we should perhaps ease up on my strict rules about foreign travel and Ferraris. And with a heavy heart I acquiesced. Mainly to demonstrate, on television, that their ideas would never work. So we started driving over the Andes and racing light aircraft across the Alps, and now up to 380 million people tune in every week, making it the most-watched television show in the world.

Very often people write to me saying we should revert to the days when we reviewed the sorts of car that people actually buy, but, armed with the power of hindsight, I can now see exactly why this doesn't work.

Because in any given week a hundred people might be interested in the new Volkswagen Golf diesel, meaning that 379,999,900 aren't. Whereas a Ferrari power-sliding through a volcano? Pretty much everyone wants to see that, pretty much all the time. Plus, of course, it means I get to live your dreams.

This morning, however, I'm living your reality, because I've gone full circle and arrived back at the aforementioned Citroën van.

It was called the Berlingo, and I said at the time that it was pretty damn good.

The idea was simple. The manufacturer had taken a van and then fitted back windows and seating in the rear to create what was a very practical, supremely comfortable and extraordinarily cheap family car.

Yes, it looked a bit ungainly, but if memory serves, it was about £9,000 – way, way less than VW was asking for a more cramped, less versatile Golf.

Since then I've been jolly busy drifting Bugattis and driving to the North Pole and hurtling across Botswana under clear blue skies, and I'm afraid I'd rather forgotten about everyone back at home trundling to the shops and back every weekend in a Berlingo. Until now.

Because I've spent the past week with the latest version and, ooh, it's grown up. To mask the fact that it was born to transport French plumbing equipment, the new car has chunky-looking body armour and chrome around the grille. '*Je suis un Range Rover*,' it seems to be saying. Yeah, right.

'*Mais oui, parce que regardez-là. J'ai un knob pour beaucoup de fonctions de traction control et donc je suis sérieux.*' You aren't. You're a van.

Now. If we apply the modern-day *Top Gear* rules to this road test, I'm afraid this van-ness is going to be a problem. Because the Berlingo understeers as though it has no steering wheel at all, and there is a great deal of driveline shunt.

However, I have a sneaking feeling that if you are interested in a car of this type, you will have no idea what understeer is and you will imagine that driveline shunt is something that is only ever encountered in *Fifty Shades of Grey*. You will also not be the slightest bit bothered to hear that there's no red line on the rev counter.

'So what? I have ears. And they will tell me when it's time to change up.' They sure will, because when you approach 4000 rpm the noise is so loud, it causes all dogs within 30 miles to faint.

Criticizing this car for its lack of ability to thrill, though, is like criticizing soup-kitchen sandwiches for lacking lemon grass.

This is emphatically not a car for enthusiasts. It would not be capable of getting around the Nürburgring, mainly because by the time you reached the Carousel corner, you'd have died from old age – 0 to 62 mph takes 12.1 seconds, for crying out loud.

However, I'm talking now to the 99.9 per cent of people who are not heading out to the Eifel mountains next weekend, and who do not have to leave half their tyres smeared all over the road to feel as though they've circumnavigated a roundabout properly.

And still, the Berlingo appears to make sense. Its styling may be writing cheques its underpinnings can't cash, but in several important areas it is superb. First of all, it is easily the most comfortable car any money can buy this side of a Rolls-Royce Phantom. Precisely because Citroën has made no attempt to make it sporty, it simply glides over potholes and speed humps.

There's more. As I'm sure you know, children tend to open the door when the car stops without bothering to see if a cyclist is coming. Well, the Berlingo has sliding side doors, which means our wizened chum in his Lycra romper suit is not inconvenienced at all.

Around the back, the massive but light boot lid opens to reveal enough space for a whole pack of hounds, though if you want, you can simply open the rear window instead.

In the cockpit there is a mass of bins and trays in which you can put all the stuff that accumulates in a family car, and lots of real-world clever thinking that makes the Berlingo more in tune with the needs of more people than just about any other car out there. It was always thus.

Yet there are one or two small problems that matter. The pillars are so thick that it's hard to see at junctions. Of course, this isn't an issue when you're in a van, because it's not yours and there's nothing of any value inside should you pull into the path of a bus. But when the van is a car, it is yours and there are children in the back.

Equally annoying is its susceptibility to crosswinds. You don't just have to hold on to the wheel on a blustery day. You have to wrestle with it, as if it's a bear and you've just trodden on one of its cubs.

And now we get to the vexed question of price. Citroën obviously knows it's vexed, which is why the actual cost of the car – the single most important fact – is not listed in the press pack. But I've managed to find it. And for the top-of-the-range 1.6-litre HDi 115 XTR that I tested, it's £16,795. You can have a normal car for that.

It gets worse, because satnav is an extra £750, air-conditioning is an extra £650 and a 'family pack' with a third row of seats is £845. And so it goes on.

I'm afraid, then, that the principal appeal of the original Berlingo is gone. This new version is soup-kitchen food at four-star restaurant prices. Yes, it's extremely practical and extremely comfortable. But that simply isn't enough.

22 July 2012

Gary the ram raider cracks Fermat's last theorem

Vauxhall Astra VXR 2.0i Turbo

It's known that people first moved to the Blackbird Leys region of what became known as Oxford about 5,300 years ago. Nothing much happened until 2,000 years later, when it's thought that someone built a circular eco-house there and someone else made a loom.

Then nothing happened again until 1991, when the younger male residents invented a new sport. They would go into wealthier areas of the region, steal a car and use it to whizz about their own estates, doing skids.

Soon news crews from all over the world were turning up to film these young men bouncing off postboxes and lampposts in front of a cheering crowd. And if the police tried to stop the mayhem, the crowd would express its displeasure by throwing stones and making cow-like lowing noises. This would cause even more film crews to turn up. Which would cause even more attention-seeking young men to steal even more cars.

The craze soon spread, and within months young men from council estates all over the land began to spend their evenings driving other people's cars around branches of Dixons and Woolworths. For a while, doing a handbrake turn in an Arndale centre was more popular than football.

Eventually, of course, the craze died down and the young men of Blackbird Leys went back to doing what they'd done for thousands of years: sitting in bus shelters chewing gum, mostly.

But they did leave two legacies. No 1: cars could no longer be stolen using a lollipop stick and a bent coathanger. And No 2: they killed off the hot hatchback.

Devised in the mid-Seventies, the recipe was very simple. You took a normal, easy-to-park, easy-to-mend family hatchback, and under the bonnet you fitted a biggish engine. It proved to be immensely popular, to the point that in the mid-Eighties 15 per cent of all Ford Escorts sold in Britain were hotted-up XR3is and 20 per cent of all Volkswagen Golfs were GTIs.

Cars such as this were classless. They were driven by Hoorays in Fulham and school-run mums in Castle Bromwich. I know someone who traded his Gordon-Keeble for a Golf GTI. They were ageless, too, and were just as popular with teenagers as they were with the elderly.

But after the ram raiders and the Twockers and that newsreel footage of Gary hooning around Blackbird Leys in someone else's turbocharged MG Maestro, the hot hatch became a byword for yobbery. A Burberry-badged back-to-front baseball cap with windscreen wipers and an out-of-date tax disc.

Now. If I were running a car firm, I'd want people back in hot hatchbacks as soon as possible because they are extremely profitable. I'd therefore be doing everything in my power to shake off the yob tag, in the same way that Stella Artois tries to shake off its wife-beater image by banging on about how it uses only hops that can speak Latin and zesty mountain spring water.

But no. Every hot Ford is festooned with trinketry that would not look out of place on Wayne and Coleen's mantelpiece. And each is painted in lime green or vivid blue or matt black. They're as subtle as being attacked by a shark while off your head on acid.

Renault is equally childish. Hot versions of the Mégane and the Clio look as if they've been lifted straight from a school playground. 'Look at me,' they seem to be saying. 'I have a mental age of nine.'

And then we get to the subject of this morning's review. The new Vauxhall Astra VXR, which was sent around to my house sporting an optional rear wing that would be dismissed by an Asian drifting champion as being a bit over the top and massive Fisher-Price 20-inch wheels.

Even if you don't specify these extras, it still has more jewellery and more tinsel than P Diddy at a rap convention. It's a car that conveys one simple message to other road users. And the message is this: 'I am extremely unintelligent.'

It's annoying, because, beneath the flotsam and jetsam, this is not just a very pretty car but also quite a clever one.

Because the turbocharged 2-litre engine develops a whopping 276 brake horsepower, making this by some margin the most powerful car in its class, much out-of-sight work has been done to ensure the front wheels don't just fall off every time you put your foot down.

Up front, it's fitted with what Vauxhall calls HiPerStrut suspension, which is designed to optimize camber during cornering and cut torque steer, and, as a further measure, a proper mechanical differential is added. Ford used pretty much the same setup on its most recent Focus RS.

But Vauxhall goes even further because the VXR comes with an adaptive ride and 'floating' front brake discs designed to reduce unsprung weight. Make no mistake: the underside of this car has been created by someone who was concentrating, and funded by a company that plainly wants to lay the ghost of the Vectra to rest and be taken seriously.

I shall oblige. The VXR is very, very good. It goes like a scalded cock, stops with an almost cartoonish suddenness and corners with absolutely no drama at all. It isn't quite as thrilling as a hot Mégane – it's much heavier – but what you lose in Stowe Corner, you gain on all the other days of the year because the ride comfort is exceptional. Even though the tyres have the profile of paint, this is a car that just glides over bumps.

There is a Sport button that firms up the suspension, and a so-called VXR button that adjusts the throttle response, adds weight to the hydraulic power steering, gives even harder suspension and makes all the dials glow red. But I don't recommend you ever use either. Because mainly what they do is add 10 per cent to the dynamism and take away 100 per cent of the comfort.

There are a few little niggles. Despite Vauxhall's best efforts, the wheel does still writhe about under harsh acceleration, and there is rather more turbo lag than I'd like.

Inside, it was the driving position that annoyed me most of all. After a while, I got cramp. And who thought it would be a good idea to fit a centre armrest that prevents the taller driver from selecting second, fourth and sixth? Also the front wheels weren't balanced properly. Grrrr.

Oh, and then there was the boot lid that wouldn't open. I'm sure there's a clever button hidden away somewhere, but finding it would have meant reading the handbook. And I can't do that because I'm a man.

None of these things, however, should prevent you from buying what is a well-engineered and well-executed car. But what might cause you to think twice is the bovine trinketry, the high price and the fact you have to tell people at parties that you've bought a Vauxhall Astra.

If these things are too much of a cross to bear, it's not the end of the world. Because, happily, Volkswagen can still sell you a hot hatch that doesn't make you look a gormless plonker. It's not as stupendous as the Astra. But it's not as stupid. It's called the Golf GTI.

29 July 2012

Kiss goodbye to your no-claims – Mr Fender-bender has a new toy

Peugeot 208 1.2 VTi Allure

It is obviously very bad when someone becomes so consumed with a project or hobby that they lose the ability to talk or even think about anything else. Hobbies are a bit like crack cocaine. You think that maybe you'll just dangle a worm in some water to see if you can catch a stickleback, and the next thing you know, you're divorced because you spent all your life savings on a carbon-fibre rod, and you're sitting by the side of a canal at five in the morning trying it out.

I've been there. Back in 1975 I became mildly interested in what we used to call hi-fis. And then, in the blink of an eye, I was very interested indeed and my girlfriend had gone off with someone who wasn't really interested in anything very much at all.

I barely noticed because my new Marsden Hall speakers had arrived. Some say Wharfedale made a better unit but I disagreed. The Marsden Halls were perfect for my slimline black Teleton amp. I caught a train all the way to London to buy that.

The deck? At the time the Garrard SP25 was popular, but I took a holiday job as a milkman so I could afford the 86SB, which I teamed with a Shure M75ED cartridge. I'm not looking any of this stuff up. It was all ingrained in my head back then and it's still there now. I actually know what sort of stylus I used, and its code name.

While my friends were out stealing traffic cones and trying to get into Annabel's bra, I was to be found at my desk, soldering an unbelievably fiddly seven-pin Din plug so I could connect my recently bought Akai tape deck to the school's PA system. I was very boring.

So you can imagine how I felt about the home-brand all-in-one 'music centres' that Currys and Comet started to sell in the 1980s. Oh, they looked all right, with all their flashing lights and damped cassette-release mechanisms. And I'm sure they were fine for listening to Dire Straits' albums at suburban dinner parties. But for someone like me, they were only a forked tongue short of being the actual devil.

And so we arrive, naturally, at the Volvo 340 DL. As we know, this was a ghastly car. Made by people in Holland who thought Jesus was coming, it was powered by rubber bands, fitted with Mr Universe steering and styled during a game of consequences.

However, it was perceived to be strong and safe, so it attracted all the people who were not very good at driving and thought they may crash. This was unbelievably useful for the rest of us. If you saw a Volvo 340 DL coming the other way, you knew to be on your guard.

Eventually, however, Volvo decided to stop making bad cars for useless drivers, so the incompetent and weak decided en masse to switch to Rover. And again, this was good: see a 45 in the left lane, indicating left, and you knew not to assume it was actually going to turn left.

But then Rover went west and the bad drivers were suddenly hard to spot. Some were in Hyundais and Kias. Some were in Volkswagen Golfs. It was a dangerous period, but luckily Peugeot rode to the rescue. For many years this French company had made excellent cars but one day it decided to make a lot of very cheap rubbish for people with hearing aids, hats and a tendency to hang something from the rear-view mirror.

The other day I saw a Peugeot upside down at the entrance to the Hanger Lane underpass in west London. It is physically impossible to roll a car here, on what is a dead-straight piece of road. But Mr Pug Driver had managed it. And I recently saw another, balanced in pretty much the same place on the Armco.

Last week I came as close as I've been for years to having a head-on with a 308 that was on completely the wrong side of the road. It is uncanny this: Peugeots are invariably driven by someone who finds every single motoring event a complete surprise. 'Oh my God, look. Those lights have just gone RED!' 'Holy cow. There's another CAR!'

If I were running the police force, I would ask my officers to pull over all Peugeot drivers just to make sure they aren't driving under the influence of Vera Lynn. Because they're sure as hell driving under the influence of something.

To find out what it might be, I've just spent a week with a 208, or to be specific, the mid-range 1.2-litre VTi Allure. It's a good-looking little thing and at £13,495 it's well priced, too, especially given the amount of equipment provided as standard.

The only slight oddness is the steering wheel. It's the size of a shirt button and it's located very low down. So low that in the event of a crash, your testes would get such a thump from the airbag you'd wish you had died.

There were many nice things, though. For a 1.2 the engine delivers a surprising lump of punch. At one stage I was doing 70 mph, and that's faster than a Peugeot has travelled for twenty years. I also liked the central command system that is used to operate everything.

The 208 is actually smaller on the outside than the car it replaces – the dreadful 207 – but inside, it's bigger. So big, in fact, that there was space in the back, with the rear seats folded, for three dogs, one of which was larger than a diplodocus. Other things? Well, it was quiet and comfortable and the visibility was good.

All the time, though, I had a nagging doubt. On the face of it all was well, but every time I started the engine there was a beat before the electric power steering woke up. It was only a moment, but it told me that behind the flashing lights and the nice design touches, the engineering wasn't quite as thorough as you might have hoped.

There's more evidence too. It's never an annoying car but it's not what you'd call delightful, either. You don't get the little shiver that you sometimes experience in a Fiat, or even a Volkswagen. This, then, to a car enthusiast is what those music centres were to me back in 1981. An attractive package with many features that is fine for playing Dire Straits as you drive to the shops. But not much else.

It is, therefore, a car for people who are not that interested in cars. And that explains everything. Because if you are not interested in something, you will be no good at it.

Perhaps that's why Peugeot says in its advertisements that the 208 is a car that lets your body drive. It does, leaving your mind free to think about stuff that matters to you: the Blitz and how it used to be all trees around here.

I suppose, however, we can draw an interesting conclusion. If you – as a good driver – do buy a 208, you will find that all the traffic parts as you motor along. They will assume you are about to crash into something.

It might, therefore, be a faster and safer way of moving around than almost anything else on the road.

5 August 2012

The nip and tuck doesn't fool anyone, Grandma

Jaguar XKR-S

A man was apprehended by the constabulary recently for turning around to admire a girl on the pavement. He'd seen her bottom as he drove by, and officers spotted him looking through his rear window to see if the front was as good as the back.

I realize, of course, that when we are behind the wheel we are expected to become robots, immune to the ringing of a telephone, the crying of children in the back and the stupidity of other motorists. We may not talk, listen to the radio, eat a sandwich or become irritated. And all of this is ridiculous. But now we discover we may not drive while under the influence of a scrotum, and that's worse.

I try not to look at pretty girls on bicycles because it is probably annoying to have half the population looking up your skirt and praying for a gust of wind. But it is not possible. I have just about trained my head to stay still but my eyes are controlled by testosterone, and as often as not I don't see the lights turn green because they've swung around so far I'm actually looking at my own frontal lobes.

I'm also distracted by roadside advertisements, new shops, the amusing driving position of shorter motorists, interesting cloud formations, work matters, idiotic signs that have no meaning, a constant fear that one of the wheels is about to fall off, the mind-numbing noise of high-power motorcycles – pretty much everything. Except other cars.

I don't turn around when I see a Lamborghini or a Ferrari going the other way, in the same way that people who work at the chocolate factory don't stand and salivate at the petrol station's confectionery counter.

That said, I can never resist a sneaky double take when I am presented with a Jaguar XK. Designed by the same man who gave us the Aston Martin DB9, and engineered by Jag when it and Aston were part of the same company, it's always been a thinking man's Bondmobile. No, really. It was nearly as quick off the mark and only half the price.

What's more, it managed to combine the rakish good looks of the Aston with more aggression. It managed therefore to be pretty and fighty at the same time. It's such a head-turner, in fact, that whenever I see one I become consumed by one of life's great mysteries: 'Why don't I have one?'

To find out, again, I've just spent a week with the newest, latest version, the super-hot, super-aggressive XKR-S convertible. And straight away I could see many problems.

There's no getting away from the fact that this is an old car now. The dials look as though they've been lifted from a thirty-year-old Peugeot, the back seats are as useful as having no seats at all, the touchscreen command system, which operates the radio and climate control, is as counterintuitive as an old twist-key sardine tin, and while an iPod connection is supplied, it won't play tunes from your iPhone – or at least it wouldn't for me. 'What!?' it says, when you try. 'Are you suggesting you can play music on your telephone? Don't be stupid.'

There's more. To distinguish this new hot model from its lesser brethren, Jaguar's stylists have seen fit to spoil the very thing that gives this car such appeal. Its looks. Up front, there's a new nose that suggests the car is frowning. Then you have two suitcase handles on the corners – I have no idea why – and at the back a large spoiler, which is fine if you are eighteen and a yob. But not if you are forty-eight and a solicitor.

What Jaguar has done is taken, say, Keira Knightley and 'improved' her looks with several nose piercings and, on her forehead, a dirty great tattoo.

And now we get to the 5-litre supercharged V8 engine. It, like the car, is old, but that hasn't stopped the engineers squeezing about 40 more brake horsepower out of it. Inside, it must look like a lemon that's been run over by a bus.

So. There we are. The looks are gone. The interior is old and the engine's a pensioner with a new pair of training shoes. And yet . . .

In two important areas, the car's age pays dividends. First, it still uses a proper automatic gearbox, not an eco-sop flappy-paddle manual. And its roof is still made from canvas rather than steel. Normally, I'm not given to camping, but somehow, in a car, it's nice to be protected from the elements by nothing more substantial than one of Bear Grylls's hats.

Not that I needed the roof up much because at the precise moment this car was delivered, the rain stopped, the sky turned blue and the temperature shot up to what felt like a million. You're going to like driving any convertible in conditions like this, and I must say, I liked driving this one a lot.

Foremost, there's the speed. It's properly fast. And for once in an XKR, the exhaust thunder is audible not just to passers-by but to occupants of the car as well. I particularly enjoyed the distant gunfire rumble it makes on the overrun. I was on the overrun a lot. In fact, I spent most of the week speeding up, just so I could slow down again.

I also liked the steering. It's not an especially light car, but it feels nimble and agile. The only thing you have to remember is that the chassis was set up by a man who likes to go sideways all the time, so you have to be a bit careful before engaging the full enchilada.

Not to worry. Unlike other XKRs I've driven in recent times, this one doesn't bang and crash over potholes. It's actually quite smooth. You can therefore cruise about the place, no problem at all.

Soon, though, you start to encounter some issues again. When, for instance, you use the paddles to override the auto

box, there's no easy way of getting it back into 'Drive' again.
Also the seats aren't very comfortable. And I'm afraid that when
you arrive at a friend's house, they will see the blingarama styling
add-ons and will not be impressed. 'Oh dear' was the most common reaction.

At this point, I must get to the price. It's £103,430. And it
doesn't matter if you squint or stand on your head or say it really
fast – that is a lot of cash. Half-price Aston DB9? Not any more.

It's easy to see what the people at Jaguar have done here. They
are busy developing the new, small sports car that many are billing as the next E-type. And over at Land Rover, finishing touches
are being put to the new Range Rover. There simply isn't the
money, or the manpower, to come up with a new XK, so, to
keep it alive, they've sprinkled a bit of mustard powder on the
old girl in the hope they can sell a few in Qatar.

Jaguar might pull that off. But here? No. It's a lovely car to
drive and it's very fast. But it's too expensive and too embarrassing when you get where you're going.

Buy a Mercedes SL instead. Or, if you've been swept up in
that 'Aren't we marvellous?' euphoria from the Olympics and
you really want an XK, look in the second-hand columns of this
paper and buy one from a period when it was new.

12 August 2012

Wuthering werewolves, a beast made for the moors

Lexus LFA

On a recent trip to America I maintained my 100 per cent record of never having driven though Nevada without being stopped by the police. Six trips. Six heartfelt roadside apologies to a selection of burly-looking men in beige trousers.

I was pulled over the first time for travelling in a Dodge Stealth at a very huge speed indeed. So huge, in fact, that they'd had to use an aeroplane to catch me. So vast that it would have needed three boxes on the official form. And only two were provided.

Seeing that bureaucracy would prevent him from recording how fast I'd actually been going, the extremely good-natured policeman said, 'Listen, son. I know and you know how fast you were going. But, hell, it's a beautiful evening. Let's call it eighty-five.' And then, after I'd said I was going back to Britain in a week, he gave me two weeks to pay the fine . . .

A few years later I was back and going even faster in a Chevrolet Corvette when, once again, Frank Cannon arrived on the scene with a stern face and a big piece. This time he was so staggered to find the communist host from that Limey motoring show on his patch, he saluted and let me go.

And so we now spool forwards to last month, when, in an attempt to show my children the real America, where real Americans live, I was taking them on the state's back roads, flashing past remote shops where signs advised us that guns were welcome on the premises. But aliens were not.

Soon we arrived in Radiator Springs. There were a few tractor carcasses, a motley collection of trailers and one police cruiser by the stop sign I hadn't noticed. He was very angry that some goddamn Limey had dared to breach the law in what was almost certainly a communist-made Range Rover and wanted to see my driving licence.

Finding it turned out to be a time-consuming affair. So time-consuming that after five minutes he harrumphed and let me go, saying he had better things to be getting on with. Quite what these 'better things' might be in a town such as his, I'm not sure. Almost certainly they would be alien-related. Or possibly something to do with communists.

Despite everything, though, I like driving in Nevada. Even the back roads are so smooth, it felt like we were on a conveyor belt. You never need to accelerate or brake or steer. Cruise control was invented for Nevada. Driving there is as tiring as taking a bath.

I also like the sense that everything is 500 miles away and that no matter how hard you try, every journey is always completed in exactly half the time quoted by locals. 'How far's Las Vegas?' you ask. 'Ooh, about eight hours,' they say. And you get there in four.

And this is even though you are forced to stop every twenty minutes because the view, which you thought couldn't possibly get any more extraordinary, just did. And then a moment later you have to stop again because you want to photograph the dashboard, which shows two things. The time is 6.30 p.m. And the outside temperature is 47°C.

I must confess that as the time came to leave, I wasn't much looking forward to driving in England, where every journey takes twice as long as you'd expected and there are mealy-mouthed Peugeot drivers who won't let you by and every road is closed so that traffic officers can safely retrieve a sweet wrapper from the carriageway and potholes are repaired by people who are being deliberately stupid and you can't reason with law enforcement because it's all done by cameras and petrol costs more than myrrh and it's raining.

But then, just twenty-four hours after leaving Nevada, I found myself on top of a moor in Yorkshire, in the drizzle, about to get inside a Lexus LFA.

A couple of years ago a friend called about this car. He'd been offered one instead of payment for a job and was wondering if it was worth it. Embarrassed to admit I had no idea what he was talking about, I said, 'Er, no.' So he took the money and bought a Ferrari instead.

So intense was my lack of interest in a Lexus sports car that when the time came to test it on *Top Gear*, I hid under the sofa and let Richard Hammond do it instead.

Why? Lexus doesn't engineer its cars for Britain. They're engineered for fatties in Texas. For the long, straight roads of Nevada. For show-off eco-mentalists in Hollywood. The SC 430 is one of the most disgusting pieces of automotive crap I've encountered. So why should I imagine for a moment that the LFA would be any different?

This was a car that took nine years to develop. Some would say that demonstrates a fastidious attention to detail. To me it demonstrates that the company hadn't a clue what it was doing. There's some evidence to suggest I'm right, because after five years, when the prototype was nearly ready, Lexus decided to scrap the aluminium body and make it instead from carbon fibre.

That took so long that by the time the finished product was ready, Formula One racing had switched to V8 engines, making the LFA's V10 look like a dinosaur. Only not a very big one.

It develops 552 brake horsepower, which is about 200 bhp less than the current going rate, and it sends this dribble of power to a flappy-paddle gearbox that has half as many clutches as, say, a Golf GTI. On paper, then, the LFA looks to be the dinner of a dog. In the flesh, however . . .

Some say it looks too similar to a Toyota Celica, or a Toyota Supra, but because it's so wide and low it actually looks like neither of those. It looks very, very special. And inside, it's even better. Unlike Ferrari, which fits buttons wherever it can find a bit of space, Lexus has thought everything out beautifully in this car. Apart from the switch that engages reverse – which is behind the mileometer – there's a Spock-like logic to everything. And when you push or pull or engage anything, there's a sense that it will continue to work for about a thousand years. It's the nicest car interior I've ever encountered. And I would never, ever, tire of the tool that moves all the dials around.

Then you fire up the engine, snick it into first, move off and . . . whoa! The noise beggars belief. This is not a car that shouts or barks or growls. It howls. Up there, on the moors, it sounded otherworldly. Like a werewolf that had put its foot in a gin trap. And while it isn't as fast as you may have been expecting, you quickly realize on damp moorland roads that 552 bhp is perfect. Any more and you're going to be picking heather out of the grille for a month.

The LFA inspires tremendous confidence. Then, up ahead, you see a dip. Gouge marks in the tarmac show clearly that, over the years, many a sump has clattered into the road, and you brace for an impact that never comes. The LFA may be lower to the ground than a worm's navel, but so successful is the suspension, it never bottoms out.

I have to say I loved it. It's an intelligent car, built by intelligent people. In some ways it's raw and visceral; in others it's a lesson in common sense. Engine at the front, two seats in the middle and a boot you can use. And yet, despite this, there's a sense that you're in a real, full-on racer.

If cruise control was invented for Nevada, Yorkshire was invented for the LFA. It's a car that reminds you every few seconds why we have corners.

The trouble is that only 500 are being made. And the reason only 500 are being made is that only 500 can be sold. And the reason for that is that each one costs £336,000. An idiotic price. Still, it's not the end of the world, because you can have a Nissan GT-R. It's nine-tenths as good. But costs almost five times less.

9 September 2012

It's certainly cheap . . . but I can't find cheerful

Skoda Octavia vRS

I spent a few days up north recently. And, at the risk of provoking howls of protest, I came home wondering if the region's love affair with value for money might be a bit overrated. In restaurants the waiter would not tell us what the food was, or how it had been cooked, or where the ingredients had come from – only how much it would cost. Up north, people like all they can eat for £2.99.

So let's take this to its logical conclusion. If I were to open a restaurant serving nothing but horse manure and grass clippings, the prices would be very low indeed. But would people eat there? No. This means that at least some emphasis must be placed on quality. And that's the problem. Quality costs. So, if dinner looks like it could be cheap, there's a reason. It's rubbish.

As I've said before, there is no such thing as cheap and cheerful. There is cheap and disgusting. Or expensive and cheerful. There is no third way.

We see this with everything. Near where I live a firm of developers recently built a row of terraced houses. They are for sale now at extremely low prices and there's a very good reason for this. I watched them being built. So I know they are made from old cardboard boxes and dust. I suspect most of the structural integrity comes from the wallpaper.

In short, there is no such thing as a bargain. Something is cheap because it's cracked, broken or hideous. If you buy cheap garden furniture, it will rot. If you buy cheap pots and pans, they will melt, and if you buy cheap antiques, you will get home to discover they were made yesterday in Korea.

However, where all of this gets blurry is when you introduce the concept of a badge. A pair of sunglasses made by Scrotum & Goldfish would sell for £14.99. Stick a Prada badge on exactly the same glasses, though, and all of a sudden the decimal point heads east.

This makes me froth with rage. I look sometimes at a T-shirt and I think, That cannot possibly cost more than 40p. But because it has a horse or a fox or some other knowing smudge on the left breast, the shopkeeper is allowed by law to charge me £40. Often I'm consumed by an uncontrollable urge to stab her.

All of which brings me neatly to the door of the swanky Audi A3. It costs more than a Volkswagen Golf and you are going to say, 'Of course it does – it's an Audi.' But, actually, it isn't. Underneath, it is virtually identical to the Golf. They just have different bodies.

You are paying more just so you can go down to the Harvester and tell your friends you have an Audi. And that in turn brings me on to the Skoda Octavia. That costs less than a Golf and you're going to nod sagely and say, 'Well, yes. Stands to reason. It's a Skoda.' But it isn't. Underneath, it is also identical to a Golf.

So why would anyone buy a Golf, or an Audi A3, when they could buy exactly the same car for less? Simple answer: badges make people stupid.

In recent years every single Skoda I've ever tested has been enormously impressive. There was the Yeti, as good an all-rounder as I've ever driven, and the Roomster, which is a blend of practicality, VW engineering and some wondrous styling details. And there was also the Octavia Scout – a perfect farmer's car with four-wheel drive, low running costs, a boot big enough for a sheep and a fine ride.

I was therefore very much looking forward to a drive last week in the Octavia vRS, because here we have a large five-door hatchback that costs £20,440. That makes it £5,210 less than the Golf GTI. Even though, at the risk of sounding like a stuck record, they're the same car.

That means you get a 16-valve turbocharged direct-injection 2-litre engine, which equates to 0 to 62 mph in 7.2 seconds and a top speed of 150 mph. You also get big brakes, lowered suspension and a six-speed gearbox.

Apart from the big wheels, though, the Octavia doesn't look especially racy. But neither does the Golf. However, it is extremely racy when you put your foot down. There's an almost diesely clatter to the engine, and a hint of lag, and then you're off in a blizzard of face ache and rush.

The ride is firm without being alarming, and the handling is neutral. There are no fireworks, just a solid, sure-footed ability to deal with any input even the most sabre-toothed driver cares to make. It put me in mind of the Golf GTI, weirdly. Only it's bigger and more practical and, as I've said, £5,210 less.

Of course, you might imagine that a car made on the wrong side of what used to be the Iron Curtain will not be crafted with the same ruthless zeal as a car made in Germany. Well, sorry, but a robot doesn't know what territory it's in. Skodas? Volkswagens? Exactly the same Taiwanese robots help to build the two.

I was feeling particularly pleased with myself at this point, and was very much looking forward to giving yet another Skoda a tip-top review. But then I started to notice a few things.

I thought at first the brakes were a bit sharp but that I'd get used to them. I didn't. Then I noticed that despite the many buttons, almost no toys – such as rear parking sensors or Bluetooth – are fitted as standard. You get cruise control, which is just about useless in Britain. And that's more or less it.

Later, in traffic, I tried to rest my arm on the door, but it wasn't possible because the seat, which is too high up, is mounted right next to the B pillar. Once I'd noticed this, it was hard to think about anything else. I began to feel as though I was sitting in the back.

Then there was a funny noise. And, as with all funny noises, once I'd heard it, I couldn't think about or hear anything else. I even forgot after a while that I might be in the back. It sounded as if a fly was in its death throes in the air-conditioning system, so I decided to put it out of my misery by turning the fan up full. This made the noise stop. Then I turned it down low and it came back. The fan was broken. So I turned it off. And heard another funny noise. A jangling sound. A rattle. Two faults? In a Volkswagen? Not possible. And I was right.

It turned out to be an empty Red Bull can in the door pocket. A door pocket that I noticed was unlined. What's the point of that? Door pockets are invariably full of stuff that rattles – coins, keys, lighters and so on. If they are made from hard plastic, the driver will quickly go mad. Buck your ideas up on that one, Skoda.

With all the noises sorted out, I started looking for other things and quickly I found one. The speedometer has no display for 90 mph. It jumps from 80 mph to 100 mph. Does this mean the car cannot do 90 mph? And if so, how does it miss it out? How would such a thing be possible? It's madness.

The Octavia vRS, then, is the exception to the rule that Skodas are the exception to the rule that everything cheap is rubbish.

16 September 2012

Ooh, it feels good to wear my superhero outfit again

Toyota GT86

In the olden days, when people had diphtheria and children were covered in soot, cars had skinny little tyres so that enthusiastic drivers could have fun making them slither about on roundabouts.

Nowadays, though, it's all about grip. Fast Fords are fitted with front differentials to ensure you can keep a tight line, even when you are doing 1,000 mph through a mountain hairpin. Then you have the Nissan GT-R, which uses the computing power of a stock exchange to make the same mountain hairpin doable at the speed of sound.

In fact, all modern cars cling to the road like a frightened toddler clings onto its mother's hand. In some ways this is no bad thing. It means the befuddled and the weak are less likely to spin off and hit a tree. And it means the helmsmen among us can post faster lap times on track days.

But is that what you want? Really? Because when the grip does run out, you will be travelling at such a rate that you will have neither the talent nor the time to get everything back in order before you slam into a telegraph pole. If you are trying to win a race, high cornering speeds are important. But if you are not, they're frightening.

For the business of going fast, a Nissan GT-R is unbeatable. But for fun – and I am not exaggerating here – you would be better off in a Morris Minor on cross-plies.

Which brings me neatly to the door of this week's test car. It's called the Toyota GT86 and it's been built in a collaboration with Subaru, which is selling an almost identical machine called the BRZ.

Unlike most coupés, such as the Ford Capri, Volkswagen Scirocco and Vauxhall Calibra, the GT86 is not a hatchback in a party frock. It is not a marketing exercise designed to relieve the style-conscious of their surplus cash. It isn't even very good-looking.

Or practical. The boot is large enough for things, but you can forget about putting anyone in the back, even children. Unless they've no legs or heads.

Power? Well, it has a 2-litre boxer engine – Subaru's contribution – which delivers 197 brake horsepower. That's not very much. But because the car weighs just 1,275 kg and the engine is so revvy, you'll hit 62 mph in 7.6 seconds and a top speed of 140 mph. It could almost be mistaken for a hot hatch.

But there's no mistaking the noise. This car is loud, and not in a particularly nice way. There's no crisp exhaust note, no induction wheezing. It's just the sound of petrol exploding in a metal box.

The interior is nothing to write home about, either. You get what you need by way of equipment – air-conditioning, stereo, cupholders and so on – but there's no sense of style or beauty. Apart from a bit of red stitching here and there, it all feels utilitarian, the product of a bean counter's lowest-bidder wet dream.

So, there is nothing about this car, either on paper or in the showroom, that is going to tickle the tickly bits of Clint Thrust, the lantern-jawed hero from the planet Oversteer. And yet there is, because, unlike most cars of its type, the GT86 is rear-wheel drive.

Rear drive in a car is like a roux in cooking. Yes, you can use cheap'n'easy cornflour front-wheel drive, but if you want the best results you have to go the extra mile. You have to fit a prop shaft. And a differential.

In a rear-drive car the front wheels are left to get on with the job of steering while those at the back handle the business of propulsion. It's expensive to make a car this way, and complicated, but the end result will be better, more balanced.

And now we get to the nub of Toyota's genius. The company fitted the GT86 with the same skinny little tyres it uses on the Prius. And what this means is that there is very little grip. You turn into a corner at what by modern standards is a pedestrian speed, and immediately you feel the tail start to slide.

So you let it go a little bit, and when the angle is just so, you find a throttle position that keeps it there. For ever. You are power-sliding, you are grinning like an ape and you are doing about 13 mph. Which means that if you do make a mess of it and you're heading for a tree, you can open the door and get out.

You won't make a mess of it, though, because the steering is perfectly weighted and full of juicy feel. I promise. The GT86 will unlock a talent you didn't know you had. It will unleash your hero gene and you will never want to drive any other sort of car ever again.

No, really. Put some cotton wool in your ears, snick the old-feeling snick-snick box down into second, stand hard on the astoundingly good brakes, wish you'd used more cotton wool as the boxer engine roars, turn the wheel, feel the back start to go and it's like being back in the time of the Mk 1 Ford Escort.

I'm sure that at this point many non-enthusiasts are wondering whether I've taken leave of my senses. Why, they will ask, would anyone want a noisy, impractical car that won't go round corners properly? Simple answer: if you're asking the question, the GT86 is not for you.

I suppose I could raise a safety question. Because, while its antics are a massive giggle on a track, I do wonder what will happen when it's raining and your head is full of other things and you try to go round a roundabout at 25 mph. There's a time and a place for oversteer and I'm not sure 5.30 p.m. in suburbia is it. Best in these circumstances, then, to turn the traction control on.

There's another issue, too. I'm willing to bet that some people will decide that the styling of the GT86 could be improved by fitting larger wheels and fatter tyres. Do not do this. Because while it may make the car more meaty to behold, it will ruin the recipe as surely as you would ruin a plate of cauliflower cheese by vomiting on it.

Frankly, I wouldn't change a thing about the GT86. Because it's so bland, it doesn't attract too much attention. You can therefore have fun without being marked out by passers-by as an anorak.

And now we get to the clincher. The GT86 costs less than £25,000 with manual transmission. That makes it cheaper than a Vauxhall Astra VXR. It makes it a Tiffany diamond for the price of a fairground lucky-dip prize.

It's strange. We thought purpose-designed coupés had gone. We thought wayward handling had gone. And we sure as hell thought genuinely good value had gone. But all three things are now back in one astonishing car. Perhaps the most interesting car to be launched since the original Mazda MX-5. I'm giving it five stars only because it's not possible to hand out more.

23 September 2012

OK, Sister Maria, try tailgating me now

Audi S6 4.0 TFSI quattro

The results of a continent-wide survey are in and it's been announced that the Italians are the worst drivers in Europe. Apparently this is largely due to a strong showing from the Italians themselves, 28 per cent of whom said, 'Yup. Nobody does it worse than us.'

Well, I'm sorry, but I'm incredulous about this, because the Italians are in fact the best drivers, not just in Europe but anywhere. They get to where they're going more quickly, and they have more fun on the way. They also look good in the process.

Just last week I was driving from Turin to Milan in a car that develops 662 horsepower. It was plainly very fast and I was plainly in a big hurry. But that didn't stop every single Italian I encountered trying to get past. The Italian driver must overtake the car in front. This is a rule. Even if the car in front is an F-15E Strike Eagle and you are in a seventeen-year-old Fiat with a two-stroke under the bonnet, you must get past or you are not a man.

They say that you never feel more alive than you do when you are staring death in the face. Which is why my drive though Rome last year in a Lamborghini Aventador was such an unparalleled joy. You can't daydream there; you can't take in the views. Many Roman drivers have no idea the Colosseum is still upright: they're so busy concentrating on getting past the car in front, they've never even noticed it.

You can't help but notice, though, the weeping American tourists, marooned on traffic islands, wondering through heaving sobs why no car will stop at the pedestrian crossing. Because they're racing, you witless idiots.

However, what makes it different from any race you've ever seen is that no one knows where the finish line is or where the other competitors are going. That's what adds to the sparkle.

I was once on the autostrada outside Pisa when the car behind indicated it wished to get past by nudging my rear bumper. It was quite a hefty nudge, if I'm honest. Which is why I was so surprised to note the vehicle in question was being driven by a nun. I promise I'm not making this up. I was rammed out of the way by a nun in an Alfa Romeo.

What fascinates me is that when you drive to Italy through France, you have mile after mile of belligerence and arrogance and big Citroëns being in both lanes at once. Then you go through a tunnel and on the other side everyone is stark, staring bonkers. Rude on one side of a hill. Mad on the other. It's strange.

Other nations to do badly in the survey are the Greeks, who drive very much like the Italians, only without the panache, skill or style, and the Germans. Ah, the Germans. I don't think they are necessarily bad, but it is the only country in the world where I sometimes feel intimidated. As if the man coming the other way really would rather die in a huge head-on smash than pull over a bit.

Apparently, the best drivers in Europe are the Finns. How can we be sure? I'm the only person I've ever met who's been to Finland. The Brits come eighth, and there can be only one reason for such a poor showing: the sheer number of Audis you see whizzing about these days.

There was a time when Audis were driven by cement sales-men, but in recent years they have become the must-have accessory for squash- and golf-playing 'winners'. And squash- and golf-playing winners don't have the time or the inclination to let you out of a side turning, that's for sure. Also, Audi drivers have it in their heads that the stopping distances in the Highway

Code are given in millimetres. You check next time you're being tailgated. I bet you any money the culprit is in an Audi.

This used to be a BMW problem but today BMWs are rather too restrained and tasteful for the world's winners. The slight flashiness of an Audi goes better with the pillars outside their houses.

That said, the new S6 is really rather good-looking. The wheels are especially handsome and overall it has the look of a BMW, the look of a car whose body has been stretched to the absolute limits to cover the wheels; the look, in short, of a car that can barely contain its muscle.

The muscle in question is actually smaller than it used to be. In the old S6 you had the Lamborghini V10 but that's gone now, a victim of the relentless drive for better emissions and improved fuel economy. So instead you get a twin-turbo V8, the same unit Bentley is using in the basic Continental GT these days.

It's a clever engine because when you are just pootling about, four of the cylinders close down – if you really, really concentrate, you can sense that happening – which means you are using far less fuel. And then at the lights everything stops, which means you are using none at all. The upshot is about 29 mpg, and that's pretty damn good for a car of this type.

Obviously the power is down a tad from the old V10. But you still get a colossal shove in the back when you floor it, and a sense that even with four-wheel drive the tyres are scrabbling for grip, like a girl in a horror film running away from the monster in the wood.

Sometimes you think there may even be too much power, because this is not a sports saloon. It may look like one, with its silver mirrors and its fancy wheels and its V8 badges on the front wings. But the steering is not sporty at all. And neither is the ride.

Gone are the days when Audis jiggled on rough roads. Some were so bad, I often thought I was going to have an aneurysm. The new models, though, even the fast ones such as this, are extremely good at isolating occupants from the slovenliness of the British roadwork Johnny. Doubtless there will be a harder, more focused RS6 in due course, but for now what we have here is a comfortable, economical, fast and good-looking cruiser.

Inside, you get quilted seats. And I'd like to say at this point how much I hate their vulgarity. But sadly I can't. Because they look great. The equipment levels are pretty impressive too. Put it this way: if there's a gadget you've seen in another car of this type, you can be assured it's available in the S6.

It's a silly John Lewis-style price-promise game played by Audi, BMW and Mercedes. When one of them introduces a new toy such as a head-up display, the others follow suit. In the S6 I notice that Audi allows you to choose how long you would like voice commands to be. The others will have that feature in a year.

Jaguar should play a joke on the Germans and say its next car will have a ski jump in the boot. Or an aquarium in the glove box.

Until then, though, the S6 is a good, well-judged car that would make a great deal of sense in Italy. Here in Britain, however, it will be bought and driven extremely badly by people you wouldn't want round for dinner. For that reason, I still slightly prefer the BMW 5-series.

30 September 2012

It's Sunday, the sun is out – let's go commando

Ferrari California 30

I suppose we all harbour a secret longing to buy a little sports car – a Triumph TR6, perhaps, that we can use for lunchtime trips to the pub on sunny Sundays. You can picture the scene, can't you, as you sit outside the Dog and Feather: a pint of Old Crusty Moorhen, a hunk of Cheddar and a car park full of people cooing over your wheels. Lovely.

Well, it'd never happen. First of all, there were no sunny Sundays, either this year or last. Which means your little TR6 would now be sitting in the garage with four flat tyres and an equally flat battery. You're going to get round to fixing it as soon as you have a spare moment. But you won't.

And even if you do, and even if next summer is lovely, your problems are far from over because you can be assured that, moments before you set off, your wife will invite a friend along. So, with a need for three seats, you'll end up taking the Vauxhall Astra instead.

Or she won't invite a friend along and you can take the little Triumph. But then you will have one too many Crusty Moorhens and you won't be able to drive it home. Which means you will have to pop round to the pub after work on Monday to pick it up, and it won't start and the tyres will be flat again. And so it'll sit there till next autumn.

There's another problem, too. On the once-in-a-blue-moon occasion when you do drive your Triumph, it will be horrible and you will hate every minute of it. You will hate the heavy steering, the useless brakes and being overtaken by vans, because what passed for power in the 1970s doesn't cut the mustard today. Driving an old car is like watching an old black-and-white television. And you wouldn't do that for fun, would you?

The truth is that none of us really drives for fun any more. The roads are too full; the cost is too painful. So keeping a car in the garage for high days and holidays is like keeping a fun pair of scissors. It's stupid and pointless.

Which brings me on to Ferraris. Over the years, I've occasionally entertained the notion that you can use a mid-engined supercar as an everyday commuter tool. But, of course, you can't. You're always worried that it'll be scratched, and it won't go over speed bumps, and it's always noisy, especially when you've had a hard day at work, and there's precious little luggage space, and in large parts of the country it makes you look as if you have a first-class honours degree in onanism, and it chews fuel, and you can never use the power, and pretty soon you are twisted into a jealous rage every time you see someone in a diesel-powered BMW 5-series.

A Ferrari is for high days and holidays. It is a special-occasion car. Which means you need another car as well. And if you have something else, that will always be more comfortable and more practical, which means your beloved Ferrari will sit in the garage for month after month, chewing its way through your finances and then not starting on the one day you decide it would be suitable. That's why the second-hand columns are always rammed full of ten-year-old Ferraris that have only ever done 650 miles. Every one of them is a shattered dream.

You sense that Ferrari is trying to address this. Its cars now come with 200-year warranties and e-zee financing for the servicing costs. Plus, except in the case of the 458, the company has stopped putting the engine in the middle. It has given up on the high-day-and-holiday supercar and is making GT cars you can actually use to take the dog to the vet when it has diarrhoea.

Or can you? Well, I've just spent a few days with what is the cheapest of all the Ferraris. It's the California and it's yours for

£152,116. Unless, of course, you decide to spend a little more on a few extras, in which case it's £258,972.

I particularly liked the 'handy' fire extinguisher that was fitted in my car. It came in a little suede fire extinguisher cosy and cost £494. I like the way Ferrari makes it so precise. If it charged £500, you'd think it was taking the mick, but because it's £494, it looks as if it has been carefully worked out. And it's the same story with the Ferrari badges on the front wings. They are not £1,000. They are £1,013. Of course they are.

Then there's the new handling pack. At £4,320, this gives you faster steering and a more aggressive feel. In theory. Mainly it's been made available to convince those who bought a California a year ago that they really should do a part-exchange deal (I wouldn't bother).

Anyway, while the price list may be daft, the car is not. There's a V8 engine at the front and a boot at the back into which the metal roof folds away. In terms of layout, it's the same as a Mercedes SL. Except that in the Ferrari the rear parcel shelf can be disguised to look like a seat. It isn't. Unless you are an amoeba.

It also goes like a Mercedes SL. Recent power upgrades mean you now get 483 brake horsepower. Which is a lot. But it's not stupid.

Inside, it's straightforward too. This is one model in the Ferrari line-up that has conventional controls for the wipers and lights. The steering wheel is used to house only the simple three-way mode selector, the starter button and, in my test car, a series of red lights warning you that it might be time for a gear change (an amusingly priced £4,321 option).

It has a satnav you can understand and a Bluetooth system that can play your music. There's no lunacy at all in the way this car works, and once you're out of town, the flappy-paddle gearbox is an utter delight.

Don't be fooled, though, because, despite everything, this is still not an everyday car. And not just because the exhaust is like a dog that has to have the last word. It never, ever, shuts up.

No. The main reason you wouldn't want to use a California every day is that it feels so incredibly special. It doesn't feel like any other car. It communicates with you in a different way. It feels . . . like a Ferrari, which means it feels lighter, more darty and more aggressive than even the lightest, dartiest and most aggressive of its rivals.

It's a mistress, not a wife. You know that it could cook and sew but you wouldn't want it to do those things. It would be all wrong. If this car knew what underwear was, it wouldn't wear any.

I loved it massively. I'd love to have one. But if I were going to buy a car I'd never use, I'd rather go the whole hog and have a mid-engined, high-day 458 Spider. Granted, it's more expensive than a California, but there's a reason for that: it's better to drive, and as you walk past it every morning to get into your Range Rover, you'll note it's quite a lot better-looking as well.

7 October 2012

Yo, bruv, check out da Poundland Bentley

Chrysler 300C Executive

Some people can go into any clothes shop and buy any item from any shelf, knowing that when they put it on, they will look good. I am not one of those people. I've never even been able to find a pair of socks that don't look ridiculous once I've put my feet into them.

It's the same story with hats. Partly because my head is the same size as a Hallowe'en pumpkin, and partly because my hair looks as though it could be used to descale a ship's boiler, it doesn't matter what titfer I select, it ends up looking like an atom on an ocean of seaweed.

Trousers, though, are the worst. Because my stomach is similar in size, colour and texture to the moon, it's difficult to know whether strides should be worn above or below the waist. Both ways look stupid.

I'm told the problem can be masked with a well-tailored jacket, but this simply isn't true. Attempting to mask my physical shortcomings with carefully cut cloth is like attempting to mask the shortcomings of a boring play by serving really nice ice cream in the interval.

This is why I have cultivated my own look over the years. It's the look of a man who has simply got dressed in whatever happened to be lying by the end of the bed that morning. I pull it off very well. Mainly because that's what I actually do.

I'm not alone, of course. Many people obviously struggle to find clothes that work, but, unlike me, they continue to make an effort. Pointlessly. That's why you see fat girls in miniskirts, and men in Pringle jumpers, and Jon Snow's socks.

We see the same problem with cars: people drive around in stuff that is really and truly wrong. Yesterday, for example, I saw a very small woman getting into an Audi RS5. And when I say small, I mean microscopic. It's entirely possible that while her mum may have been diminutive, her dad was an amoeba. And she was getting into a super-fast Audi. A car that only really works if you look like the chisel-jawed centrepiece of a watch advert.

Let's take Nicholas Soames as a case in point. He is a somewhat large – and larger-than-life – Conservative MP with very little time for . . . anyone, really. Can you see him in a Nissan Micra? Or even a Volkswagen Golf? No. It would be all wrong.

Can you see Stella McCartney in a Kia Rio or Mick Jagger in a van? James May drives around in a Ferrari, and I'm sorry, but that's as hysterical as the notion of Prince Philip turning up to open a community centre in a Mazda MX-5. With Jay-Z on the stereo.

It's strange, isn't it? We all pretend that we pay attention to the cost of running a car and how much fuel it will consume. We tell friends that we made our choice on our particular needs and the needs of our family. But the truth is that we buy a car as we buy clothing. With scant regard for how it was made, or by whom, because we're too busy looking in a mirror thinking, Would this suit me?

I, for example, like small sports cars. But I know that driving around in such a thing would be like driving around in a PVC catsuit. It would be absurd. I also like the BMW M3. As a car, it ticks every box that I can think of. But I could not have one because Beemers have not quite managed to shake off an image that is at odds with the one I'd like to portray.

Ever wondered why you see so few big Jaguar XJs on the road? Is there something wrong with them? Not that I know of. Except that, among the people old enough to be interested in such a car, the memory of Arthur Daley is still vivid. They don't want a Jag for the same reason they don't want a sheepskin coat.

And all of this brings me nicely to the door of the Chrysler 300C that is parked outside my house. And my neighbour's house, too. And the one after that. Other places that it is parked include Hammersmith, Swindon, Bristol and the eastern bits of Cardiff. It is very big. More than 16½ feet long and almost 6 feet 3 inches wide, in fact.

Naturally, it is also extremely large on the inside, which would make it ideal for anyone with many children. A Catholic, for example. Or with children who are very fat. An American, perhaps.

And yet, despite its size, prices currently start at less than £30,000. About half what you'd be asked to pay for a similarly large and well-equipped Mercedes S-class.

A bargain, then? Well, yes, but every single thing you touch in a Mercedes feels as if it has been hewn from rock and assembled in such a way that it will last for a thousand years. Whereas everything you touch in the Chrysler feels like a bin bag full of discarded packaging.

It feels American. But actually it's made in Canada, and on the Continent, thanks to a flip-chart-and-PowerPoint meeting somewhere, it's sold as a Lancia Thema.

Under the bonnet you have a choice of engines. One is a 3-litre V6 diesel and the other is a 3-litre V6 diesel. They produce the same power, though it's not what you'd call quite enough. And you can't expect eye-widening fuel economy either – not from a car that weighs more than Wales.

The old 300C handled terribly. And I'm here to report that the new one handles terribly as well. It responds to all inputs from the wheel with what feels like casual indifference. Imagine asking a French policeman in a rural town for help. Can you picture his uninterested face? His nonchalant shrug? Well, that's how the big Chrysler responds when you ask it to go round a corner.

But there is an upside to this. Because it sits on tall, non-sporty tyres, it is extremely comfortable. And despite the diesel flowing

through its arteries, it's very quiet as well. And it is fitted as standard with every single thing you could dream of.

I can think of hundreds of people – probably thousands –
who would love a car such as this. People who are not bothered
about handling or driving along as though they are on fire.
People who just want a quiet, comfortable, gadget-laden cruiser.
At an amazingly low price.

But there is a problem that takes me right back to the gentlemen's changing rooms. It's a very showy car, very brash. And
who would that suit?

I have seen several people in American gangster movies who
could pull it off, but here in Britain? Hmmm. A rapper, perhaps,
but in my experience most like a bit of Bentley & Gabbana.
They like a brand name, and Chrysler's a bit Poundland. Beyond
rap-land . . . I can't think of anyone.

It is, then, like the perfect pair of trousers. They are keenly
priced and made ethically and well by adults in a clean factory
with many fire escapes and wheelchair ramps. They are exactly
what you need and they fit like a glove. Lovely. Except they are
purple.

14 October 2012

Out with the flower power, in with the toothbrush moustache

VW Beetle 1.4 TSI Sport

Enzo Ferrari once described the E-type Jaguar as the most beautiful car ever made. And even today, fifty-one years after it first sent a fizz down everyone's trousers, you can still turn more heads by driving down the street in an E-type than you could if you rode into town on the back of a diplodocus.

The E-type has transcended fashion, and even Marilyn Monroe hasn't been able to pull that one off. Back in the day she was considered to be the most beautiful woman in the world and she died before age wearied her. Today, though, most young boys would describe her as 'a bit fat'.

Buildings? Nope. I guarantee that all of those über-modern, über-cool houses you see on Grand Designs will, in twenty years' time, look absolutely ridiculous. As stupid then as a 1970s house looks now. But a hundred years after that they will all be listed and revered and people will come from Japan to photograph them.

Our taste changes constantly. Sunglasses have to be round. Then they don't. Trousers are worn high. Then they are not. But through loon pants and punk and new Labour, the E-type has soldiered on, winning every single poll to find the best-looking car ever made. It was considered pretty at launch. It was still thought to be pretty when it went out of production. And it still causes people to swoon and faint today.

So it must have at least crossed the mind of Jaguar's board to make next year's F-type look like some kind of modern interpretation. An E-type with a 21st-century twist. But no. It's not curvy or small. It doesn't have especially pronounced haunches or an oval radiator grille. As I see it, there is not one single detail that's been carried over, not a single nod of acknowledgment. And I think that is very, very weird.

The new car is good-looking, make no mistake about that. I don't doubt it will be fast, and will oversteer so controllably that the helmsmen at *Autocar* magazine will be tempted to rub warm oils into the gentleman bag of its chassis engineer.

We hear that in terms of size it fits neatly between the Porsche Boxster and the 911. We hear that prices will start at around £55,000, that there's a choice of V6 or V8 power plant and that, while it will be launched as a convertible, a coupé is in the pipeline. It all sounds very well thought out and lovely.

But what in the name of all that's holy caused Jaguar's board to say, 'Yes. We have the legal and moral right to make it look like an E-type. But we won't'?

It's not as though there isn't a taste for retro designs right now. Fiat has the 500, which, in London at least, seems to have taken a 75 per cent share of the market. You also have the Mini, which is bought by everyone else. Ford has unveiled images of what it thinks a modern-day Cortina might look like, and then, of course, until recently we had the Chrysler PT Cruiser . . . which shows, I suppose, that things don't always work out as well as the company hoped.

Speaking of which. The Volkswagen Beetle. When I was first introduced to the reincarnated version, I was much taken by it. I thought it was a great idea to clothe the VW Golf in some Wehrmacht clobber and fit a vase. I even considered buying one.

I'm glad I didn't, because quite quickly it became clear VW had somehow missed the mark. Part of the problem is the Bug looked a bit too friendly. And friendly-looking cars – the Nissan Micra is another – always seem a bit gormless. Cars need to have at least a hint of aggression – and the Beetle didn't.

I thought VW would give up, but it hasn't. It has come back with another Beetle, which is lower and more purposeful. It has spooky wheels, a menacing spoiler and a hint of the night about it. Think of it as Herbie's bank-robber cousin.

Inside, the flower-power vase is gone, and in its place you have a dash to match the colour you've chosen for the body, an odd glovebox, modern-day electronic equipment and the biggest fuel gauge ever fitted to any car in all of history. It's the size of the moon. It's so big you get the sense that you could drive for 3,000 miles in the red zone.

So – drum roll – has it worked?

Well, I hate to be a party pooper but I don't think it has. It looks like a hippie in a Rambo suit. The Beetle may have been a by-product of war but it became a symbol of peace. And the new aggression? I don't know. Imagine a CND symbol picked out in razor wire. A dove with machineguns.

I suppose you could soften things up with some stickers. The Mini and the Fiat 500 are available with a great many adhesive options designed to conjure up images of Mary Quant or Rome on a sunny day. But what stickers would you put on a Beetle to put you in mind of its origins? Second thoughts, best leave the stickers out of it.

And, anyway, there is no doubt some people like the design as it is. So let's move on to see what it's like as a car.

Not bad, is the answer. There's a 2-litre 197-brake-horsepower version on the market, but I tried the clever 1.4 Sport, which has good fuel economy and, thanks to two turbochargers, 158 bhp as well. You have to work the six-speed gearbox hard to get at the power but it's nice to know it's there if you can be bothered.

The handling's good, which is due in no small part to what is basically an electronic differential that tames the driven front wheels and allows you to drive like an ape, stamping on the throttle when really you shouldn't.

Mostly, though, it is a comfortable and quiet, easy and relaxing cruising machine with some genuinely nice touches. The big glass sunroof is one, and the optional Fender sound system is another, partly because it glows at night, partly because the sound is good but mostly because it's a Fender. And who wouldn't want that in their life?

In essence it feels very like a Golf, and that's not surprising because, of course, under the Hitler suit that's what it is. You don't get the practicality of a Golf – it's quite cramped in the back – but at least the boot is bigger than it was on the last Beetle. You do, however, get a much higher standard of fit and finish, because the Golf is made in slovenly Germany, while the Beetle is made in Mexico – a byword for fastidious attention to detail, as we know.

So. The nub. The price. The Beetle costs about the same as the VW Scirocco, which, of course, has the same engine. And that seems to be rather clever. You choose. Retro or modern? Maybe that's what Jag should have done with its new car. F-type or G-spot?

28 October 2012

You can keep your schnapps, Heidi – I'll have cider with Rosie

Mercedes A 250 AMG

When my dad announced that he'd become engaged to a girl from the next village, his parents were mortified. 'What's the matter with the girls from our village?' they cried.

Psychologists don't call this limited-horizon thinking 'Nissan Almera syndrome', but they should. The Almera was just some car. White goods you bought by the pound or the foot. It did nothing badly, but it did nothing well, either. It was for people who saw no need to eat fancy food or to holiday outside Britain. It was a bucket of beige, a non-car for those frightened of the exotic.

Of course, it was not alone. There was also the Toyota Corolla. A fridge with windscreen wipers. A car for people who daren't look at the sunset lest they become aroused. Chicken korma people.

Happily today in Britain both the Almera and the Corolla are gone, buried with the ghosts of Terry and June in a cemetery on a bypass, under a perpetually grey sky, beneath a headstone that no one will ever visit. We've moved on. We all want Range Rover Evoques these days. Or mini MPVs or maybe a swashbuckling coupé. The meat-and-potato hatchback is dead.

Except it isn't. It's lower than it used to be and more sleek. It's replete with styling details to arouse the curiosity. It's no longer the girl from down the street. It's an internet bride, a brogue with scarlet laces. The Ford Escort has become the Focus, all independent rear suspension and tricksy diff. The Vauxhall Astra has stepped out of its mackintosh and slipped into a pair of open-crotch panties. Even the new Volkswagen Golf looks as if it knows where Tate Modern is.

And now we get to the Mercedes A-class, the latest frumpy-dumpy hatch to have been force-fed a diet of vodka and Red Bull. The original had two floors, one a few inches above the other. With straight faces, Merc's engineers explained that in the event of a crash, the engine would slide into the gap and thus would not turn the occupants into paste. And I don't doubt this was true.

So why does the new car not have such a feature? If it was such a bonzer idea, why drop it? Could it, I wonder – a bit rhetorically – have something to do with the fact that the real reason the original had two floors is that it had been conceived as an electric car and needed somewhere to store the battery?

Happily Mercedes has now realized electric cars have no future and, as a result, one floor is enough. It has also realized that it can't just sell a packing case with wheels any more. Today we live in a skinny latte world and instant coffee won't do. A hatchback, therefore, has to have some zing.

So the new A-class has all sorts of styling creases down the flanks, a titchy rear window and a massive bulbous nose with the grille from what appears to be a truck stuck on the front. It now looks like the sort of car they might have used on the moon base in Space 1999.

And I tested the 250 AMG version, which has massive wheels as well. I want to tell you it looked a bit silly, a bit garish, a bit overstyled. But I can't because, actually, it looked tremendous. Many others also thought so.

Inside, it's good, too, chiefly because it feels like a much bigger Mercedes. However, there were a couple of issues. I have new shoes. They are Dr Martens and I like them very much but they were too wide for the gap between the wheelarch and the brake pedal. This meant that every time I pressed the accelerator, I slowed down.

And there's more. When you push the driver's seat fully back, your shoulder is adjacent to the B pillar. This means you can't drive with your arm resting on the window ledge. I'm surprised by how annoying that was.

There was another surprise as well. This is an AMG-badged car, and that is the same as a three-chilli warning on the menu at your local Indian restaurant. You expect, if you turn your foot sideways to press the throttle, to have your eyes moved round to the side of your head so you end up looking like a pigeon. But no.

The turbocharged 2-litre engine spools up nicely enough and the rev counter charges towards the red zone, but the speedo confirms what your peripheral vision has been suggesting: you aren't picking up speed at anything like the rate you were expecting.

A quick glance at the technical specifications reveals the reason. There's no shortage of power but most of it is used to move the excess weight. This is a heavy car. You feel that weight in the corners, too. No AMG Mercedes is built to generate 6 g on roundabouts – you need a BMW for that – but this one feels inert and out of its depth. So it's not that fast in a straight line. And it's not that exciting in the corners. And the gearbox isn't much cop, either.

Perhaps the AMG badge is to blame. Perhaps it's writing cheques the car isn't even designed to cash. Perhaps, beneath it all, it's designed to be a quiet and unruffled cruiser. On a smooth road, that's certainly the case. But introduce even the slightest ripple and you'd better be sitting on a cushion at the time because the ride in this car is terrible.

I'm told that on standard wheels, with normal suspension, the new A-class is pretty good. But in the AMG trim it is – and I'm choosing my words carefully here – effing unpleasant. Fast Mercs in the recent past have got quite close to the line in terms of unacceptable stiffness. This one crosses it.

But towering above the ride in the big bag of mistakes is the fuel tank. It may be large enough if the engine under the bonnet is a diesel, but when it's a turbo nutter petrol bastard, you can't even get from London to Sheffield and back without filling up. God knows what it will be like when the 350-bhp four-wheel-drive version arrives next year. That won't be able to get from 0 to 62 mph without spluttering to a halt.

The standard car, I don't doubt for a moment, is all right. It's certainly getting rave notices from all quarters. But this hot one? No. It's surprisingly poor in too many areas.

And it's not like you're short of alternatives. If you want a prestigious badge, Audi will sell you a fast A3 that won't break your back or cause you to spend half your life putting petrol in the tank. But my recommendation is that you forget the badge and buy an Astra. I drove the VXR recently, and while it may have only three doors, I was extremely surprised by how good it was. And how comfortable.

Strange, isn't it? The Astra. It used to be a byword for everything we thought we'd left behind. But after a bit of a makeover, the girl from your own village is better than the generously breasted temptress from Stuttgart.

4 November 2012

A real stinker from Silvio, the lav attendant

Chrysler Ypsilon

Many years ago I saw a magnificently idiotic film in which Sylvester Stallone played the part of a tough cop who was cryonically frozen for a crime he had not committed. Then, at some point in the future, he was defrosted so that he could rush about punching people in the face.

Every single thing about it was idiotic, especially the director's vision of what the future might look like. People drove around in cars that were satellite-controlled to keep them at the speed limit. The radio stations only played silly little ditties from television commercials. It was illegal to swear or make fun of anyone because of their colour or their creed or the state of their mental health. It all seemed bonkers. And yet here we are in 2012 and it's pretty much the Labour party manifesto.

There was something else as well. Capitalism had run amok to the point where there was really only one company that controlled everything. I seem to recall it was Taco Bell. That isn't in Ed Millipede's head, of course. But it's almost certainly coming anyway. In fact, in the car world you could be forgiven for thinking it's already here.

You might think when you buy a Seat that you are buying something with a bit of Spanish flair but, actually, you are buying a Volkswagen Golf. You may think when you buy a Skoda that you are buying 15 feet of sturdy Czech ingenuity. Nope. That's a Golf too. Audi A3? Golf as well, I'm afraid.

So what about an Aston Martin Cygnet? Surely, you're thinking, that can't be a Golf. You're right. It isn't. It's actually a Toyota. The Subaru BRZ? That's a Toyota also.

I am particularly excited at the moment about the new Alfa Romeo 4C. Which is a Mazda MX-5. Then you have the Ford Ka, which is a Fiat Panda. The Fiat 500 is also a Panda. And the subject of this morning's missive is a Panda as well, even though it doesn't say so on the back. It doesn't say Lancia, either, which is strange because that's the company that made it. And that's why I've been trying for two straight years to get my hands on one.

I firmly believe that in the past hundred years Lancia has made more truly great cars than any other brand. Ford gets close. So does Ferrari. But Lancia edges it, thanks to the Stratos, the Fulvia, the 037 – the last two-wheel-drive car ever to win the world rally championship – the Delta Integrale and other, more elderly models with running boards that exist now only in the minds and garages of people who played the drums with Pink Floyd.

Even when Lancia was not very good, it was still rather brilliant. The Gamma was a classic case in point. We all knew that on full left lock, a design fault meant the pistons could meet the valves in a head-on collision, causing the engine to explode. But we didn't care because it was so very, very pretty to look at.

Then you had the supercharged HPE. Made from steel so thin you could use it as tracing paper, and sold as an estate even though it was no such thing, this was a triumph of style over absolutely everything else that matters and I loved it.

Some say that Lancias were unreliable and while this is almost certainly true, it's hard to be sure because they had usually rusted away long before any of the mechanical components had the chance to malfunction. Fans didn't mind, though, because of those bite-the-back-of-your-hand-and-faint looks.

The trouble is that in the 1980s the Italians handed the styling department over to someone who plainly went to work with a box on his head. The result was a range of cars that oxidized and blew up. And didn't look very nice in the process. With hindsight this was not a good idea. We can tolerate bad-tempered lunatic girlfriends if they are pretty. But not if they look like the Beta saloon. Or the Dedra.

The result was disastrous. Sales plummeted and Fiat, which owns Lancia, decided to pull its problem child out of Britain. And that, we thought, was that. Only now, almost twenty years later, Lancia is back. The Ypsilon.

Let's look at the obvious problems first of all. Number 1: the man with the box on his head is plainly still in charge of styling because the list of things I'd rather look at includes every single thing in the world.

Then there's the name: Ypsilon. The company may argue that this is a Greek letter but it sounds like another Greek letter, epsilon, and as anyone who has read Aldous Huxley's *Brave New World* will know, an epsilon is synonymous with idiot. It means lavatory attendant. It means loser. The manufacturer may as well have fitted a swastika badge, arguing that it's an ancient Buddhist symbol. It is, but . . .

I'm afraid things get much, much worse. This is a horrible car to drive. The 1.3-litre diesel engine feels as if it's running on gravel. The driving position is suitable only for an animal that doesn't exist. The dials are so far away from where you sit you can't read them, the handbrake sounds like it's been made from bits of a 1971 roof rack and, as a result of the materials used to line the interior, it feels like you are sitting in a wheelie bin. Still, at least it's slow and devoid of any excitement whatsoever. And fitted with a gear lever that has been shaped specifically to make it extremely unpleasant to hold.

There are two settings for the steering. Nasty. And Very Nasty. The latter makes the system so light that you daren't open the window for fear the resultant breeze would cause you to do a U-turn. And Nasty means you drive along, suffering from a nagging doubt that the wheel has nothing at all to do with your direction of travel.

Ride? That's dreadful. Noise? Awful as well. And then we get to the brakes. You get what looks like a pedal but actually it's a switch. So you are either not braking, or braking so violently that you are going through the windscreen.

Other stuff? Well, it's got back seats that fold down, a boot and a big button on the A pillar that, so far as I can tell, does nothing except distract you from the rest of the terribleness. It also has cruise control, for no reason that I can fathom. Still, you might be thinking, at least you can go to parties and tell everyone that you have a Lancia. Well, yes, I agree, that would be good. Except you can't because this car is actually sold here as a Chrysler.

This is because Fiat recently bought Chrysler and reckons that in Britain that badge is better than the Lancia one used on the other side of the Channel. That's the sort of thinking that resulted in a car this bad being made in the first place.

Still, at least there's a solution. You simply buy a Fiat 500 or a Fiat Panda or a Ford Ka instead. They're all exactly the same as the Ypsilon. But much better.

11 November 2012

Ask nicely and it'll probably cook you dinner underwater

BMW M135i

When we buy a really fast car, the last thing we want is a really fast car. We may think we do. But we don't. The top speed of a car matters when you're a child. My dad's car is faster than yours. And it matters when you are a teenager.

I bought a Volkswagen Scirocco when I was twenty because *What Car?* magazine said it accelerated from 0 to 60 mph a little bit faster than my mate's Vauxhall Chevette. But when you are an adult you realize that you will never accelerate from 0 to 60 mph as fast as possible because a) people will think you are an imbecile and b) you will need a new clutch afterwards.

Nor can you ever indulge in the 1970s pastime of proving to other motorists that you have a faster car than they do, because these days all cars can do 120 mph. This means you have to do 140 mph to make your point, and when you're at that speed, someone's going to put you in a prison.

Let's get to the point. If all you want from a car is speed, you should buy a Nissan GT-R. If you use its launch control, it will leave the line as though a comet has crashed into the back of it. And it will keep on accelerating until stark, naked fear causes you to remove your foot from the pedal. And we haven't got to its party piece yet: its all-wheel-drive ability to get round any corner at any speed of your choosing. With the exception of a few silly track-day specials, the Nissan GT-R is the fastest car money can buy.

But you didn't buy one, did you? Because it's a bit ugly. And it's a Nissan. And you thought your friends and neighbours might laugh at you.

My colleague James May recently bought a really fast car. It's a Ferrari 458 Italia and with a fair wind it will zoom along at 200 mph. But he will never drive it at anything like that speed. Ever. And even if he did take it to ten-tenths on a track – unlikely, I know – he'd still get overtaken by a GT-R.

You buy a Ferrari because you think it makes you look interesting, rich and attractive. You buy one because you like the feel of the thing, or the styling, or the cut of the salesman's jib. You buy one so, at night, when it's dark and you're feeling worthless, you can say to yourself, 'But I have a Ferrari.' And you will feel better. I know. I've been there.

Another friend recently bought a Mercedes C 63 AMG Black Series. And within days he was sending me texts saying it was a bit scary on full throttle. Wouldn't know, mate. I've never used full throttle on my Black, the CLK, because there's a big difference between admiring a slumbering crocodile and running up and poking it with a stick.

I have a Black for all sorts of reasons. I like the pillarless doors. I like the flared wheelarches. I like the body-hugging seats. And I like the noise it makes. Unfortunately, in order to make its tremendous sound, the engine has to be very powerful, which, as a by-product, makes the car very fast. But it's not fast not in the way that a GT-R is fast. You can use the speed in the Nissan. If you try to use the speed in a Mercedes Black it will put you in a tree.

Every human being on the planet, with the possible exception of Ed Miliband, likes the feeling of being a little bit out of control. Push a child high on a swing and it will squeal with delight. But when the big kids start pushing the roundabout too fast, the sound it makes tends to change somewhat.

Which brings me to the new BMW 1-series. The top-of-the-range M135i has been winning rave reviews because, unlike the hot hatches made by every other company, it has rear-wheel drive. This means you can 'hang the tail out in a corner'.

Indeed you can, but there is a price to pay for this. Because the car has rear-wheel drive, the big six-cylinder engine is mounted longitudinally. Also there is a prop shaft running under the cabin, and at the back, beneath the boot, are many components that aren't necessary in a front-wheel-drive car. Net result: you have less space inside than you do in, say, a Ford Focus or a Vauxhall Astra. So you pay more and get less space, simply so that you have the ability to power-slide through roundabouts. Something you will never, ever, do.

However, here's the thing. I have a watch that will still work 3,000 feet underwater. I have plumbing that can deliver water so hot it can remove skin. And I often eat in restaurants that serve food so complex that it's way beyond the limited range of my smoke-addled palate. Also, as we know, I have a car that can go 80 mph faster than I will ever drive.

And that's what gives the BMW M135i such massive appeal. You will never go round a corner trailing smoke from its out-of-shape rear . . . but it's nice to know you could.

There is a lot more to commend this car as well. It has a supremely comfortable driver's seat, an excellent steering wheel, impossibly Germanic controls and a perfect driving position. Get in and, no matter what age has done to your frame, you will immediately feel at one with the machine.

Then there's the engine. To appease those of a tree-hugging disposition, it is fitted with a compound turbocharger, which means that, after a hint of lag, there is a never-ending stream of bassy, gutsy power. In the real world, where there are other motorists and lampposts and policemen, this car is as fast as you would ever want.

And because it's rear-wheel drive, the front wheels don't have to multitask. They have only to worry about steering, which means the car feels balanced. It's fantastic – as good as the Mercedes A 250 AMG I tested recently was bad.

There's more, too. While it's better-looking than its predecessor, which had the appearance of a bread van, it's still no beauty. But, unlike all its rivals, it's free of bling. Like all modern BMWs, it's understated and tasteful. Yes, rivals have more space inside, but we're talking about a few centimetres here and a bit of an inch there. And if you truly like cars and truly like driving, that is a price well worth paying.

One thing, though. I do wish BMW would reserve that M badge for cars that have come from its motor sport division, rather than sticking it on anything that's a bit faster than usual. The M135i may say M on the back. But if you look underneath, there's no limited-slip diff, so it isn't an M car really. Unless the M here stands for marketing.

That, however, is my only gripe. And it isn't enough to warrant a lost star. Because the M135i is so lovely to drive and because it's available with a proper automatic gearbox and because it has pillarless doors and because it's only £3,000 more than a similarly powerful Vauxhall, it gets full marks from me.

18 November 2012

The pretty panzer parks on Jurgen's golf links

Volvo V40 D4 SE Nav

In essence there are three peninsulas that stick out into the Mediterranean: Greece, Spain and Italy. And choosing which is best for a summer holiday is a no-brainer. It doesn't matter what you're looking for – heat, landscape, wine, culture, food, history or architecture – Italy wins. By miles.

When I see people on holiday in Greece, I always think, Why have you come to a country where they grow vines, eat the leaves and throw the grapes away, choosing instead to make their wine out of creosote? Of course, Spain is more civilized than that, but it doesn't have a proper word for 'beer' and the food seems mostly to have come from the nearest bin.

It's the same story with supermarkets. If you have a choice of outlets within easy reach of where you live or work – and most people do – why would you not go to Waitrose?

There's more. When you are in need of a refreshing soft drink, why would you not have a glass of Robinsons lemon barley water? Why do people buy BlackBerrys when they could have iPhones? And, conversely, why have a Mac, which has no right-click, when you could have a PC that does?

In almost every sphere of life – baked beans, cola, television channels – there is a bewildering choice on offer but actually no choice at all. Because one product is almost always head and shoulders above the rest. I'm trying my hardest at this point not to mention *Fifth Gear*.

It certainly applies in the world of cars. If you want a big off-roader, you can waste your time test-driving the Toyota Land

Cruiser if you like, but it simply isn't as good as the Range Rover. And that's the end of it.

Supercars? Yup. By all means buy a McLaren MP4-12C or a Lamborghini Gallardo, but you must know that because you didn't buy a Ferrari 458 Italia your life will not be quite as good as it could have been.

You may imagine that the theory gets a little blurred in the risk-averse world of the humble hatchback. These are the bread-and-butter cars and any attempt to do something risky or interesting might put buyers off. Car makers know this, so they stick to four wheels and a fold-down back seat. And yet . . .

What is it you want? Economy? Value? Speed? Comfort? Reliability? Handling? Space? A blend of all those? It doesn't really matter because the Volkswagen Golf does more things more betterer than all of its rivals. It is the Italy of hatchbacks. The Heinz baked bean. The iPhone. The Waitrose. The best.

I recently drove a Vauxhall Astra VXR and it was deeply impressive, fast like you would not believe yet blessed with a level of comfort that you could not reasonably expect. But too flashy, really. So you're better off with a Golf. Which isn't flashy at all.

The Ford Focus ST? Great fun. But not as good as a Golf. Mercedes A-class? Well, the model I tested was the 250 AMG Sport and it was flawed in many ways. But it does at least have that Germanic quality. Much like the Golf, which isn't flawed in many ways.

The BMW M135i? This is a fabulous car. I loved it. It's better to drive than any Golf I've ever experienced, but the payback is a slightly cramped interior. A Golf doesn't have that problem.

However, in recent months I've been seeing a new boy on the block. It's so pretty that I've found myself hoping its undersides can cash the cheques its body is writing. Because if this is as good to drive as it is to look at, Johnny Golf may have finally met his match. I'm talking about the Volvo V40.

The interior is just as good as the exterior. Great seats, a good driving position and a 'floating' centre console that's festooned with cool Scandinavian buttonry. With prices starting at a whisker under £20,000, I thought I might be on to something . . .

The washer nozzles are particularly impressive. You get six. This means that when you pull the stalk, it's like driving through a car wash. Sure, you're temporarily blinded, but when the spray has gone, it's as if the glass has been burnished clean. I think, however, that I could make do with fewer. So I'd angle five to clean the windscreen and aim the sixth so I could wash the faces of passing cyclists. I think they'd like that.

So far, then, all is good. But I'm afraid there are a few problems. First of all, Volvo diesel engines are not the most refined you can buy, and the five-cylinder in my test car was no exception. On start-up, it sounded as though the cylinders were full of pebbles.

Plus, when you compare it with the similarly sized engine BMW offers in a 120d, it's not as powerful and, despite being hardly any more economical, takes nearly 1½ seconds longer to propel the car to 62 mph. I don't understand this. Why build an engine that you know straight away is not quite good enough?

I fear there's more, because while the floating dash may look nice, it is almost impossible to use. I couldn't turn the radio up or down, couldn't operate the satnav, couldn't turn the seat heater off, couldn't find the button that switched the car from Performance to Eco setting and couldn't work out how to engage the system that parks the car for you. It's bewildering and hopeless.

It's not a spacious car, either. Realistically, you're only ever going to get two adults in the back, headroom is at a premium and while the boot does all sorts of funky things, it's not very commodious. It looks, then, like a car that's been designed and engineered by a company that didn't quite have enough money to design and engineer a new car. Which is probably the case. And as a result, despite the looks, it's not as good as a Golf. The end.

Except it's not the end because this car scored a whopping 98 per cent in independent safety tests for adult occupancy. The highest score of any car in history. And I'm not surprised because it comes with a vast range of devices to warn you of impending doom as well as many features to ensure you're OK even if the worst happens.

And it's not just good at protecting those inside, it also comes as standard with an airbag that inflates to protect any pedestrians that get in its way. This is all part of the company's mission to ensure that by 2020 no one should ever be killed or injured in a Volvo.

It's an ambitious target, and in all probability it's completely unrealistic. But the aim is noble, nonetheless, and for that reason I would completely understand why you might buy a V40 rather than a Golf.

All things considered, the VW is a vastly superior car. But in an accident you will probably be better off in the Volvo. Think of it, then, as Cuba. In terms of Caribbean islands, Mustique is much better. The food, the beaches, the crime, everything. But if you become ill, you can't get round the fact that Havana has better hospitals.

2 December 2012

I ordered a full English but ended up with bubble and squeak

Aston Martin Vanquish

I'm a fiddler. Whenever I get the furniture in a room arranged just as I like it, I sit down and decide immediately that I don't like it at all. It all started in my study at school. I had to share it with five younger boys, so that meant five tables and five chairs. And pretty much every day I'd try something a bit different.

Usually this meant four boys sharing one table and me having the others for my various hi-fi components that had to be laid out horizontally for aesthetic reasons one day and then vertically the next.

I think, therefore, I'd make a good fist of running Aston Martin. It is a small company with limited resources and no big-boy owner to help out with the economies of scale when buying components. Bentley can get its masters at Volkswagen to make noises when negotiating a deal on a new supply of brakes. Rolls-Royce can turn to BMW. Ferrari can look to Fiat. But Aston has to go and see ZF, the German gearbox manufacturer, and say, 'Please, sir, can we have some more?' And usually the answer is, '*Nein*, Englander.' So it has to produce a range of cars using nothing but what it's got. And what it's got is two engines. And one basic design.

It started with the DB9. An excellent, graceful and fast car. It heralded a departure for the company and many were sold. So a new car was launched, the V8 Vantage. To the untrained eye, it looked pretty similar to the DB9 but to start with, it had a different engine and was a bit more sporty. Then it was fitted with the same engine as the DB9. And then the DB9 was given some new sills to become the DBS, which was very brilliant but quite expensive. So some new sills were invented to create the Virage. And you could also buy most of these cars as convertibles or coupés. In essence, then, Aston's engineers were in my study, endlessly rearranging the same bits of furniture.

And now they've rearranged them again. The DBS and the Virage are gone and in their place we have this car. A car that doesn't even get its own name. Instead they've rummaged around in the company tuck box and found an old moniker – Vanquish.

You may remember the first effort. I do. Mainly because it was so very poor. Yes, it had a big price tag and a big V12 engine – made from mating two Ford V6s together. And yes, it was very fast, but only in theory. In practice it went from 0 to 6000 rpm in one clutch.

So the new car uses an old name and the same basic engine that's been in the DB9, the DBS, the Virage and the V12 Vantage. The same basic styling, the same construction techniques and a six-speed automatic gearbox from ZF's end-of-season everything-must-go discount bin. And the price tag for this re-arrangement? Well, before you start with options, it's a whopping £189,995.

There are other issues too. This is a car made from a clever, glued-together aluminium and carbon-fibre tub. It has aluminium side-impact beams. It should be so light that it needs mooring ropes rather than a handbrake to stop it floating away. And yet somehow it weighs more than 1.7 tons. Perhaps that's why it's so expensive – because the seats are filled with gold ingots.

On paper, then, this looks like a bubble-and-squeak car. At first appealing, but when you stop and think, you realize you're eating leftovers. And yet . . .

While the profile may be familiar, there is no doubting the fact that the styling tweaks are extremely successful. This is a car that moons you with its beauty. The new Ferrari F12 is good-looking enough to cause a grown man to faint but the

Vanquish is better still. It's delightful on the inside too. The old Volvo satnav has been replaced with a system that tells you where you're going rather than where you've been and the fiddly little buttons that plagued the short-sighted in previous models have been replaced with bigger 'haptic feedback' knobs that buzz slightly when you touch them. Why? Not a clue, but it's nice.

Thanks to a smaller transmission tunnel, there's more space than in previous models too. So much, in fact, that if you specify the optional rear seats, you can actually put people in them. People with heads, if you want, and legs.

And then there's the upholstery. It looks like leather bubble wrap and it's wonderful. So when it comes to practicality – it even has a big boot – and styling, this car is world class. It's an iPhone in a sea of Bakelite.

And there's more. The engine is a masterpiece. You get almost 11 per cent more power than you did from the DBS, and that means 565 bhp. To remind you that they are there, all of them bark every time you go anywhere near the throttle.

It's more than just the aural effect, though; it's the sense that this engine is not a thousand parts whizzing about under the watchful eye of an electronic overlord. It feels like one big muscle – a mountain of loud but lazy torque.

Normally I worry about V12s. I always fear that a V8 does pretty much the same job with much less complexity and much less opportunity to go wrong. But the V12 in the Vanquish is a thing of such unparalleled brilliance, I'd be prepared to forgo my worries.

It suits the gearbox too. Yes, almost all of the Vanquish's rivals are now fitted with eight-speed manual gearboxes that are operated by flappy paddles. This is very good for the environment, as they use less fuel, and it's also good when you are on a racetrack. But it's not so good when you are in town and you have a millionth of a second to exploit a gap in the traffic. Because at low speeds they are dim-witted, jerky and hesitant.

That's where the Aston scores. Yes, it has flappy paddles, but they are connected to an automatic gearbox. That may be old-fashioned, but in town it works much, much better.

But, you may be wondering, what about out of town? Can the bubble-and-squeak car really hold its own against its more modern rivals? The short answer is no. You feel the weight and that makes the car feel bigger and more intimidating than it really is. It doesn't flow and it doesn't really matter whether you engage Sport mode or put the suspension in Nutter Bastard mode, the Vanquish doesn't boggle the mind quite as effectively as, say, a McLaren MP4-12C or a Ferrari California. It doesn't feel special.

However, since it was a pre-production car, it's probably unfair to say its boot lid broke and its passenger-side electric window was wonky.

If we assume these issues are addressed before the car goes on sale – that was never a certainty with Aston Martin in the past – then what do we have here? Well, it's a thing of immense beauty and it does have a fabulous engine. In many ways you could call it a modern-day British take on the muscle-car idea. And that would be great.

But I fear Aston is at the point where we can all see this car for what it really is. A new car that's not really new at all.

9 December 2012

The cocaine chintz has been kept in check

Range Rover Vogue SDV8 4.4L V8 Vogue

Helsinki airport is vast. You walk for miles and miles past hundreds of shops and thousands of commuters, through a line of passport inspection booths that stretches across six time zones. And then you are told that your luggage will be arriving at baggage carousel No. 36.

That's what gave it away – thirty-six baggage carousels. Do me a favour. Why would you need that many in Finland? Well, I did some investigating and it turns out there are, in fact, only six. It's just that the numbering starts at thirty-one. And once you notice this, the whole charade falls apart. All the people? Actors, plainly, employed by the government to make the country look busy and industrious.

The shops? Well, I didn't check but I bet that behind the Dior and the Jack Daniel's advertising, all you can actually buy are half-hunter watches, No6 cigarettes and confectionery items such as Spangles and Opal Fruits.

Lots of smaller countries do this. They know that their airport is the nation's porch, so they make it enormous to give the six annual visitors a sense that they have arrived in a country that's going places. 'Look at us. We make mobile phones and cars you haven't heard of, and Santa lives here and that's why we need an airport that is eighteen times bigger than LAX in Los Angeles. Because we are important.'

It's all very lovely, I'm sure, but the problem is that by the time the visitor makes it to the taxi rank – and has hopefully not noted they're all Singer Gazelles full of shop window mannequins – he feels like he's done an Ironman triathlon and is knackered.

This is an especially big problem for those who visit Helsinki in the winter because unless you arrive between 12.03 p.m. and 12.07 p.m. it will be black dark. And if you do manage to arrive between these times, it will be dark grey. It's a lovely country, Finland, but the dimmer switch appears to be broken.

That's why, when I emerged from Helsinki airport last week, exhausted, with my stomach demanding brunch and my eyes telling me it was time for a mug of cocoa, I thought I was getting into a normal Range Rover.

Only when I arrived at the hotel and went to retrieve my luggage from the boot did I notice something was afoot. I own a Range Rover. We use a fleet of them to make *Top Gear*. There is no car I know better. I certainly know where the boot-release catch is. But on the car in Finland it was in a different place.

Now I know some car manufacturers make small changes to tailor their models for various markets. Cars sold in China, for example, have a longer wheelbase than cars sold elsewhere because Chinese motorists like a lot of room in the back. I'm also aware that the Americans and the French insist on having their steering wheels mounted on the left.

But changing a car in this way is expensive. And I couldn't for the life of me work out why Land Rover would agree to move the boot-release catch just for the Finns. So I did some journalism and realized after no more than fifteen minutes that I'd just been driven from the airport to the centre of Helsinki in the new Range Rover. One of the most eagerly awaited cars of recent times. And so at dawn the next day, I broke off from breakfast to give it a closer look.

The Range Rover has to walk a fine line. Yes, the vast majority are sold within the M25, but only because the city folk are buying into the country dream. They have to know that Lord Fotherington-Sorbet has one as well. So first and foremost, the Range Rover has to appeal to him.

The most recent examples of the old model were definitely getting too chintzy. Land Rover was listening to the people who were buying their cars – footballers and drug dealers – and was losing sight of the reason why. Lord Fotherington-Sorbet, for instance, does not like a chrome grille. Or piped upholstery.

I was fearful that with the new model the company would go berserk and give up on the countryside altogether. But it has not done this. The chintz is kept in check. Yes, my car was black (a town colour) and had a silver roof (a Cheshire option) but in the dim Finnish light, it looked very good.

My only complaint: on the last model, the heat-extracting gills used to be on the front wings and therefore seemed to have a purpose. But on the new one, they've been moved onto the doors where they just look stupid because they're obviously fake.

Inside, though, I had no complaints at all. The car I drove was a pure four-seater, with a box of electronic goodies separating the back seats. This was nice but you wouldn't actually buy this option unless you were mad. Up front, I was amazed how similar it all felt to the last model. You have the same split-opening glovebox, the same controls. All the company has really done is redesign the buttons and fit the gear lever from a Jaguar. And thank God for that.

There are some new things, though. The stereo is quite simply the best I've ever encountered. The seats are sublime. And now you can choose what colour you would like the interior lighting. And I'm not talking about blue or red. I'm talking about a full Dulux colour chart. I liked the purply blue best. It made up for Helsinki's broken dimmer switch.

Apart from this, though, it looked and felt like a Range Rover. And then the door handle fell off, which means it's probably built like a Range Rover too.

To find out for sure, I opened and closed the door twenty times. In the last model, this would have flattened the battery because each time the car was unlocked, the computer thought, 'Oh, we are going somewhere. I'll power myself up.' Then when the door closed, it would power itself down. But Land Rover has obviously fixed the problem now because the battery was fine.

On the road? Well, now this is the really clever bit. The new car may be bigger than the old one, but some versions of it are almost a staggering half a ton lighter. I was fearful this would make it feel less substantial, more Japanesey. But it doesn't. You now get better acceleration and much better fuel consumption, and it still feels as solid and as regal and as comfortable and as imperious as ever.

Off road? I didn't have a chance to find out but all of the features you found on the last model are still fitted to the new one. So it should be about the same. Fine on winter tyres. A bit slithery if not.

Three engines are on offer. There's the 5-litre supercharged V8, which is fine if you are a bit unhinged, and then there's a 3-litre V6 diesel, which offers extraordinary fuel consumption for a car of this size. And in the middle is a 4.4-litre V8 diesel. That would be my choice. Will be my choice, in fact . . .

My main emotion after driving this car was much the same as my main emotion during the Olympics opening ceremony. Relief that they hadn't cocked it up. Then as time went by I started to realize that, like Danny Boyle's effort, it's more than not a cock-up. It's actually brilliant. Expensive, yes. But worth it.

The company spent a billion quid on designing the new lightweight chassis. And then clothed it in a modern-day interpretation of what made the last car such a massive hit, not just with people who wear nylon shorts at work but also people who wear tweed shorts at play. It is a fantastic car. Not just the best off-roader in the world, but one of the best cars full stop.

16 December 2012

Thanks, guys, from the heart of my bottom

Audi RS 4 Avant 4.2 FSI quattro

From a road-tester's perspective, the good thing about Audi's RS cars is that you never quite know what you're going to get. Some are nearly as good as their rivals from BMW. Some are forgettable. Some are dire. And then we get to the new RS 4 Avant, which has just provided me with one of the worst weeks of my entire motoring life.

Bad is a small word that doesn't even begin to cover the misery. Misery that was so all-consuming that, given the choice of using this car or taking the Tube, I would head straight for the escalator. Not just would. Did.

Day one involved a trip to a place called Stoke Newington, which pretends to be in London but is, in reality, an hour north of the capital, just outside Hull. And straight away I knew there was something terribly wrong.

I have driven bumpy cars in the past. My own Mercedes is extremely firm. But the RS 4 was in a different league. It was like sitting in a spin-dryer that was not only on its final frantic cycle but also falling down a very long, boulder-strewn escarpment. I couldn't begin to imagine what Audi's engineers had been thinking of. The interior was typical of the breed. It had all the toys. All the features. So it didn't look like a stripped-out racer. But that's what it felt like every time I ran over a pothole or a catseye or a sweet wrapper. It was, in short, a nicely finished brogue – with a drawing pin poking up through the sole. I hated it.

I tried to test some of the features, but so vigorous was the shaking that I gave up. I'd aim my finger at a button, but by the time it got there I'd have run over a bit of discarded chewing gum so it'd bounce off course and hit something else. Usually a bit of carbon fibre that had been added . . . to save weight.

That's another issue with the RS 4. It felt like it was set up to scythe round Druids with no roll at all, and yet it weighs more than 1Ç tons. Some of that is muscle from the big V8. But most of it is fat.

And then the brakes started misbehaving. This meant that every time I pulled up, they made a sound exactly like I was running a wetted finger around a wine glass. This made passers-by look at me very crossly.

But the worst thing, by miles, was the steering. There's absolutely no feel at all when you are going in a straight line. It's so floppy, you actually begin to think, as you bump along, that it may be broken. And then, when you get to a corner, it suddenly becomes extremely heavy.

That's why on day two I used the Tube and taxis to get around. But on day three I had to go to Luton, and then Chipping Norton. So I slipped into the same sort of padded underpants I'd wear when being beaten at school and headed north.

You may imagine I'm going to say things got better. But they didn't. So violent was the bumping and so alarming was the steering that I stuck to 55 mph on the M1. Listening to Radio 3. I'd wanted Radio 2 but my finger had cannoned off the roof, the wiper switch and various other bits and bobs before alighting in completely the wrong place.

What's interesting is that wherever I went, squash-playing lunatics in other Audis and people on the street would bound over to ask what it was like – I can't remember a car attracting so much interest – and all looked terribly deflated when I explained that it was utter, utter crap.

All weekend I didn't drive it at all. Why would you? But then on Sunday night, with a heavy heart, I climbed back on board and set off back to London. As I drew near, it started to rain, so I reached for the wiper switch. Unfortunately I ran over a white line as I did this, and as a result my arm boinged into a button on the dash marked 'drive select'.

And everything changed.

The steering suddenly developed some feel. The ride settled down. The revs dropped. The RS 4 stopped being a wild animal and became a car. It turns out that 'drive select' alters the entire character of the machine. It changes the engine, the steering, the suspension and even the noises that come out of the tailpipe.

It's there so you can tailor your car to suit your mood. Which does raise a question: what sort of mood was the delivery driver in when he left it at my house, set up to achieve a new lap record in the Saturn V engineering shop?

I have never met anyone who would want, ever, to put their RS 4 in what's called Dynamic mode. I can't imagine such a creature exists. Because on all the other settings it's a good car. So good that I rode into town on a wave of guilt and shame, remembering what I'd been telling people about it. And how they'd be better off with a diesel BMW. Or a pogo stick. Or some new shoes.

In Comfort mode it's quiet, the steering is light, the seats are seductive and the double-clutch gearbox creamy and smooth. And because it's so relaxing, you can sit back and enjoy the firepower from that big V8.

Unlike the engine in most modern performance cars, this one is not turbocharged. The upside of that is crispness and lots of hectic goings-on at the top of the rev band. The downside is that a polar bear could get a bit of asthma at some point in the very distant future. This was one of the world's best engines when it was introduced six years ago. And nothing's changed.

Handling? Well, in the past all Audis have been determined understeerers, partly because they were nose-heavy and partly because of the four-wheel-drive system. In this one much work has been done to shift some of the weight aft. And at the back there's a locking diff. So now you have the grip, but when you get to the end of its tenacity, it's the rear that starts to feel light, not the front. This, of course, is better, because if you go backwards into a tree, you don't see it coming.

You could buy one of these cars, and, provided you never, ever, put it in Dynamic mode, you'd be very happy. Your dog would also like it because in the boot there's a bit of equipment designed, in my mind, to stop him falling over. It works for shopping too.

However, I thought pretty much the same thing when I drove the original RS 4. I liked it very much indeed. But it was not quite as good as BMW's M3. It lacked the Beemer's liveliness and, ultimately, its speed.

The new RS 4 bridges the gap and is therefore quite a tempting proposition. But I can pretty much guarantee that as soon as you take delivery, BMW will launch a new version of the M3 and that will once again surge ahead. It was always thus, I'm afraid.

23 December 2012

Just like Anne Boleyn, there's no magic with the head off

Volkswagen Golf GTI cabriolet 2.0 TSi

We may all harbour the dream of driving down a glorious stretch of road, on a jasmine-scented day, in a fast and beautiful sports car. But if you buy such a thing, there's going to come a time, possibly in the middle of February, when there's a hint of sleet in the air, and it's cold, and you're parked outside a hardware store and the timber you've just bought won't fit in the boot.

Then you're going to wonder, fervently and out loud, and with many expletives, why you didn't buy something a bit more sensible. Something a bit more like the Volkswagen Golf GTI cabriolet. A car that gives you top-down sporty motoring and a boot and four seats.

In theory, this is the chocolate pudding that won't make you fat. It's the cigarette that repairs your lungs. It's the gravy-resistant silk tie. And it has been that way since the first incarnation burst into the world of tanning salons way back in the early days of Mrs Thatcher.

However, do not imagine for a moment that the Golf will suit all your requirements. Because as I discovered in the run-up to Christmas, it sure as hell didn't meet a single one of mine.

I was in London packing for a range of engagements over the festive period. I'd need a suit, my shooting things, some T-shirts for a quick trip to the Caribbean, an Elvis outfit for New Year's Eve, a gun, some ammunition and all the usual undergarments. Plus, of course, all my Christmas shopping.

Other things that needed to fit into the car were two teenage children and their shopping. Along with all the electronic equipment needed to keep people of this age from dying of boredom on the long and arduous fifty-five-minute trip to the country.

I'll be honest, I'm not the best packer in the world. But compared with a teenager, I'm a professor of getting a square peg into a round hole. The pair of them made no effort, did everything wrong and there was not much festive spirit as we huffed and puffed and swore at one another. Until eventually the car was loaded.

Unfortunately the driver's seat had to be pulled so far forwards that I looked rather like Mr Incredible. Still, we made it to the Cotswolds with nothing to report, apart from a lot of complaining, some light arthritis and a healthy dollop of cramp in my left leg.

The next day I was due to go shooting. And there's a rule here. Your car must be the colour of mud, equipped with four-wheel drive and have good ground clearance so that it can handle all the rough tracks and slippery grass. A white Golf GTI convertible fits the bill in no way at all. I may as well have turned up in a motorized mankini.

Nor did it work for taking the dogs out on a Christmas Eve walk, or going to the garden centre to pick up a tree, or for making sure the pheasant feeders were full. Collecting logs? Running a family to the station? Taking rubbish to the dump? All of these simple, mundane things were beyond its cramped rear seat, its tiny boot and its low-profile tyres.

So what about the roof, which slides away electrically in just 9.5 seconds? Nope. Useless. Because, as you may remember, the run-up to Christmas was marked by some of the wettest weather Britain has seen. On the roads, only submarine commanders felt completely at home.

The car didn't really work as a style statement either. Because there was a time when a white convertible was just the thing. But now? Turning up in a car of this type is like turning up with a Wham! days George Michael blow wave.

There's a similar problem on the inside because it has the sort of upholstery used to make jackets for German newsreaders. VW will tell you it's a nod to traditionalism since it's the same material used to line the seats in the original Golf GTI. But what's the point of resurrecting something no one remembers or cares about?

What we have here, then, is a car that appears to be a mass of fun but still has one eye pointed in the beige-infused, drip-dry direction of common sense. It sounds good but actually it's like a leather-soled training shoe. The worst of both worlds . . .

Because as a sports car it's not much cop either. I have always been a fan of the Golf GTI. And I'm very much looking forward to trying the new version, which will be the same as all the others; a no-nonsense blend of hard-charging speed and dog-in-the-boot practicality. It doesn't really matter who you are or what you do, a Golf GTI is the answer.

Take off its roof, though, and it all goes a bit pear-shaped. Not only do you lose a lot of interior space but also you have quite a lot of scuttle shake; a sense that the front of the car and the back aren't connected quite as well as they should be. Which is true, of course, because whereas in a hard top, you have a sheet of steel joining them together, in a cabrio you only have a bit of canvas.

And to make up for the lack of rigidity, engineers are forced to fit many strengthening beams. Which a) add weight and b) don't really mask the problem. So the Golf cabrio is like a normal Golf in the same way that Anne Boleyn after her beheading was like Anne Boleyn when all her bits were still joined together.

And it gets worse, because although the roof has many layers of fabric and fits with Germanic precision, you still get a fair bit of extra noise. Which is annoying when you just want to listen to the radio and go home.

In various other ways, the GTI cabriolet is very good at this. It has a delightfully easy-to-use command and control setup. The seats are very well thought out, offering both comfort and support in all the right places. And its ability to deal with speed bumps and potholes is exemplary. Plus it is fast.

But I couldn't live with the drawbacks and that is neither the end of the story, nor the end of the world. Because the Golf I tested was £30,765, and for around £2,500 less you can have exactly the same car with far fewer problems. It has the same 207-bhp engine, the same suspension, and the same seats, comfort and ease of use. It's made by the same company and it's called the Eos.

This comes with a folding metal roof, which is more rigid and better able to protect you from the elements and all their noises. Sure, it's not as practical as the normal Golf, but for walking the fine line between sports-car motoring and everyday usability it's not bad at all.

But here's the clincher. The Golf cabrio really does have a whiff of fake tan about it. I think it may even have the ghost of Duran Duran in its genes. But the Eos does not. The Eos is exactly what you want and expect from VW. A simple, clean blend of well-made anonymity.

Plus, with an Eos, you don't get a sense that it's only been made to use up the last of the Golf bits and bobs before the new model arrives next year.

So there we are. If you want a Golf convertible, buy the Golf convertible that isn't actually called a Golf.

30 December 2012

Come on, caravanners, see if it will tackle the quicksand

Hyundai Santa-Fe Premium 7-seat

Motoring journalism: someone brings a car to your house on a Monday morning. It's clean, full of fuel and insured. You have it for a week and then you say whether you like it or not. Couldn't be easier.

That was certainly the case when I began in the job, because adjusting the vehicle to suit my requirements was a doddle. I simply tweaked the reins and jiggled the saddle around, and all was well. Even as recently as five years ago things were a piece of cake. The car was dropped off and all I needed to do to get comfortable was move the seat back.

Now, though, it's often Saturday afternoon before I've got it set up just so. Because every single thing is adjustable. Not just the seat but the components inside it: the lumbar support, the massage facility, the headrest. All have been set to suit the chap who dropped it off and they need to be reset to suit me.

When you've done that – and I'm well aware these are First World problems – you have to waste more valuable time finding the button that adjusts the suspension. Or, as I discovered with the Audi RS 4 that I wrote about here recently, you spend the whole time being vibrated so badly that your skeleton turns to dust.

Then we get to the climate control. You used to have a choice: warm or cold. Now you can select a temperature – to within half a degree – for each person in the car. This takes about a week. And you don't have a week because you are way too busy reconfiguring the satnav.

Most press-fleet delivery drivers like to have the map constantly spinning round so it's pointing in the direction of travel. I prefer north to be up. So I have to find the buttons that make this possible and work out the sequence in which they have to be pressed – hard when the lumbar support is digging into your back and the temperature is set at absolute zero. And you've got the suspension set on Rock. And then we get to the voice guidance. I cannot imagine for the life of me why delivery drivers like to have their chosen radio station interrupted every few seconds by a woman barking orders when there's a perfectly good map on the dash. But most do. Which means I have to work out how she may be silenced.

If I designed a satnav system, there would be a massive red button in the middle of the steering wheel marked 'Silence the Nazi'. But I haven't. So there isn't. And in the Hyundai Santa Fe I was driving last week that was a problem.

I tried every single thing I could think of. I even resorted to pulling over and reaching into the glovebox for the handbook . . . which wasn't there. So eventually I had to turn to Twitter. And it worked. I was told that while the woman was speaking, I had to turn the volume knob to zero. Doing so at any other time would simply silence the stereo. Not that this would have been a bad thing, as it had been left on Radio 1.

Small wonder the satellite recently launched by the Korean rocket went out of control once it got into orbit. It had probably been driven mad by the constant stream of spoken instructions about where it had to go next.

So, anyway, the first impressions of the Hyundai were not good. And the second weren't much cop either. Because it's all a bit rubbishy.

Cleverly, the company has fitted a soft-touch leather steering wheel, so the first thing you touch when you get inside feels expensive and luggzurious. But don't be fooled, because everything else feels cheap and nasty. The box between the front seats, for instance, has the quality of a Third World bucket. Johnny Hyundai knew a box was necessary and fitted one with no thought at all about how it felt to the touch. If he'd thought for a moment that it could be made from cardboard, it would have been.

Then we have the leather that covers the seats. It is leather. It must have come from a cow. But most cows I've seen are made from meat. The cows Hyundai uses are plainly a bit more synthetic.

This, then, is not a car you can love, because you sense all the time that it was made using bottom-line engineering by a gigantic Korean corporation that produces cars only to make money.

Small wonder this car is so popular with caravannists. They choose to go on what nobody else in the world would call a holiday. So it stands to reason that they like what I can't really call a car.

However, the Santa Fe is cheap. The high-end, seven-seat, four-wheel-drive version that I tested is £30,195, way less than you'd have to pay for a European seven-seat, four-wheel-drive car.

It's also cheap to run, though only if you go for a version with a manual gearbox. The automatic will send your fuel bills through the roof.

The options list, however, will not. You get, as standard, ABS, BAS, DBC, EPB, ESP, ESS, HAC, TSA and VSM. This thing has more abbreviations than the British Army. And more airbags – seven, to be precise. In fact, it's hard to think of anything you'd get on a Volvo XC90 or a Land Rover Discovery that you don't get on the Santa Fe. Apart from a sense of style, wellbeing and oneness with yourself.

That said, the Hyundai's not a bad looker and it drives pretty well too. Again, the people who set up the suspension were plainly dancing to a tune conducted by the company's accounts department, so they haven't gone the extra mile. Or even the extra inch. It's not a rewarding car to drive in any way, but it goes round most corners at most speeds without crashing.

Can it go off road? Yes, but not very far. With a part-time

four-wheel-drive system, it'll get your caravan into a field. But it probably won't get it out again. Which is a good thing for the rest of us.

And so, as we approach the end, I have to start thinking about a conclusion. It's tricky, because the petrol in my veins dislikes cars of this type in the way that a restaurant critic would dislike a McDonald's Filet-o-Fish.

However, in the real world where people live, where a quail's egg's a bit daft, the Filet-o-Fish is very popular. And there's my problem. In the real world the Santa Fe is cheap, it doesn't drink a lot of diesel, it's well equipped, it's good-looking, and the 2.2-litre engine is torquey enough to pull your caravan. Even if it's a Sterling Europa 565. So who cares if the seats are a bit plasticky?

It's not a car for the silly world in which I live. But elsewhere it's bloody brilliant.

13 January 2013

No one can reinvent the wheel quite like you, Fritz

VW Golf 1.4 TSI ACT GT

It's funny, isn't it, how people argue about which mobile phone is best. Honestly and truthfully, to the vast majority of people, they are indistinguishable. It's the same with wine. Of course there is a handful of enthusiasts who in a blind tasting really can tell white from red, but to the rest of us a £4.99 bottle of Château d'Asda tastes the same, and has the same effect, as a £45,000 bottle of Pétrus.

In fact, this is true of absolutely everything. Cheese. Pizzas. Caribbean islands. I spoke with a famous rock god the other day, who agreed that the whole debate about guitars is nonsense because they're all identical.

And so are cities, really. Brummies will argue that Birmingham is better than Manchester or Liverpool or Sheffield. But to the casual observer they're as different as milk bottles.

Then there's music. To those who were born under the influence of Stanley Baldwin, the Rolling Stones sound exactly the same as N-Dubz. It's all just boom, boom, boom, as the elderly are fond of saying. And I know what they mean because I simply cannot tell one piece of classical music from another. Unless it's been used in an advert on the television, it's all just one endless parade of girls sitting with their legs wide apart, sawing a cello in half with a bit of horse, and men blowing in tubes.

Yes, there are people who can tell not just Bach from Chopin, but also what orchestra they're listening to and even what conductor is in charge. But for people with jobs and friends? No. It's all just bars and tone.

You know where this is going and, of course, you're right.

Cars are all the same too. They're all Volkswagen Golfs. There are fast Golfs and big Golfs and cheap Golfs. There are Japanese Golfs and V12 Golfs and American Golfs. But they're all Golfs.

I can tell the difference between a Ford Focus and a Vauxhall Astra, but that's because I'm a nerd. However, I'm not such a nerd that I don't realize both are actually Golfs. You could put my mother in a Lincoln Town Car and she would be incapable of telling it apart from her own car. Which is a Golf. She thinks my Range Rover is a Golf, too, albeit one that is idiotically hard to park.

She's right, of course. I sometimes wonder why anyone ever buys anything else. You want a fast car? Buy a Golf GTI. You want an economical car? Buy a Golf diesel. You want a cheap car? Buy a Golf from the second-hand columns. You want a big car? Buy a Golf Plus. You want a convertible car? Don't buy a Golf convertible. It's terrible. But do buy a Volkswagen Eos. Which is a Golf.

You can fit five people in a Golf, the same number as you can get in a Rolls-Royce Phantom. A Golf will cruise easily at 95 mph, the same as a Bugatti Veyron. It is as reliable as Switzerland, as comfortable as your favourite armchair, as parsimonious as a Methodist's auntie and, all things considered, good value too.

I'll tell you how brilliant it is. Volkswagen has spent the past five years working round the clock on an all-new model. The company started with a clean sheet of paper and an open mind. And what it has ended up with is a Golf.

If you set out to rethink the concept of a table, you'll end up with a table. And if you set out to rethink the concept of a car, you'll end up with a Golf.

The boot is a little bigger than it was before. There's a little crease running down the side. It's cheaper as well. And though it's longer and wider, it's 100 kg lighter. Which means the new model is as parsimonious as a Methodist auntie's lapdog. Or indeed any dog, because they're all the same too. In fact all pets are the same. Some stand in a field. Some live in a tank. Some purr, and some have a shell. But they all need feeding and housing and . . . I'm digressing.

Economy is probably the big news with the new Golf. And rightly so. Because if all you want is 'a car', then you want to spend as little as is humanly possible on fuelling it. Other car makers are fitting their Golfs with all sorts of stuff – independent rear suspension, for instance – which is fine for the tiny number of connoisseurs. But for everyone else? Independent rear suspension and the benefits it brings are less interesting and important than the result of a village cricket match in Pakistan.

No. When it comes to cars these days, the top 10 things that matter are: economy, economy, economy, economy, economy, economy, economy, economy, economy and safety.

My son is about to turn seventeen and wants a car. Does he want it to be fast? Good-looking? Comfy? Nope. All he is bothered about is economy. And he's not alone.

VW has obviously realized this, which is why the 1.4-litre TSI GT model I tested is fitted with a four-cylinder engine that switches to two cylinders when you're pootling along. And then shuts down altogether at the lights. What's more, messages flash up on the dash, giving helpful hints on how to get the most miles from each gallon. One tells you, for example, not to disengage the clutch until the revs have dropped to 1300 rpm.

This makes sense because most people don't realize how modern engines work. They think that if they put their foot on the clutch and freewheel when, say, going downhill, it saves fuel. Not so. An engine not in gear quickly slows to idle and needs fuel to keep it ticking over. But if you slow down without disengaging the clutch, the engine uses absolutely no fuel at all.

So, the new Golf. It's light. It can run on one lung. And you get hints on how to maximize the mpg. That means almost 60 mpg, which makes this one of the most economical family cars on sale today. So that covers the nine most important things. For the tenth, it comes with a forward-facing radar system that applies the brakes if it thinks you haven't noticed you're about to crash.

So what's it like to drive? Well, you can choose from a variety of settings to suit your mood. All of them were good. Speed? That was good too. And so were the handling, the comfort and the quality. It was all good. Everything was good. It even has an optional self-park system. That was good as well.

Honestly, reviewing this car is like reviewing a floorboard. It's impossible to say anything other than: 'It does what it's supposed to do.' Which is why, for a summary of what the new Golf is like, I've turned to the *Oxford English Dictionary*. Golf. Noun. A road vehicle, typically with four wheels, powered by an internal combustion engine and able to carry a small number of people.

27 January 2013

Great at a shooting party – for gangsters

Mercedes CLS63 AMG Shooting Brake

THERE is almost nothing harder in this business than reviewing an AMG Mercedes. Because the engine dominates everything. It doesn't matter whether you are in a two-seat sports car or a four-wheel-drive off-roader or a large executive saloon, all you can think about, and hear and feel, is that massive, throbbing, tumescent muscle under the bonnet.

AMG-powered cars all handle in pretty much the same way as well. In that they don't really handle at all. You push the accelerator and you are going sideways, immediately. And I'm sorry, but when you are on the Oxford ring road, in a cloud of your own tyre smoke, being deafened by the jackhammer soundtrack, it is pretty difficult to think about seat comfort or visibility or indeed anything much at all.

In an AMG the car is simply the knife and fork. The engine is what you're eating. And in a restaurant it's the food that matters. Not the utensils needed to deliver it to your mouth.

However, in recent months things have changed. When AMG started making mainstream cars, it had supercharged and non-supercharged 5.4-litre V8s in various states of tune. And then it did a range of non-supercharged 6.2-litre V8s that, for reasons we don't understand, were claimed to have a capacity of 6.3 litres.

All these engines were exactly the same. Choosing which you liked best was like choosing a favourite type of thunderclap. Some were a bit more powerful than others, but when you were screaming and praying and sweat was flooding out of every pore, you really couldn't tell.

Now, though, things have changed. All the new AMG engines are twin-turbocharged 5.5-litre V8s, and they're different. The bellow has gone, and with it the terror. You can now drive an AMG without ear defenders and nappies. And this means you can concentrate on what the utensils are like.

As a diesel-powered estate car, the CLS Shooting Brake doesn't appear to be that good. It's less spacious – at least with the seats down – slower from 0 to 62 mph and considerably more expensive than rivals from Jaguar, Audi and BMW. But that isn't the case with the AMG 63 version. Because it doesn't really have any rivals at all.

The CLS saloon was, it's said, designed by a Mercedes stylist almost as a doodle. He wanted to know what a Jaguar would look like if it were done by his company. His bosses liked what they saw, his squiggles made it into production and now there is an estate version called the Shooting Brake.

Strange name. A 'brake' was the name given in the olden days to a carriage that was used to break spirited horses. Later, British landowners would modify them so they could be used to carry people, guns and dogs on shooting parties.

So, in the early days of the motorcar anything designed to carry stuff in the back was called a shooting brake, and when these became common on various estates around the country, the generic name changed to 'estate car'. Think about that the next time you fire up the five-door Ford Escort.

Today, to get away from the rather toffish imagery, car makers call their estate cars by other names. BMW has the Touring and Audi the Avant. And the Americans, of course, the station wagon. Because over there large, practical cars evolved at railway stations, hauling goods from train to train, rather than on grouse moors.

But Mercedes has ignored this egalitarianism and resurrected the shooting brake handle. On perhaps the least practical estate car in human history. For a shooting party it would be hopeless. Unless, of course, the shooting in question were a drive-by in Los Angeles. Then, I suspect, it would be excellent.

The 63 AMG version is very, very fast. And because the his-trionics are mostly gone, sometimes you barely even notice. There's a momentary strain on your neck muscles when you first plant the throttle, and the next thing you know, the white lines have become a blur and you're in Arbroath.

It even does well on more testing roads. You always felt in old AMG cars that you were having to manhandle them through the bends. That's something I like. But I also like the way the Shooting Brake seems to be dancing. It feels agile and light, which it emphatically is not.

And yet somehow, despite the power and the corpulence, it seems to have grip. You no longer have to countersteer every time you pull over to pick up a pint of milk. You can even get from one side of Guildford to the other without spinning.

It's comfortable too. On even the most potholed roads it would make a tremendously stable gun platform. It's a bit like an ageing rock god. The anarchy is still there in its DNA. But these days it prefers cocoa to cocaine.

Problems? Yes. The traditional automatic gearbox with its torque converter has been replaced by a multi-clutch system. Doubtless this is very good news for Johnny Polar Bear, but, like all modern flappy-paddle solutions, it is annoying.

In Comfort mode, things are just about acceptable, but if you switch the system to Sport, it becomes dim-witted. Especially around town, where it has a habit of responding to requests for more speed exactly half a second after the gap you were aiming for has disappeared.

Also, my test car had so many toys, the dashboard was almost indecipherable. And if you dive into a submenu to see what's what, or maybe turn off the lane-keeping assistance, you could be stranded in a world of electronic gobbledygook for a year.

Further back we find that the boot lid opens and closes elec-trically. Why? Who thought that customers would enjoy standing in the rain for an age, looking like dorks, while the motors went about their business? I have arms. I can open a boot perfectly well, thanks.

Not that there's much point, in the CLS, because this really isn't designed for the antiques dealer in a hurry. You're certainly not going to get a wardrobe in there, that's for sure. And that's because this car is like its four-door brother. It's mostly about style.

And I must say that on that front it does rather well. Today a lot of modern Mercs are overdesigned. They have too many creases in their flanks and too many unnecessary details. But this one is bob-on. It's tremendous-looking, which leads me to a peculiar conclusion.

If you want a car to carry school trunks and horse paraphernalia, you can do better with the offerings from Jaguar, Audi and especially BMW. I think the 530d Touring is one of the best cars made.

And yet even that doesn't have quite the appeal, somehow, of Merc's new drive-by shooting brake. It's not a sports car. It's not really an estate. But, unlike anything else, it is a little bit of both.

I said earlier that it has no rivals, and that's true. But only because these days you can no longer buy a Reliant Scimitar.

10 February 2013

Yippee! It's OK to be a Bentley boy again

Bentley Continental GT Speed

I don't understand why people get so cross when politicians do a U-turn. Would you rather they were so consumed with towering self-belief that they ploughed on regardless? Because I wouldn't. Take Michael Gove as an example here. The education secretary expressed a desire to change the way exams were run. There was much brouhaha from interested parties. And, having listened to their objections, he's decided to abandon his plans. What's wrong with that?

If I were a politician I would constantly express a desire to invade and conquer France. I'd explain that it simply isn't morally correct for such a lovely country to be in the hands of the French and, as a result, I'd ask the armed forces to work up some kind of plan.

And what they'd do is talk me out of it. They'd explain that we simply don't have the muscle, and that even if we did, the United Nations would impose all kinds of unpleasant sanctions. So that eventually, and much to the relief of just about everyone, I'd announce that we would be invading and conquering Spain instead. Technically that would be a U-turn. But it would also be an example of common sense.

It'd be the same story with cats. I'm afraid that if I were prime minister I'd announce that they must all be executed. This would prompt a great deal of debate and eventually I'd be forced to say, 'Oh, OK. Your cat can live. Just so long as it doesn't leap into my lap and show me its anus.'

This is because I am wise. Unlike His Tonyness, who was not. He was so wrapped up in his own self-importance that after he'd decided to invade the Middle East, no amount of reasoning would cause him to back down. And look where that got us.

So, no. We need to stop criticizing politicians when they make U-turns and start congratulating them for being open-minded and flexible. Which brings me neatly to the Bentley Continental GT.

I was first shown this car shortly before it went on sale in 2003 and I'm afraid I was at a loss. The designer was standing there, all wide-eyed and expectant, but I simply couldn't find the right word. 'Amazing' is what I usually use when presented with a hostess's terrible pudding or an actor's disastrous new play. But 'amazing' didn't really work with the Conti. 'It's awful,' I said.

And it was. I'd never seen a car that managed to be both bland and ostentatious at the same time. And there were some details that looked plain wrong. But still, the styling was a triumph compared with the way it drove. It may have had a mighty W12 engine, but there was absolutely no sense that it was doing anything that a much less wasteful V8 couldn't achieve. It was just there, like a big piece of Georgian plumbing, turning fuel into absolutely nothing interesting at all.

What's more, the four-wheel-drive system removed any sense of finesse and the suspension settings had plainly been chosen by June Whitfield. Well, not necessarily by June herself, but someone of her ilk: someone with no interest in cars whatsoever. As a result, the Continental GT felt like a big, useless, thirsty waste of leather and aluminium. I hated it and said so.

Then came the customers. Bentley gave one to Mrs Queen and was probably hoping for James Bond. But what it ended up with was Jay-Z, Mario Balotelli, Paris Hilton, Xzibit and Steven Gerrard. A shocking array of people you would not like to have round for dinner.

Small wonder the depreciation was so bad. A new Bentley would look good in your exclusive gated community. But the entire appeal of this car was wanton consumption, so nobody wanted to be seen dead in a used one. Which is why you could, and still can, pick up barely run-in examples for about 6p.

As the years crawled by, Bentley took the cash from its increasingly terrible customer base and spent it on a raft of technical improvements. None of which did anything to change my mind.

There was, for instance, the Continental Supersports version, which was intended to be a pointy racer. But it was no such thing. Because you can't make a go-kart if you start out with a lorry. It was, frankly, an idiotic car.

But still Bentley kept on going. And last year announced a new, lighter, more economical V8 GT. It was just as fast as the W12 but did 10 more miles to the gallon. And it made a dirty sound when you accelerated. And I liked it.

Of course, the customer base didn't. Because footballers, rappers and the stars of various sex tapes are not given to saying, 'Please may I have the second-most-expensive Rolex?' Or, 'Yes, I like this hideous house very much. Especially the fake pillars. But do you have anything that is easier to heat?'

So, to keep them happy, Bentley has fiddled with the W12 version to create what it is calling the GT Speed. The suspension is tweaked, as is the twin-turbo W12, so that now, with a flat-out maximum of 205 mph, it is the fastest Bentley yet. Plus, according to my televisual colleague James May, who recently took one rallying on *Top Gear*, it's the best.

He's wrong, of course. It isn't as good as the V8. But I will admit he does have a point. First, unlike almost any other fast car, it rides properly, steering a hitherto unexplored path between the wallowy detachment of a Rolls-Royce and the youthful and sometimes misjudged harshness of a BMW, Jaguar or Aston Martin.

Then there's the gearbox. Other manufacturers are moving at breakneck pace towards the double-clutch systems that make it easier to pass EU emissions legislation. But Bentley has stuck with a traditional automatic, albeit with eight speeds.

Don't think, however, that because it's soft and fitted with a slushmatic that it's a slouch. I put my foot down on day one and, so savage was the acceleration, I never put it all the way down again. This is a blindingly quick car.

And it's good value too. The new Aston Martin Vanquish is nudging £200,000. The Bentley is just £151,100. Not cheap. But for what you get, not bad at all. Certainly it comes with a lot of toys. And while many of the buttons and knobs are handcrafted and chromed and very Bentleyish, the fact is that behind the scenes, it's pure Volkswagen. Which is another way of saying, 'It'll all work.'

Sprinkle into the mix four seats, a commodious boot and a body that isn't as big as it seems and you end up with a pretty compelling car. So here comes the U-turn. In the past you would see someone cruise by in a Continental and you would think, You bought that because you know nothing about cars, you are not interested in driving and you wanted simply to tell your friends at the golf club you have a Bentley.

Now, though, things are different, because if I see someone cruise by in a Continental GT Speed, I shall think, There's a chance you bought that because it's a bloody good car.

17 February 2013

Thrusters on, Iron Man, this'll cut through the congestion

Audi R8 5.2 FSI quattro S tronic

I wonder. Does AA Gill review a restaurant when he has a cold? Because surely, when your eyes are streaming and your head is full of hot mercury and your nose is like a leaky tap, it must be very hard to tell whether you're eating fish or chicken.

Also, when you are feeling low and miserable, there's no waitress in the world who will pass muster. She may be smiley and lovely and knowledgeable, but because you are consumed by an overwhelming need to be in bed, with your teddy bear and some warm milk, you will sit there, as she runs you through the specials of the day, wondering how she might look without a head.

I have a similar problem when I have a cold and I have to review a car. Because no matter how comfortable it is, it's not as comfortable as where I want to be: in bed. And even if it can get from 0 to 60 mph in two seconds, I will fume because that's not good enough. I want it to get from where I am to where my bedroom is . . . immediately.

Which brings another problem into sharp focus. We are not allowed to drive a car after we have consumed alcohol or if we are using a mobile telephone or if we are eating a sandwich. But we are allowed to drive while we have a cold. And I think that's odd. Because – and I'm sure I'm not alone in this – when I have a snotty nose and a heavy head, I am a madman.

You know those turtles that lay eggs so far up a beach that there is no way their young will make it back into the sea without being fried or eaten? Well, they would make better drivers than me when I'm poorly.

Take a recent a case in point. Shepherd's Bush Green in west

London was closed because of roadworks, which meant every-one in the world was using the Hammersmith flyover. So obviously some bright spark closed that too.

Normally I'd simply sit in the resultant jam, accepting that the people in charge are morons. But because I had a cold, I needed to be at home. So I set off on a hate-fuelled charge through the back streets, hurling insults at absolutely everyone and everything.

Had the dithering minicab driver in the Toyota Camry actually heard what I was saying as he sat there for an age, making no effort to turn right, I'd be in prison now for breaking all sorts of modern-day laws.

Eventually I squeezed by on his left, which meant I may have accidentally popped a couple of wheels on the pavement. And moments later a police officer knocked on my window. I don't know why, I'm afraid, because before he had much of a chance to speak, I let rip, telling him that the pavement was too wide, that the minicab driver was a stupid idiot and that if he wanted to speak to someone, he should talk to the halfwit who'd shut both main roads into west London at the same time. I then drove off.

The next day the A1 was shut, preventing me from making a meeting on time. So I pulled over and rang various contacts in my phone to shout at them. And that night, on the M1, while running late for my daughter's school play, I didn't do as normal and sit behind the stream of Peugeots doing 50 mph in the out-side lane. I just overtook them all on the inside, muttering and chuntering like a homeless American drunk.

In short, there was much mayhem and rage and misery. But there was at least one crumb of comfort: the car I was using. An Audi R8 V10.

Normally, when the roads are full of idiots in Peugeots and your head is full of mucus, the last place you want to be is in a low-slung, super-wide, Lamborghini-engined two-seat supercar.

Supercars are tricky. All the things that make them great on sunny days in the Tuscan hills make them utter pigs in Shepherd's Bush on a dark, wet Wednesday. They steam up. They pop. They bang. They growl. They won't fit through gaps. You can't see out properly. And they are uncomfortable.

The R8, though, has always been different. It feels normal. The cockpit is big. Everything is where you expect it to be and it all works. It can traverse speed humps without leaving 40 per cent of its undersides behind. It's quiet. It's unruffled. It is the Loyd Grossman cook-in sauce of supercars. An easy, fuss-free alternative to the complexity of the real thing.

Unless you are at an oblique junction where visibility is a bit restricted, you forget when you are driving an R8 that it is a supercar. It lulls you, cossets you, soothes you. Which is why you are always surprised when you put your foot down and it takes off like a bloody Ferrari. 'Whoa,' you say, 'I thought I was in an Audi TT.'

It grips well too. You might imagine a car that could comfortably handle a farm track – it can and did – would fall over in a bend, but the R8 doesn't. I'm not suggesting for a moment that it has either the tenacity or the controllability of the (much more expensive) Ferrari 458 Italia, but when it's raining, and the motorway slip road has tightened unexpectedly, and you haven't noticed because you've been sneezing non-stop for the past 30 miles, you'll be glad of the four-wheel-drive system, that's for sure.

There's something else as well. Thanks entirely to its starring role in *Iron Man*, the R8 is a popular car. People like it, and by association they will like you for driving it. But there has always been one problem: the gearbox. You could have either a manual that came with a lever from a Victorian's signal box or a semi-automatic that was as refined as a South Yorkshire hen night.

Well, now you can have a seven-speed dual-clutch system. Many car makers are going down this route, saying it brings Formula One-style driving to the road. While not explaining the real reason.

The EU is demanding that engines produce fewer and fewer 'harmful' emissions, and gearboxes such as this help a lot. They're fitted to tick a bureaucrat's box, not because they make life better for the driver. This is a point you will note at low speed in town. Go on. See a gap and try to exploit it. You can't. Because the box is either too slow or too jerky – put your foot down and the R8 sets off like you've never been in a car before. Really, flappy-paddle gearboxes should come with P-plates.

Of course, out of town they're great. The gear changes are fast and smooth and can be done either manually with flappy paddles or automatically. On balance, then, I'd say the new box in the R8 paints over the only bit Mamma Achilles missed. It's a damn good car. It really is. And that from a man who's spent a week hating everything else.

24 February 2012

They'll be flying off the shelves at Poundland

Porsche 911 Carrera 4S

The weekly motoring magazine *Autocar* recently published a truly remarkable road test of Jaguar's new baby sports car, the F-type. Apparently it has been engineered by a man whose superhuman abilities as a boffin are matched only by his unparalleled excellence behind the wheel. He was described, almost homoerotically, as a blend of Jackie Stewart, Mother Teresa and Wernher von Braun, the German rocket scientist. So not surprisingly his creation is apparently rather special.

We were told that the transmission is silky and beautifully smooth. We learned that the V8 engine produces a tidal wave of torque and that the V6 'has the soul of a responsive, agile sports car'. The eulogies went on for page after page, and what made this truly astonishing piece of journalism so extraordinary is that the man who wrote it has never actually driven the car.

He admits that he determined everything from the passenger seat. And the ability to do this, I must say, is a rare gift. He demonstrated his talents the following week by telling his readers that there are subtle differences between the Subaru BRZ and the Toyota GT86. This is true. There are. Both may be made using exactly the same parts, by exactly the same people, in exactly the same factory, but we are told in the official blurb that each car has slightly different suspension settings.

I'll be honest with you – I drove both cars back to back on the *Top Gear* test track and could not tell them apart. It was the same story with the Stig. I told him that they had differently tuned suspensions, but after many laps he came back in to report that they both felt exactly the same.

Not so, it seems. Because according to our friend on *Autocar*, the Toyota feels slightly softer up front while the Subaru feels a shade more 'planted' in faster corners. I felt deeply ashamed not to have spotted this and made a silent vow to redouble my road-testing efforts in the future.

Which brings us on to the new Porsche 911. It's longer than before and has a wider track. And although 90 per cent of the components are new, it is still very definitely a 911. Except for one thing. It isn't.

My colleague Richard Hammond is a demented fool, of course, but I'll give him one thing. He knows the 911 better than anyone and he reckons the new car is a huge step backwards. He tells us that all of the strange little quirks that set this squashed Hitlermobile apart from other cars have been erased in the new model. And that it now feels just like anything else.

He points with special venom at the new electric power steering, saying that this has completely ruined the unique feel. He accepts that electric power steering is necessary for cars to meet new emissions legislation – hydraulic systems take more power from the engine and therefore use more fuel – but he says that as a result of the change, the 911 is George Clooney without the twinkle, Cindy Crawford without the mole.

Naturally the gifted helmsmen on *Autocar* disagree. They say we should not worry about the new electric power steering setup and tell us that, 'information, rich and abundant, comes streaming through' the car's wheel rim. And that all is well.

Allow me to be the judge in this matter. Ready? Here goes. The steering system in the new Porsche works extremely well. If you turn the wheel anticlockwise, you go left. And if you turn it clockwise, you go right. If you do nothing at all, it goes in a straight line.

There's more, too. If you push the seat back, you end up further from the windscreen. If you turn the heater up, you will be a bit warmer. And if you turn the radio on, music comes out of the speakers. Unless it's tuned to Radio 4, in which case you get some Brummies talking about cows.

This is what the 911 has always been about. A taste of the exotic in an-easy-to-use, everything-works package. It's bloody good value as well. Most interesting cars these days cost upwards of £200,000. They're so expensive that I was surprised and amazed the other day to find the new Bentley Continental GT Speed is 'only' £151,100. But the 394 bhp sequential automatic Porsche 911 Carrera 4S I tested last week is a mere £90,346. And this, remember, is a 184-mph car. So that's proper performance with a Poundland price tag.

There are some lovely details. It has a satellite navigation system that has been loaded with a current map of Britain. Unlike, say, the Audi R8, which has been given a map not used since your farmer had oxen and all it said over Somerset was: 'There be witches.' In one journey in the Audi it simply drove me into a field and gave up. The exact same journey in the Porsche was a breeze.

I also like the way that, in the Porsche, you can choose which sporty element you'd like at any given moment. In most cars, you push the Sport button and it alters everything; the gearbox, the exhaust noise, the steering and the engine response. In the 911 each one of these can be adjusted individually.

I should explain at this point that I am not a 911 enthusiast. But in the same way that a Chelsea fan will admit that behind the gum and the bad temper, Sir Ferguson is a bloody good manager, I can see this is a bloody good car.

There are some niggles, though. When the differentials are cold and you are manoeuvring in a tight space, the front wheels just plough straight on. In this respect, it's a bit like my P45 micro-car. And on the move, the accelerator pedal feels mushy. There is no instant response from the engine, partly because it's fed in an old-fashioned way and partly because the new seven-speed gearbox is a bit dim-witted.

There's something else too. The four-wheel-drive system is a complete waste of time, unless you live in Val d'Isère. Which you don't. Or are a road tester on *Autocar*. They say the all-wheel-drive grip makes the car slightly more sure-footed than a two-wheel-drive version. But I'm not sure. I think that on a normal road, it just adds weight, chews fuel and offers absolutely no benefit.

I have more advice on the 911 range. As well as a new cabriolet, in the fullness of time there will be a Turbo and all sorts of hunkered-down, uncomfortable track-day monsters. Forget all of them. A 911 is a sports car, pure and simple. If you add a turbo, it becomes a supercar, something it is not. If you take off the roof, or fit a roll cage or do anything at all, you are adding or subtracting ingredients to a recipe that was fine in the first place.

You buy a 911 like you buy wine in a restaurant. You go for the second cheapest. And that's the two-wheel-drive Carrera S. *Autocar* agrees with me on this. It says the two-wheel-drive versions are purer and more involving. Doubtless this is so. But more importantly, some of them are more than £17,000 cheaper and make you look less of a plonker.

3 March 2013

So awful I wouldn't even give it to my son

Alfa Romeo MiTo 875cc TwinAir Distinctive

One day, when I'm done with shouting and driving round corners too quickly, I want to be the managing director, chief executive officer and Obergruppenführer of Alfa Romeo. Because I want to sort the damn thing out. The first car to be launched under my completely autonomous dictatorship will be a two-seat sports car. It will be a little larger than a Mazda MX-5, it will be rear-wheel drive, it will have a manual gearbox and it will be called the Spider.

Under the bonnet there will be a rorty little 2-litre twin-cam four-cylinder engine that will snuffle and pop on the overrun. It will be fed with carburettors that will cause the EU's faceless emissions people to write me a strong letter about the need to preserve the world's polar bears.

I will write back, in extremely strong terms, explaining that I am not interested in polar bears, or glaciers, or how much carbon dioxide there is in the upper atmosphere, because the carburettor is a thing of exquisite delicacy and magnificent simplicity. I will then attach this letter to a brick and throw it through their window.

Afterwards, we have to decide how the car will look. And again it's not hard. I will walk into the company's design centre, with a gun, and announce that it is to create the best-looking car made by anyone, ever. Failure to do so, I will announce, while glancing at my AK-47, will not be tolerated.

The end result will look a bit like a scaled-down Ferrari 275 GTS. There will be hints of the Fiat 124 Spider in there too, with notes, perhaps, of the Maserati 3500 GT. It will have wire wheels. And maybe pop-up headlamps. And if anyone writes to say that these are bad for pedestrian safety, I will find out if they are correct by running them down.

The end result will be beautiful to behold, lovely to drive, as characterful as Jack Reacher and as sought-after as pictures of a naked royal. Money will cascade into the company's coffers and it will all be spent on a new hatchback in the mould of the old 'Sud. And on bringing back the 159, which was recently dropped for no reason. Because there was nothing wrong with it.

I recently went to see James McAvoy's new film, *Welcome to the Punch*. In it he drives a 159 and it looked so sensational that I lost control of what was going on. I love that car. I wish it was still with us. I miss it.

I love Alfa Romeos in the same way as old ladies love their cats. I know they are unreliable and stupid and mad. But that's what makes them seem human. I love Alfas in the same way as Arsenal fans love their team. But, like Arsenal fans, I'm being forced to sit and watch the love of my life being ruined.

My son turns seventeen this month and will be getting a car. But what? Perhaps because he is my son he is about as passionately uninterested in all things automotive as is humanly possible. He knows less about cars than June Whitfield. Actually, come to think of it, he knows less than King Herod.

All he knows is that he doesn't want a Ford Fiesta because that's what his big sister drives. And he wants to be different. So I suggested an Alfa. The MiTo, perhaps. The new TwinAir version with the two-cylinder 0.9-litre . . . his eyes started to glaze over. 'It's Italian,' I said. He liked the sound of that. He likes Italy.

So I borrowed one for a test drive and – Holy Mother of God. It was shocking. Bad almost beyond belief. The idea, though, is quite good. There are plenty of people who would love to be propelled from place to place by this barnstorming little engine with its promise of 60 mpg-plus. But originally it was available only in the Fiat 500, which for many is just too small.

So here it is in the MiTo and it's as good as ever. It revs like a terrier and makes the sort of noise that causes you to smile. But there's no getting away from the fact that it's very tiny and the MiTo is quite heavy. So while you have the hysterical revs and the comedy noise, you're still doing only 4 mph.

Time and time again I'd pull over to overtake a Peugeot, pleased with the initial surge of power, but then it would wither and die like a grape in a furnace, and something would be coming the other way, and its lights would be flashing, and I'd have to brake and get back behind the Peugeot again. Fast, this car is not.

It sounds fast. It even looks fast. And it's an Alfa Romeo so it should be fast. But it isn't, no matter what you do with the DNA. There is a switch that lets you choose from three settings: Dynamic, which give you nowhere near enough power; Natural, which gives you less; and All-weather, which is irrelevant.

I was happy with all of this. Because, while my son is not interested in cars, he is a teenager. And I very much like the idea of him driving around in something that has exactly the same top speed and acceleration as the Queen.

However, there were some things that were not so good. The gear change was woeful, the ride was catastrophic and I can only assume the driving position was set up for that orang-utan that used to hang around with Clint Eastwood. The pedals are virtually underneath the steering wheel, which means the only people who can get comfortable are those whose arms and legs are exactly the same length.

And guess how much it costs. Nope. You're wrong. But don't feel bad, because I was wrong too. I thought that it would be £13,000. But in fact the car I tested, which had one or two small extras fitted, was £16,500. Way too much for any car this size. But for a car this size and this bad it's a joke.

Of course, you're paying for the badge. Which is worth a lot, I admit. There's a lot of symbolism in there: serpents and Milanese legends and a red cross that harks back to the first crusader to scale the walls of Jerusalem. There's more, too. Alfa Romeo once won every single race in the Formula One grand prix calendar. It's where Enzo Ferrari began his career. There's more heritage and more raw emotion in that symbol than there is in the rest of the car industry put together.

And yet the company has the cheek and the barefaced effrontery to put it on the MiTo TwinAir. It's like putting the Rolls-Royce badge on a corner shop. No. It's worse. It's like Princess Anne appearing in a television commercial for payday loans. It's simply not on.

So when I take over Alfa Romeo — and it may have to be an armed coup — the MiTo TwinAir will be dropped and any library or internet site that contains a mention of it will be burnt.

In the meantime, though, I think my son will get a Volkswagen Polo. I'll tell him it's Italian. He won't know.

10 March 2013

Off to save the planet with my African queen

BMW528i Touring SE (1999, T-reg)

When I spend a night at my flat in London, I like to cook myself some supper. And since my culinary skills are a little bit northern, I don't overreach myself with exotic spices, pestles or mortars. It's just pasta, some pork and various bits and bobs to liven things up.

Then, the next day, I have to call a skip-hire company and a forklift truck to take away all the packaging. This makes me so angry that my nose swells up and my teeth move about. I mean why, for instance, do spring onions have to be sold with a rubber band?

And it's not just food, either. If you buy a Gillette razor, it comes in a hardened shell that is impervious to any known form of attack. Scissors, fire and even atomic weapons are useless. And pretty soon there is blood everywhere, which requires a sticking plaster. And therefore another skip to take away the packaging in which that was sold.

Bought a USB dangly thing recently? It occupies less than a square inch but comes in a clear plastic container that could be used by a family of six as a fallout shelter. And that's before we get to Easter eggs. They come in boxes the size of B-52 bombers and are wrapped in Cellophane and tinfoil and paper and about a million things the world can ill afford to waste. And it makes me seethe.

I'm not, as you probably know, given to environmental thinking. I do not care about the composition of gases in the upper atmosphere, and when it's chilly outside I will gladly fire up the patio heater. But I have a profound dislike of waste, and there is nothing on God's decreasingly green earth as wasteful as pack-aging. When I come to power, people will be allowed, by law, to murder any shopkeeper who sells them anything that does not need to be in a bag, in a bag.

We shall start with the *Daily Mail*. Because it runs a campaign about the idiocy of plastic bags, but on a Sunday its supplements are delivered in . . . a plastic bag. Which I have to throw away.

Then we shall move on to the makers of anything that is billed as disposable. Why would you want to throw away a camera, for crying out loud? Or cutlery? Or anything at all?

This nonsense is even spreading these days to the world of cars. You may have noticed recently that Fiat is advertising what it calls the spring/summer collection of 500s. And I know exactly what is going on. It wants people to think of their car as a skirt, something that should be discarded after six months because white is so last year.

I'm afraid, however, that Fiat is by no means the only offender. We all are. Because why do we ever sell the car we have now?

Just the other day my wife said we should change our Range Rover for the new model because it keeps getting punctures. That's like saying, 'Oh, no. We need to change the children because they've got measles.' But she is not alone. The world is full of people who get rid of their cars because they imagine they're about to start costing money, because they appear to be on their last legs. But that ain't necessarily so.

Top Gear is often criticized – by people with terrible shoes, usually – for concentrating too vigorously on expensive cars that no one can afford. This, of course, is rubbish. Elton John could easily afford anything and everything we feature. It's also not true. We put far more effort, money and time into featuring old crocks.

You see, we have a message. When you think a car is done for and washed up, we will demonstrate that, actually, it can still get across Botswana, or India, or Bolivia, or the Middle East. Deep down, *Top Gear* is the most environmentally sound programme on earth. Or it would be if only more people had got the message. And since they haven't, I've decided this morning to tell you a little bit more about the BMW I recently drove on a 1,000-mile odyssey through Uganda, Rwanda and Tanzania.

It was a T-registered 528i Touring that had covered 150,000 miles, and there were a number of features that would cause a wary buyer to shy away. It had a manual gearbox, suggesting the original owner had been enthusiastic. It had mismatched tyres, indicating it had been run on a shoestring. And it had a towbar, implying it had been used to haul heavy stuff. It was, in other words, one breath away from the skip. And that's why we were able to buy it for just shy of £1,500.

Yes, the throttle was calibrated all wrong and, yes, the electronic boot lid was broken, but mechanically it was in fine fettle. The gearbox was sweet, the engine was strong and even the air-conditioning still worked. And so, as I cruised across Uganda on smooth, Chinese-built roads, I found myself thinking: why would you not want this car?

Of course, it came from a time before satellite navigation, but that could be sorted with a TomTom. And, yes, it wasn't equipped with other modern features such as parking sensors, but I solved that when manoeuvring by simply looking out of the windows.

Anyone who saw the show will have noticed that some of its underfloor wiring became damaged. That's true. It did. But only because in one edited-out moment I decided to see if doughnuts went in different directions on either side of the equator. And hit a rock. Not the car's fault. Mine. And it was the same story with the rear window.

When the tarmac ended and the road became a quagmire, logic dictated that I should simply give up. BMWs do not have the best reputation for longevity and I was asking it to climb a track that, half the time, was flummoxing the crew's Toyota Land Cruisers. But even though that car had been owned by a penniless enthusiast with a trailer, nothing broke. Nothing.

It didn't even suffer unduly when the going became extremely rough. Yes, two of the airbags deployed over one nasty jolt, but unlike the estate cars chosen by my colleagues – a Volvo 850 R and a Subaru Impreza WRX – it arrived at the finish line with all its wheels still attached.

For nearly two weeks it had been driven on washboard gravel, through mud and, some of the time, on no kind of track at all. And yet I could quite happily have driven it back to England afterwards. And, despite all the hardships and all the torture, it would have made it.

So bear that in mind if you are looking at your own car now. You may think it's on its last legs, but I'd like to take a bet that it isn't.

Jonathon Porritt, George Monbiot, Greenpeace and Friends of the Earth will tell you that to save the planet you must dispose of your old car and use the bus or a bicycle instead.

But I've got a better idea. If you really want to save the planet, and a fortune too, do not buy a new car. Follow the teachings of *Top Gear* and simply carry on using the one you've got now.

17 March 2013

Oh, I hate the noise you make in 'wounded cow' mode

Toyota Corolla GX (aka the Auris but GX model not sold in UK)

Qualifying for last weekend's Australian Grand Prix, the first race in the Formula One calendar, was to be held in the middle of the night. But that was just fine by me because I had big, saucer-eyed, loneliness-of-the-night jet lag.

Annoyingly, though, the start of the action was delayed because of rain. And then it was delayed some more, and then a lot more. And then it was abandoned. Various serious-faced men in short-sleeved shirts came on the television to explain that there was standing water on the track, and on Twitter all sorts of people were sympathizing with the decision because it's dangerous to drive a car into a puddle at 160 mph. They're right. It is.

But here's the point that everyone missed. The drivers would not be forced to drive into a puddle at 160 mph. Some would, and of those, a few would spin and crash in an exciting explosion of noise and carbon fibre. Others would choose to slow down. So the puddles would, in fact, be a test of the driver's bravery. And isn't that why we watch?

Well, it was in the olden days, when the men who took part wanted to win at all costs because then they'd get more sex. They'd bash wheels and do four-wheel drifts, and as often as not they were still nursing a hangover from the night before. It was all very excellent.

Now, though, the sport is run by people who don't really think it's a sport at all. They think it's a science. And they don't want to run their aerodynamically honed, electronically measured instruments through a puddle any more than the boffins at Cern would want to study their Higgs boson in a children's ball pool.

For these people the cars are not cars at all. They are carefully considered probes into the world of advanced maths and the laws of physics. And the drivers? Robots, really, programmed to do as they're told. Sky Sports interviewed the five rookies who this year have joined what's laughably called 'the circus'. And they were like FIA puppets, saying exactly what they'd been programmed to say by someone in a branded shirt. Dead people would be more interesting.

I like to think that if I'd been one of them, sitting in my pit in Australia, and I'd been told that someone in the health-and-safety vehicle had abandoned the qualifying session, I'd have fired up my car and driven round the track in a roar of barely contained power-sliding fury to show that they were talking nonsense.

Occasionally you hear about a driver insisting a race should be stopped because of bad weather — Alain Prost in Monaco and Niki Lauda in Japan — but for the most part, and in private, they'd be happy to race even if it was snowing. Quite right too. It's the men in the monogrammed headphones. The geeks with the laptops. And the finger-wagging stewards. They're the killjoys who are turning F1 into a dreary blend of computer science, corporate public relations and cricket.

When we watch an F1 event, we crave the merest hint of humanity or passion or emotion. But instead it's Martin Whitmarsh's hair and those shudderingly awful branded shirts and all the lorries parked exactly in line. It's a televised obsessive-compulsive disorder.

I wish I ran a team. I'd turn up late and a bit drunk. I'd park my lorry at an angle and send out a car with a giant cock and balls painted on the side. I'd goose the drivers' girlfriends, over the radio, while they were racing, and if I won, I'd run up and down the pit lane making the loser sign at Christian Horner.

But that's the trouble. I wouldn't win. My cars would break down and explode and come last. And I'd be a laughing stock. And that, of course, brings me to the Toyota Corolla.

This has more in common with an F1 racer than any other car

on the market because it too was built without emotion or passion. It was built only to be logical and ordered, and as a result it has been humongously successful.

Ford shifted more than 15 million Model Ts. Volkswagen smashed that record with its original Beetle, which sold more than 21 million. But those are Zager & Evans compared with Toyota, which, to date, has sold about 40 million Corollas. This means there are more of them in the world than there are Canadians.

Except, of course, there aren't. Toyota makes Corollas that don't last for ever. After a period of time – let's say about eight years – they are likely to implode and their owners will have to buy a replacement.

I suppose I should point out that in Britain today the Corolla is actually sold as the Auris. No idea why. Seems to me like Apple changing its name to Pazizzle. But I was interested to find out what it might be like driving around in a car that was deliberately designed to be as uninteresting as possible.

It's so uninteresting that on the whole of the worldwide web there is not a single review of this car. Not one. You can read about Koenigseggs and Gilbern Invaders and the Peel P50. But not a single journalist has thought, Hmmm. I wonder what the world's bestselling car is like.

Well, let me tell you here and now that it is extremely uninteresting because you know when you turn the key, the starter motor will whirr and the 1.8-litre engine will fire up. You know that if you turn the wheel, the car will go round the corner, and that if you press the middle pedal, it will slow down.

It handles well. It rides well. It is about as economical as you could reasonably expect and . . . I'm struggling to stay awake here.

You think a Volkswagen Golf is reliable and predictable? Well, I laugh in your face. A Golf is an offbeat German arthouse film featuring laughing clowns and naked women fighting with deranged crows. Whereas an Auris is a glass of tap water.

However, there is one thing. The model I drove had a continuously variable transmission. Baffled? Let's see if I can help. Imagine the gears on the back wheel of your racing bicycle. There are five cogs, yes? Well, think of CVT as a cone. This means there's just one gear with an infinite number of ratios. Sounds great. It isn't.

Because when you put your foot down to accelerate away from a junction, the revs rise first and then your speed increases to match them. The noise is terrible. It sounds as though there's a wounded cow under the bonnet.

And on the Auris I drove – which was badged as a Corolla because I was in New Zealand and it's still 1952 there – a Sport button had been fitted. This meant the cone now had steps, as with a normal gearbox. Which rendered it not just unpleasant and noisy but useless too.

I was very grateful for this feature, though, because on a twelve-hour drive it gave me something to think about, something to hate. And that kept me alive. If I'd been driving a version with a normal manual or a traditional automatic, I'd have done what I do after ten minutes of the lights going out in an F1 race: fallen into a deep and dreamless sleep.

I'm not saying you shouldn't buy an Auris. That would be silly. But if you are a driving enthusiast, and you find yourself in such a thing, you will be enraged by the scientific approach. You'll desperately want it to do something – anything – out of the ordinary. Something excellent or mad or bad.

In the same way as when you watch an F1 race you hanker after the days when the drivers had oil on their faces, wore chinos and didn't sit in the bloody pits all day because it was a bit rainy.

24 March 2013

That puts paid to my theory on the ascent of manual

Aston Martin Vantage V12 roadster

Show a man a gun, and if he has anything at all in his underpants, he will start to play with it. He will hold it to his shoulder and look down the barrel. He will want to take the safety catch off, and if it's loaded, he will want to pull the trigger.

A gun is designed for one purpose: to kill things. We should find it abhorrent. We should shy away and cower. And yet we don't. Because beneath the cashmere outer layers and the frontal lobes and the ability to make a lovely supper, there is the root of our brain, the old bit. And that is consumed by two things: sex and violence.

It's the bit that draws us to the gun and makes us want to fire it. Because it knows that when we have an AK-47 in our hands, we are better at hunting and killing than someone who doesn't. And that makes us more attractive. Which means more sex.

I'm not making any of that up. I once made a television programme about the history of the gun and I was stumped for a conclusion. I simply couldn't explain why I liked guns and why, on the shoot, the all-male crew liked them, too. So I spoke to some brain experts and they told me.

It's because, deep down, we are all penises and teeth. It's fight, flight, eat and shag. And we know that given the choice of an unarmed George Clooney or Nicholas Witchell with an Uzi, every woman in the world is going to ignore the Hollywood superstar and tear off the royal correspondent's trousers.

But, of course, today you can't go about your daily business with a machinegun slung casually over your left shoulder. Which means that you need a substitute. Some say it's regular trips to the gym. Others say it's wit, or money, or a kitchen full of cookbooks. But, actually, it's a manual gearbox.

If you have an automatic, you are telling the world that you are too lazy to change gear yourself. Which means you'll be hopeless at killing antelopes. You're the sort of person who watches a lot of television and has warts.

And it's the same story with the new-fangled flappy-paddle gearboxes. In essence, these are the left and right keys on a laptop. They're switches. You are just asking a computer to do something, and half the time it will say no. Which means you are a slave to software. A journeyman. An epsilon.

A manual is different because you can't just sit there while the car drives along. You have to wrestle with it, tame it. You have to take charge, be the boss. Taking a car to the red line and then pulling back on a big metal lever is exactly the same as sprinting across the Serengeti and wrestling a wildebeest to the ground. A manual gearbox makes you a man.

It's a thing of joy too. Charging up to a corner, braking and downshifting with a smooth double declutch. This is poetry for the petrolhead. And unlike with flappy-paddle systems, you can change down when you feel like it. There are no fail-safe systems on hand to suggest that the shift into second will over-rev the engine. It knows that it's your engine and they are your valves and if you want to ping them through the bonnet, that's up to you. In a manual, you are master of your own destiny.

But there are some problems with this, as I discovered over a few days with the Aston Martin V12 Vantage roadster. On cold mornings the big metal lever was jolly chilly. So chilly, in fact, that you needed to think about wearing mittens before setting off. And in the fight-or-flight, hunter-killer world of male pride, you really want to be in a loincloth, not mittens.

There's more, I'm afraid. Because if you have a can of soft drink in either of the cupholders, it is quite tricky to engage second, fourth, or sixth. Then there's the bothersome business of traffic. After a while your clutch leg starts to ache. And I'm sorry, but when the traffic thins, a car with a manual gearbox simply isn't as fast as a car with flappy paddles.

Last year, on a deserted Romanian motorway and with the blessing of the local constabulary, I raced a manually equipped Aston Martin DBS against a paddle-shift Ferrari California. In terms of acceleration they were almost identical. But each time I changed up, I lost maybe five yards. I was quick. I was smooth. I was very manly. But I lost.

So while the manual gearbox may cause you to have more sex, it makes you a bit hurty, it's hard to use, it's not as efficient as flappy paddles and it feels surprisingly old-fashioned. As if you've switched from an iPad to a typewriter.

There are other things in the Vantage roadster that feel old-fashioned too. The buttons, for instance, are extremely small, and none of them does what you expect. And behind the wheel you do feel cramped.

Strangely, though, the Vantage roadster is new. Well, newish. Because what we have here is the company's venerable 6-litre V12 shoehorned into the V8 Vantage's drop-top body.

You can tell it isn't the normal V8 version because the bonnet is festooned with many air intakes and vents. Some say that fitting these to an Aston Martin is a bit like fitting a nose stud to the Duke of Edinburgh. But I like them. I think they give what is a very pretty car some edge. They hint at some savagery.

And there is savagery. It isn't as powerful as the new Vanquish, which has a differently tuned version of the same engine, but somehow it feels way faster. It feels so quick that I became concerned that the chassis wouldn't be able to cope. Sure, there have been a couple of modifications to the suspension and there's a restyled rear end designed to keep its backside more firmly glued to the road. But tinkering in the face of such grunt is a bit like trying to direct lava, using hay bales.

Yet, curiously, it's OK. The ride is compliant and the handling's fine. At the *Top Gear* test track I ended up going backwards only once. The rest of the time it was all fun and games and controllability. Until, as is normal in all Astons, the front tyres lost their bite and I was rewarded with yards of game-over juddering understeer.

It's a hard car to sum up, this, because – let's be honest, shall we? – an Audi R8 Spyder is a better machine in almost every way. But you're not interested. You're going to be happy to put up with the problems and the old-fashioned feel and the ice-cold gearknob because it's an Aston Martin and because it is one of the best-looking cars ever made.

That's fine. But I will say one thing. It costs £150,000. And for £39,300 less, you can have the V8 Vantage S roadster, which is also an Aston Martin and which looks pretty much exactly the same.

31 March 2013

Oh, how you'll giggle while strangling that polar bear

Ford Fiesta ST 1.6T EcoBoost

I went for tea at a London restaurant the other day and when we'd finished, I said I'd pick up the tab. Generous, I know. But that's the kind of guy I am. So, anyway. Go on. How much would you expect the bill to be? Four people. A pot of tea. And a round of cucumber sandwiches.

Nope. You're wrong. It was £78.42.

Now if it had been a nice round £80, I'd have summoned the manager and inserted the teapot in him. But because it was £78.42, it gave the impression that they hadn't simply said, 'Look at that rich bastard. Let's charge him eighty quid.' It looked like they'd done some workings out. And that the actual price in that part of London of two tea bags, a pint of tap water, eight slices of bread and a quarter of a cucumber really does add up to £78.42.

Apparently this is a well-known trick. When someone phones and asks you to quote for a job, you should never say, 'Oh, £500 should do it,' because that looks like you've simply plucked a number from the ether. Which is an open invitation for the customer to engage in that most barbaric of things: haggling. For the British, haggling is like talking in lifts. It's disgusting.

What you must do when asked for a quote is mumble to yourself as though you are doing some long multiplication on the back of an envelope and then, after twenty or so seconds, say £512.63.

That looks like it is the real cost of doing the job, and as a result, haggling is not possible. So instead of settling on £450, you get paid £512.63.

This brings me on to the retail habit of pricing everything at something and 99p. We know why they do this. Because £4.99 sounds less than £5. Plus it makes the customer feel all warm and well disposed towards the notion of a return visit if he walks out of the door clutching a penny change.

Yes. But am I the only person who thinks that rather than knocking a penny off, the shopkeeper has simply added 99p? Or is that just my tight-fisted Yorkshire genes?

Probably not. Because if there's one thing I despise, it's a bargain. If someone offers me some goods, or a service, that costs less than I was expecting, I automatically assume there's something wrong with it. Usually I'm right. Cheap cola tastes less good than expensive cola. A cheap vacuum cleaner will not do quite such a good job as an expensive vacuum cleaner. A cheap holiday will be rubbish. An expensive one will be great. Unless you're on a cruise, obviously.

However, in the world of cars, this is not necessarily so. An Aston Martin Vanquish, for instance, costs £189,995. This tells us that Aston has done its costings (yeah, right) and that this is how much the car costs to make. Plus a bit of profit added on.

It's the same story with the Ferrari 458 Italia. In basic form, this costs £178,491, because that's how much the metal, glass, plastic and carbon fibre cost to assemble. No, really. They do. Ferrari metal is more expensive than Ford metal because, er, it just is.

I could go on. The McLaren MP4-12C. The Mercedes SLS AMG. The Lamborghini Gallardo Superleggera. All cars of this type cost between £165,000 and £200,000. It's the price you pay for having something a bit different. Something a bit out of the ordinary.

So why, then, does an Audi R8 cost £91,575? Why is a Porsche 911 £71,449? Why is a Bentley Continental V8 GT £123,850? These are exotic cars, too, but they are half price.

Does that mean they are rubbish? Virgin Cola cars? Well, that's the thing. I don't think so. If you reduce a Porsche 911 and a Ferrari 458 to their component parts, you'd struggle to see why one costs more than twice as much as the other.

The truth is, the 911 is a bargain. So's the Bentley and the R8. And that brings me on a wave of beige-tinted common sense to the door of the Ford Fiesta ST.

Cracking car, the Fiesta. It's good-looking, spacious, safe, economical and, if you avoid the base models, nippy as well. But I've always felt that the chassis was so good, it could easily handle a bit more oomph. Which is where the ST comes in.

I do not know why Ford continues to name its faster cars after lady towels, and I'm not sure either why it says the engine under the bonnet is from the EcoBoost range. EcoBoost gives the impression that it runs on armpit hair and produces about as much power as Luxembourg.

That ain't so. Because, actually, it's a turbocharged 1.6-litre that produces 180 bhp. This means you go from 0 to 62 mph in a polar bear-strangling 6.9 seconds and onwards to a top speed of 139 mph. That's quick.

It's probable the standard suspension setup could handle the extra grunt, but to be on the safe side, and to make it look sportier, the Lady Towel is 15 mm nearer to the ground than standard Fiestas and sits on fat 17-inch wheels. There's a roof spoiler, too, that does absolutely nothing at all. But it looks nice.

Does it all work? Yes, it does. And some. Like the Ford Focus ST, it all feels loose and light and, if you reduce the traction-control system, or turn it off altogether, a bit wayward. You have understeer and liftoff oversteer and patches of cling-on-for-dear-life grip, and the upshot is: it's bloody good fun.

You can feel the electronic limited-slip diff doing its best to keep things orderly and neat, but you get the impression it's like a not very good teacher, trying to organize a class full of unruly seven-year-olds.

I like that. And I like the noise as well. Ford has pinched an idea first seen on the £336,000 Lexus LFA and has fitted a 'sound symposer', a tube that feeds the induction roar directly into the cabin. It makes you feel like you're actually sitting in the engine bay. The interior is good, too. You get enormous body-hugging Recaro seats, and even in the base model lots of toys as well.

I liked this car. A lot. It has all the qualities I look for in a hot hatchback. There's everyday practicality. There's comfort. There's the sense that each of its body panels will cost no more to repair than it would on a cooking model. And yet despite all this Terry and June down-to-earthness, there is also lots and lots of juicy speed and joie de vivre.

And here's the best bit. It costs just £16,995, which is £2,000 less than Peugeot or Renault charge for their latest hot hatches. In fact, £16,995 is as near as dammit what Alfa Romeo charges for some versions of its woeful two-cylinder TwinAir MiTo. And you don't even get ripped off with extras. I visited Ford's online configurator and once you've selected the model, pretty much the only choice you have is the colour.

The Ford Fiesta Lady Towel, then. It manages to be something that's quite rare these days: cheap and cheerful.

7 April 2013

Another bad dream in a caravan of horrors

Honda CR-V 2.2 I-D TEC EX

In 1990 there was the NSX, a mid-engined two-seater that waded into battle with Lamborghini and Ferrari, sporting a small V6 engine. While it couldn't sting like a bee, it could dance the dance of even the most zippy butterfly. And the induction roar was a noise that stirred the soul.

It's been said many times that Japanese cars have the character of a washing machine. And it's true. Japanese car makers always designed their cars to keep everyone happy: retired postmasters in Swansea, east African taxi drivers, soccer moms in Houston . . . everyone.

But, as modern politicians know, if you try to keep everyone happy, you end up looking a bit boring. If you make a stand, stamp some character into the mix, you get a big funeral that shuts half of London. That's what the NSX did. It was the Boris Johnson of supercars. Only a bit lighter.

There was also the CRX, a small 1.5-litre coupé that served no purpose at all. The rear seat was a birdbath. It had the grunt of a mouse. And the comfort of an acacia tree. But because it was so unusual I bought one the moment it went on sale.

I've always harboured a soft spot for the Prelude, too. Most coupés at this time were based on normal saloons. The Capri was a Cortina with a comedy nose. The Scirocco was a Golf. The Calibra was a Cavalier, and so on. But the Prelude was a Prelude. Honda didn't try to save money by sharing parts. It made it to be as good as it could be, and then clothed it in a very pretty body and gave it pop-up headlamps. That was the Honda way. Different. Better.

But, despite the unusualness, the company never lost sight of its origins. It started out making piston rings for Toyota and knew that quality was the beginning, the middle and the end of everything.

Honda knew that its own cars had to be just as reliable. And they were. The variable valve timing system is a complex blend of mechanical and electronic engineering. So you'd have to expect some failures. It would be only natural. But after the company had made 15 million units, guess how many warranty claims there'd been. Nope. You're wrong. The correct answer is zero.

Hondas, then, were iPhones that didn't jam. They were style icons that worked. They were the embodiment of what Charles Babbage was on about – the unerring certainty of machinery. Or, to put it another way, Alfa Romeos that started.

Remember the Honda Civic Type R? What a machine that was. Or the Integra. Or the original Insight. Oh, and I've just remembered the S2000, which was a beefed-up Mazda MX-5. Slightly bigger, and slightly more butch to behold, this two-seater soft-top had an engine that screamed up to 9000 rpm and would sit there all day. Every day.

It wasn't just cars either. There were motorcycles and generators and marine engines and lawnmowers and water pumps and mopeds that were so important to the world that they were immortalized in a Beach Boys song. There were leaf blowers and quad bikes and hydrogen fuel cells and Formula One powerplants that won the world constructors' championship six times on the trot.

All of these achievements were immortalized in one of the greatest television commercials of all time. Set to Andy Williams singing 'The Impossible Dream', it showed a balding man charging across New Zealand in a range of everything Honda had made over the years. You watched it and you wanted to have a go on every single thing.

But what Honda would you want to drive today? A Jazz?

A Civic? An Accord? A hybrid? I suspect that, in the best traditions of multiple-choice questions, the correct answer is E – none of the above.

The range of cars sold by Honda in Britain is about as dreary as a Victorian tea set. The reliability is still there, but the flair, the innovation, the genius? All gone. Honda's demise is like the Rolling Stones deciding to start recording hymns. Or the hotel chain Raffles deciding that Formule 1-style bathroom cubicles are quite good enough.

I've just spent a week in the Honda CR-V diesel EX, which is a big and quite expensive bucket of nothing at all. Honda tells us the latest model is rammed with significant improvements but then struggles a bit when they are listed. It has, for instance, daytime running lights. Just like every other car on the market.

It has a powered tailgate, which means you have to stand in the rain while electric motors take five seconds to do a job you could have done in one. It has a 12 per cent reduction in CO_2 emissions, which is irrelevant unless you are a bear. And the four-wheel-drive system is now electronic rather than hydraulic. Which is another way of saying 'worse'.

Under the bonnet of my test car was a diesel engine. Honda was one of the last big car manufacturers to make such a thing, and it is nowhere near as good as the ones made by everyone else. It's noisy, rough and, compared with, say, BMW's effort, way down on power.

Inside the CR-V are no features you cannot find in cars that cost less and a few that are annoying. The satnav screen is surrounded by buttons so small, you can't see what they all do. And there's another screen that tells you a raft of stuff you don't care about. Such as how many hours you've driven since you accidentally set the trip meter. It'd be more interesting to know when high water was due on the Solomon Islands.

The rear seats, apparently, are 38 mm lower than in the previous model. I mention this simply because I'm running out of things to say. As the miles droned by, I began to wonder who on earth would spend more than £31,000 on the model I was driving. It has no more seats than a Vauxhall Astra, and if you really need part-time four-wheel drive and a tall boot, Ford, Hyundai, Kia and many others can sell you something similar and better for less.

I suspect the answer is caravanists. People who enjoy this type of holiday tend to be the sort who vote UKIP and therefore like the fact that the CR-V is made in Swindon by British people, not by a sausage jockey or a garlic-munching surrender dog.

They also like the promise of great reliability and the sense that four-wheel drive is on hand to help out should the site be on a bit of a slope. Plus, of course, the boot is capable of taking all the paraphernalia they need for a summer holiday in Britain: umbrellas, windbreaks, cagoules, wellies and so on.

I still don't get it, though. Buying a car because it suits your requirements for two weeks in the summer surely is like wearing ski boots all year round because you go to Verbier every February.

There is no reason for buying this type of car. And even if you can think of one, there is no reason for choosing the Honda. The Land Rover Freelander is much better. So's the Nissan X-Trail.

Honda needs to buck up its ideas. I realize that there will be a new version of the NSX, and I have high hopes for that. But it needs a bigger range of other stuff, too. It needs to get different again. It needs to get better. Because until it does, there's no reason for you or me to get out our chequebook.

28 April 2013

Ooh, you make me go weak at the knees . . . and the hips and the spine

Jaguar F-Type S

Legend has it that when Frank Sinatra first clapped eyes on the then new E-type Jaguar at the New York motor show, he said, 'I want that car. And I want it now.' The E-type had that kind of effect. Pedants were saying that the actual cars would not achieve the claimed top speed of 150 mph, and back then everyone knew that the 'Made in England' tag was another way of saying you'd arrive everywhere in a cloud of oily steam.

But the E-type was so bite-the-back-of-your-hand pretty that grown men lost their capacity for reason. They simply didn't care how much it cost or how uncomfortable it might be, or even if it did only six miles to the gallon . . . of myrrh. They had to have one.

And now history has repeated itself with the F-type. I'd seen early spy shots of this car, and I'd even spotted camouflaged mules being tested around the Cotswolds. And I'd got it into my head that it would be nothing more than an XK that had been boiled for a bit too long in the washing machine. I was very, very wrong.

It is a spectacularly good-looking car. The bulge of the wheelarches, the length of the bonnet, the flatness of the boot lid, the angle of the rear window – every single little thing is completely flawless. There must have been a temptation to fit a retro nod to the E-type. But apart from the slender rear-light clusters, there's not one. So it's perfect and it's modern, and as I stood there gawping, I had a Frank Sinatra moment. I have to have this car. And I have to have it now. A few days later, however, when the test drive was over, I was starting to have some doubts . . .

When I was growing up, Jags were driven by people who had sheepskin car coats and they were very soft to sit in. They were also very quiet. This is because they'd usually broken down. All over the world they were known for these two qualities: quietness and comfort.

The F-type offers neither of those things. I tested the mid-range, V6-engined S version and it is a bite-sized parcel of barely controlled fury; the angriest, most brutal and loudest car I've encountered in some time. Then I discovered a button on the dash that makes the exhausts louder still.

Even on the overrun, as you cruise up to a junction, a clever system of flaps causes the back boxes to spit and crackle and bang. It sounds like distant artillery fire. Then you pull out from the junction and press the accelerator and, whoa, now it feels as if you have actually become an artillery shell.

God knows what the 5-litre V8 is like because the firepower from the 3-litre V6 is plenty savage enough. One minute you're balancing the throttle against the interruption of the traction control system, and the next you're in Arbroath, doing a million, there's blood streaming from your shattered eardrums and your hair's on fire. Exciting doesn't even begin to cover the thrill of taking this car by the scruff of its neck and giving it a damn good spanking.

The steering, the brakes, the grip. No idea what they're like. Concentrating on details in a car such as this is like trying to solve a Rubik's cube while being machinegunned.

And yet, behind the veneer of bloodcurdling violence, the cockpit is all very civilized. There are cupholders, for a kickoff, and there's a satnav system and blue teeth. And while there are flappy paddles behind the steering wheel, they are connected to a proper automatic gearbox – not a scrappily converted manual unit that is sold to misguided enthusiasts as a must-have Formula One-style racing accessory, but is, in fact, a fuel-saving sop to petty, polar-bear-minded Eurocrats.

And, what's more, the interior styling is every bit as successful as the exterior. Maybe the graphics on the dials are a bit 1977, but I did like the bronze-coloured controls and I loved the Range Rover-style facility for changing the colour of the interior lighting. Very Reykjavik vodka bar, that.

There were one or two little niggles, though. The warning lights in the door mirrors are designed to illuminate whenever a vehicle is in your blind spot. But, in reality, they come on to alert you to pretty much everything: trees, crash barriers, signposts you've just passed, churches, apples, the lot.

And accessing some of the day-to-day functions does mean you have to go through quite a few layers of computer submenus. You can press one button that makes the car more sporty – as if such a thing were possible – but if you want to keep the steering and gearbox in comfy mode, you need to be Bill Gates, really.

I could, and would, live with this, and the epilepsy-inducing door mirrors. I could even live with the price, which is extremely high. But there are two things that cause me problems.

First of all, my test car had no boot. Well, it had a boot, but it was filled, completely, with a space-saver spare wheel. There was not even enough room left over for my briefcase. Many of the road tests of the F-type so far have not mentioned this, as though it doesn't matter. But I think that in the real world you would occasionally wish to transport something other than just a passenger. And you can't.

And then there's the biggest problem of them all. Jaguars have been getting firmer and firmer in recent years. The engineers seem obsessed with setting the cars up to be good at the Nürburgring and nowhere else. But with the F-type they have gone completely bonkers.

It's just about bearable at 70 mph on a motorway, but around town, on urban potholes, it is a joke. I thought my Mercedes CLK Black was firm and stiff, but the F-type is in another league. I know of no other car that deals with bumps so badly.

When you see a manhole cover approaching, you try to

prepare yourself for the jolt in the way that a boxer prepares when a punch is coming. But you can't. And as you feel your skeleton creaking and snapping, you can't help wincing. And there's no need for it.

When the F-type went back to Jaguar, I took the trouble of borrowing a Porsche Boxster for a couple of days. This is also a two-seat soft-top and no one has said that it drives badly in any way. Quite the reverse. Its grip and handling have been much praised. And yet despite this it can go over speed bumps, man-hole covers and bits of grit without causing the driver to suffer pain.

This, however, is irrelevant. Because comparing the F-type and the Boxster is like comparing the latest supermodel with Angela Merkel. I could sit here all day and tell you that the super-model has a screeching, hideous accent, a virulent STD and a propensity for extreme violence, and that Mrs Merkel is very clever and comfy, but I know which you'd prefer to take to your bed.

So it goes with the Jag and the Porsche. And I get that. I respect the Porsche immensely. But I simply couldn't walk away from the Jag's looks, or the noises it makes, or the absolute right-ness of its cockpit. I'd buy the Jag. And I suspect you would too.

It's just a terrible shame that the only real winners will be the nation's osteopaths.

12 May 2013

Mirror, signal, skedaddle – Mr Bump's been turbocharged

Peugeot 208 GTi

Many, many years ago, during an economics lesson at school, I spotted a picture in the *Melody Maker*. It was an aerial shot of Emerson Lake & Palmer's tour trucks thundering down some remote motorway. One had the word 'Emerson' picked out on the roof. One had 'Lake'. And one had 'Palmer'.

It was pretty cool. And so, while moving the *Top Gear Live* tour from a gig in Amsterdam to the next night's venue in Antwerp, we saw no reason why we shouldn't do something similar.

Actually, to be fair, the promoter did see a reason. 'There are only three of you,' he said. 'We don't need three coaches. We could go in a van.' But we were most insistent. One would have 'Top' on the side. One would have 'Gear'. And one would have 'Live'.

And so it came to pass. And we were jolly pleased with ourselves as we trundled down the motorway. They were great coaches, with tea and coffee-making facilities and light snacks provided free of charge. This was rock-star living.

It was only when we arrived in Antwerp that we realized we'd driven the entire way with our buses in the wrong order. 'Live Gear Top', said the message. And that was just the start of the problem.

Antwerp is a small city with many narrow streets and cobbles so bumpy that even the prostitutes have to walk around in flat shoes. It was designed for pedestrians in sandals, or maybe the odd horse. Not a convoy of three fifty-seat coaches, which soon became as stuck as Winnie-the-Pooh in Rabbit's hole. This caused a bit of a traffic jam.

Within a minute, one of the inconvenienced motorists expressed his displeasure by blowing his horn. And this unleashed a torrent as all the others joined in. The noise was incredible.

The drivers said we'd have to get out and finish the journey on foot, but for two reasons we didn't want to do that. One, we'd be lynched, and two, if we continued to sit on board, everyone would assume they were being held up by a party of Japanese tourists.

'Unlikely,' said the promoter. 'English is not the mother tongue in Belgium but they speak it well enough to work out the anagram of "Live Gear Top".' And so, with heavy hearts, James Emerson, Richard Lake and Jeremy Palmer emerged into the cacophony.

I've never heard anything so pointless. What did these people hope to achieve by sitting there, leaning on their hooters? Did they imagine that the coaches would simply disappear, crumbling to dust in the barrage of decibels? Or did they think that the noise would cause the offending drivers to think more clearly about how a solution might be found?

Fat chance of that. The car horn is like the screech of a frightened vervet monkey. It's a warning. A precursor to impending doom. One blast is enough to make anyone look around. When you have a hundred horns, though, sounding continuously, it causes the brain to go into a sort of panicked mush.

You can't think straight. You don't know whether to go left or right, whether to run or stand your ground. So you just wait there, ghostly white in the headlight beams, looking as if every ounce of gorm you've ever had has suddenly migrated from your face to your feet. For all I know, the buses are still there. Because I didn't just walk to the hotel. I ran.

I do not understand why the modern car has a hooter. Because, as we've established, it is useless at clearing an obstacle. And if you do have to use it to sound an alert to another driver, it means you weren't anticipating the road ahead and what might happen next. The horn, then, is an admission on your part that you're a crap driver.

I blew mine the other day. I was making my way out of a petrol station and had spotted the hatchback pulling out of a parking space just ahead. Of course the driver had seen me. Cars are pretty big, and mine was a Ferrari so it was pretty loud and pretty red as well. You can't sneak up on things in a Ferrari, so with that in mind, I kept a watchful eye on the hatchback, and kept going.

Sadly, though, it did too, and eventually I had no choice. I had to pip. What makes this incident even more harrowing is that the hatchback in question was a Peugeot. I should have known. I warned my daughter when she was learning to drive that she could forget all about mirror, signal, manoeuvre. The most important thing on the road is to be aware that Peugeots never, ever, do what you're expecting them to do.

Just because they're in the left lane and indicating left, that doesn't mean they are actually going to turn left. Last summer, on these pages, I said I'd seen two accidents in one week. Both featured Peugeots. And still it goes on. You look, when you're out and about, and I can pretty much guarantee that the next car you see upside down in a hedge or parked in a florist's window display will bear the prancing lion on its nose.

I don't know why the police bother with measuring tapes and photo evidence. In any two-car smash, it's dead easy to work out who was to blame: the chap in the Peugeot.

I was driving one last week. It was the new 208 GTi, and I must say that, like many other cars in the range, it's rather good. Peugeot is at pains to say that it's not designed to be a word-for-word replica of the original 205 GTI. On the upside, this means you don't get incredibly heavy steering and woeful unreliability. On the downside, you also don't get riotously exuberant at-the-limit handling.

What you do get is the same turbocharged 1.6-litre engine as Mini uses in the Cooper S, four seats, a usable boot, a steering wheel the size of a vicar's saucer and the sort of character that causes you to drive along as if your hair's on fire.

It's not perfect. The gear change feels loose, and when you hook a tyre into the potholed apex of a bend, you can feel it banging and scrabbling through the steering wheel. But this doesn't really put you off. In some ways it's nice to feel that you're part of the action and that the gearbox is rather more than a switch. It seemed a very human car. It's also fairly well priced and well equipped. The satnav on my test car was very good, and the glass roof was a joy too.

Problems? Well, the seat gave me a bit of backache after an hour. But at my age a Vietnamese massage does that as well, so I can't grumble too loudly. Worse was the 208 GTi's tendency to settle to a 110 mph cruise. The Mini does this as well. It must be a characteristic of that engine and its acoustic qualities. But whatever; on a motorway you have to keep a constant watch on the speedo.

Overall I did like this car, but when all is said and done, the Ford Fiesta ST is better. However, the Peugeot does come with one feature that the Ford cannot match. When you are driving it, everyone gives you a wide berth.

18 May 2013

Not now, Cato – keep turning the egg whisk while I push

MG6 Magnette 1.9 DTi-Tech

According to the promotional material, the MG6 was conceived, designed, engineered and built in Britain. It's a great British name, and a great British car. A slice of Jerusalem among the dark, satanic mills of Germanic nonsense. A UKIP pin-up girl with windscreen wipers. Brown beer with a tax disc.

This is a car for people who grew up dreaming of driving a sporty B but who are now to be found at home, in their wing-backs, flicking through the channels and muttering about how there's nothing to watch on television these days. It's a car that takes them back to their youth. A car that reminds them that Britain was, and still is, the greatest country on earth. It's the Spitfire, the hovercraft and Nelson. It's Churchill. It's Elgar. It's Wordsworth and Shakespeare and Brunel.

Except it isn't. Because it turns out that the MG6 is actually built in China by SAIC Motor – the company that now owns MG Rover (though not the name Rover). It is then shipped over to Birmingham, where a small team inserts the engine. Claiming that this car is British is like claiming that an Airfix model was built in your front room. It wasn't. It was merely assembled there.

Yes, the car we buy here was styled in Britain, and some of the chassis work was done here too, with – whisper it – German components. But in essence, while this car may be pretending to be Kenneth More, it's as Chinese as a chopstick.

This is not necessarily a bad thing. Because, typically, what happens when a Far Eastern country starts exporting cars to the West is you get a substandard product with a price that's so low, no one really cares that it's held together with wallpaper paste and has an engine that sounds as if it's running on gravel.

Then, in an alarmingly short time, the cars are suddenly as good as their European rivals. Toyota went from the Toyopet to the Lexus LFA in about five weeks. One minute Kia was making the woefully awful Rio, and the next it had the bloody good Cee'd.

The company behind the MG6 started with an advantage. It didn't begin with a jungle clearing and a workforce that thought it was making dragons: it began with the underpinnings of the Rover 75 and employed people who navigated to work with their iPhone 5. In theory, then, the MG6 would drive quite well and come with a DFS everything-must-go price tag. Which would make it a tempting proposition even without the nonsensical *Last Night of the Proms*-style marketing.

So why, I wondered, was it so hard to book a test drive? Excuses were always made. Other phones were ringing. Other priorities had to be addressed.

Well, last week I sneaked behind the wheel for a short drive, and very quickly the reason became obvious. This car is not bad at all. It's hysterically terrible.

Let's start with the ignition key. You know those cheap electronic toys that you buy children from the gift shop on a cross-Channel ferry? Well, this has the quality of the wrapping in which they are sold. And naturally it didn't work.

I learnt this outside the police station in Ladbroke Grove in west London. The traffic lights went green and I set off. But I didn't because the car stalled and it would not restart. So I pushed it to the side of the road where, after several attempts, the diesel engine finally clattered into life.

At the next set of lights exactly the same thing happened again. And so at the third set I made sure it didn't stall by summoning 3,000 revs and setting off nice and gently. This made the whole of Notting Hill smell of frazzled clutch.

There are some other interesting faults as well. This is not a small car. It's a little larger than a Ford Focus and a little smaller than a Mondeo. But inside it has the headroom of a coffin. Speaking of which, it didn't do especially well in its Euro NCAP safety tests. The airbag didn't inflate sufficiently well to stop the dummy driver's head hitting the steering wheel, and while the feet and neck were well looked after, protection for the thighs and genitals was only 'marginal'. I make no observation about that. Yet. Of course, as it's a Chinese car that's assembled in Longbridge, you would not expect much in the way of quality. And it doesn't disappoint . . .

It's a widely held belief that mass-produced plastic was developed around the turn of last century. Well, the dashboard on the MG6 appears to be fabricated from a plastic that pre-dates that. I think it may follow a recipe laid down in the Middle Ages, when villagers would use cattle horns to make rudimentary windows.

Naturally there are many sharp edges. There's one in particular on the steering wheel that could probably give you an elegant paper cut on that sensitive bit of webbing between your index finger and thumb.

Then there's the kung-fu cupholder. It's not damped, as it would be in a normal European car, so when you push the button your drink leaps out onto your passenger's leg like Cato from the Pink Panther films. And it is a struggle to get any can I've ever seen to fit in it.

I shall talk now about the steering. It's electric. But only literally. It feels as though the steering wheel is connected to an egg whisk of some kind. Spin it fast enough and the blades turn, causing a vat of creamy milk to start thickening. After this happens it begins to revolve v-e-r-y s-l-o-w-l-y and that action produces a centrifugal force that turns the front wheels. It's a neat idea but I'm not sure it works very well.

As a boy, I used to look at my dad driving and wonder how he knew how much to turn the wheel when going round a corner. Alarmingly, in the MG6 you don't.

Last weekend in Scotland I encountered many members of the MG Owners' Club, driving from breakdown to breakdown with dirty fingernails and big grins on their faces. They had their roofs down, despite the cold, and it all looked very hearty and rorty and James May-ish.

The MG6 offers an experience that is nothing like that. It may say MG on the rump but it is as far removed from its predecessors as you are from an amoeba. It's a carrier bag with a Coco Chanel badge. And I think that's rotten.

The whole car's rotten, really, and here's the clincher. It's not that cheap. The Magnette model I drove is £21,195. And for that you can have a normal car that doesn't lacerate your fingers, stall, refuse to start, bash your head in every time you go over a bump and ruin your gentleman sausage if you have a crash.

In the whole of April the new MG operation sold thirteen cars throughout the whole of the UK. I'm surprised it was that many.

26 May 2013

No grid girls, no red trousers – it's formula school run

Mazda CX-5 2WD SE-L

Monaco bills itself as a glittering jewel in the south of France. But in reality it's a mostly overcast collection of people who choose to live far from their friends and family, in a 1960s council tower block, under the control of an extremely weird royal family, among a squadron of arms dealers and prostitutes.

And all so they can save a pound in tax. This makes it the world's largest open prison for lunatics. And then, once a year, the grand prix circus rolls into town – and it all gets worse.

I stayed on a giant boat on what's called the T-jetty. That's pole position for the gin palaces, and you probably think that this would be heaven. Hot and cold running waitresses dropping tasty morsels into your mouth whenever you are breathing in the right direction. And some Formula One whizz-kid and his almost completely naked girlfriend waiting next door for you to nip round and chew the fat. That's the message you get from the television pictures.

The reality is somewhat different. You hear of a party on a neighbouring boat, so you think you'll pop by for a drink. Alas, every single person in Monaco has heard of the party also and has a similar plan. So, to prevent them all from getting on board, the boat's captain has hired a French security team that stands about with curly-wurly earpieces making sure nobody gets on board at all.

You watch the men pleading and explaining that they are personal friends with the boat's owner, but this is no good because he's not at his own party. That's the key to being a proper billionaire. Throw a party and then have dinner somewhere else.

Then you have the women, who are selected for admission purely, it seemed, on the basis of how naked they are. Amazingly the party does somehow happen, although everyone on board spends their entire evening making sure that they are talking to the most important person in the room. An example. I thought I'd introduce myself to Martin Whitmarsh, McLaren's boss. But he was chatting to an Indian chap who was more important than me, so I was ignored. In fact, I was ignored so spectacularly by everyone that I ended up talking to the cabin boy for most of the night.

The next day you wake with a sore head. And to make everything more terrible, someone has pushed a microphone into a beehive and is blasting the resulting sound through the Grateful Dead's speaker system across the whole principality. So you stagger about looking for Nurofen, eventually sourcing something appropriate from someone who'd crashed on a sofa. She was a nice girl. Apart from her Adam's apple.

You then think it would be nice to go over to the paddock. But between your boat and the vast F1 motor homes is a 15-yard strip of water. And to cross it in a knackered dinghy with a Kenwood mixer on the back costs €20. That works out at more than £1 a yard. I think it would have been cheaper to use a private jet.

And it's pointless anyway, because to collect the passes that have been supplied by Bernie Ecclestone, you need to go through a security barrier for which you need your pass. '*Non*,' said the security guard.

This is one of the most important things about 'doing' Monaco for the grand prix. Yes, you need to spend all day smoking cigars the size of telegraph poles and wearing red trousers. That's important, of course. But mostly you must be festooned with so many passes that you are in danger of slipping a disc. A lot of passes shows a lot of connections. None means you are paying another €20 to go back to your boat.

Then the race starts. And even though the boat on which I was staying was about the height of Nelson's Column, and even though I climbed right up to the radar mast, all I could see was the top of the cars' air intakes, momentarily, as they sped past the swimming pool.

So I went into the cabin to watch it on TV, which was fine except I couldn't hear what Martin Brundle was saying because of the din outside. It's strange. Most sports are perfectly watchable without someone explaining what's going on. But with motor racing you really do need Mr Brundle to tell you why no one is attempting to overtake the car in front. Or else it just looks like twenty-two thin young men driving around a town.

I'll let you into a secret. Not one of the people on any of those boats saw the race. Nobody in the council blocks did either. In fact, the only people who could see more than a few feet of track were the real fans who'd arrived by train that morning in their branded Vodafone shirts and climbed the hill beneath the palace. They may have found it wasn't worth the effort.

And increasingly that's what I'm starting to think about F1. I love the idea of watching men race cars. But more and more I sense that, really, F1 is now merely televised science. It's just earnest chaps staring at laptops. And then lodging protests against one another for the tiniest of things. And in Monaco, which is supposed to be the highlight of the season, it's simply science in a big, daft wedding cake. I'd prefer to see a street race in Wakefield.

And there's more. There was a time when we were told that F1 was the launch pad for new technology and new ideas that one day would filter down into our road cars. But I suspect it doesn't even do that any more.

Which brings me to the Mazda CX-5. There are plenty of cars such as this on the market today. They're called crossovers or soft-roaders and they are very popular with school-run mums and caravanners. And I struggle to think of a single thing they have in common with F1 racers.

In Britain the bestseller of the breed is the Nissan Kumquat – it's not actually called that and I can't be bothered to look up its real name. But the only reason it's the biggest seller here is . . . that it's built here.

On paper the best is the Mazda. It's cheaper to buy, cheaper to insure, cheaper to fuel and cheaper to tax than most of its main rivals. It's faster and more powerful than lots of them, too. On paper it's a no-brainer. The winner.

And it's not bad on the road either. The ride is very good. The steering wheel is connected to the front wheels, and when you change up, the 2.2-litre diesel engine becomes a little more quiet. This, I suppose, is its chief drawback. It's a little boring. Actually, scratch that. It's catastrophically boring. It's Jane Austen with cruise control.

There's no pomp at all. There's no kinetic energy recovery system. No paddle-operated gearbox. No carbon fibre. No aero. It's a thing for the real world. The Monaco Grand Prix, on the other hand, really, really isn't.

2 June 2013

Where does Farmer Giles eat his pork pie?

Range Rover Sport SDV6 Autobiography

In the olden days, when it was possible to make a few shillings from cultivating the land, farmers could afford two cars. They had a Land Rover, on which they could lean while gnawing on a pork pie. And perhaps a Humber for nights out at the Berni Inn.

But then consumers got it into their heads that the correct price for a pint of milk was about 6p less than it costs to make. They also reckoned that bread should be 1p a loaf, and meat should be pretty much free. This meant farmers couldn't afford two cars any more.

Happily, Land Rover came to the rescue in 1970 with something called the Range Rover. It was quite brilliant. The world's first dual-purpose vehicle. For just £2,000 you had a car that would bumble about quite happily with some pigs in the back, and then, after work, you could simply hose down the interior and use it as a comfortable limousine for a trip with Mrs Farmer to the theatre.

The last Range Rover Sport was also a dual-purpose vehicle. It worked in Wilmslow just as well as it worked in Alderley Edge. It suited both footballers and their wives. You could use it to distribute pharmaceuticals during the day, and then in the evening it was a stable gun platform should a drive-by shooting be necessary.

I know lots of people who have what I call the proper Range Rover. In fact, I'm struggling to think of any friend in the country who doesn't have one. But I know only one chap who has a Sport. And there's no other way of putting this: he's called Gary. A point made plain by his registration plate.

The main problem, however, with the Range Rover Sport was that it wasn't a Range Rover. Underneath, it was actually based on the Discovery, which meant it came with a complex double chassis system. That made it heavy. And that meant it wasn't a Sport either.

Then there was the problem of its tailgate. In a proper Range Rover it splits, and you can use the bottom half as a seat when you are at a point-to-point. The Sport didn't have this feature. It just had a normal hatchback, like a Volkswagen Golf.

I was never a fan. I thought it was more of a marketing exercise than a genuine piece of engineering. And I harboured similar worries about the new model, which has just gone on sale. The problem is simple. For forty-three years Land Rover has been demonstrating that you can make a car that works extremely well off road and is still comfortable and quiet and refined on the road. But if you try to make it into a triple-purpose car by attempting to put some sportiness into the mix as well, you're going to come a cropper.

Think of it as a stout brogue. You can use such a thing on a ruddy-faced country walk. And you can use it while window shopping in St James's. But you cannot use it in a 100-metre race, unless you want to lose. And that's what the Range Rover Sport is attempting to be: a brogue that works on the moors and in central London . . . and on a squash court.

A sports car must have direct, quick steering. But if you do that with a Range Rover, you will find the steering wheel bucks and writhes about on rough ground. A sports car must have firm suspension too, but that's precisely what you don't want on a ploughed field. Or on the M40, actually.

I therefore approached the new Range Rover Sport with a sense of dread. But I emerged a bit astounded because somehow Land Rover's engineers seem to have pulled off the impossible.

It's not sporty. Let's be very clear about that. The throttle response is too slow, and the engine in my test car was too dieselly, and the steering, though quicker than I was expecting, is not as quick as it is on, say, a Ferrari F12. But it does have a sporty feel, which is quite good.

In Dynamic mode the ride comfort is seriously compromised, but I have to say, for a big car, you really can hustle it very, very hard. How hard? Well, through the Craner Curves at Donington, how does 100 mph sound? Sure, you could probably get a proper Range Rover to achieve a similar speed, but it would be extremely scary.

Later I went to have a look round the charitable institution I laughably call a farm, and here, I'll be honest, it felt pretty much identical to its proper brother. It had the same push-button system that lets you tell the car what sort of tricky terrain lies ahead, so that it can work out which differential should be locked and what range the automatic gearbox should select.

Then, afterwards, it was back to London, where the sporting brogue became as comfortable and as quiet as your favourite armchair. Some of the fixtures and fittings are not quite as satisfying as they are on its proper brother, but the architecture is great: the high centre console put me in mind of a Porsche 928, and there's no getting away from the fact that there are many toys to play with. Possibly because I had an Autobiography-spec car.

DAB radio was one of those toys, and I'm sorry but it's about time people stopped jumping up and down with excitement about the quality of the sound, because most of the time there isn't any. When the signal is a bit weak, normal radio goes hissy for a moment or two. But when the digital signal is a bit weak, you get silence. For mile after mile after mile. It may work in your kitchen, but in a car you would be better off with a record player.

Still, because there was no radio, I did notice the fuel gauge, which, after many miles, was still resolutely stuck on full. And this really is the ace up the new Sport's sleeve. You see, underneath, this is not a Discovery. It shares much of its basic architecture with the new Range Rover, and that makes it light. And that in turn means massively improved fuel consumption. I swear my car wasn't using any at all.

It's where the sportiness comes from too, and the sometimes vivid acceleration. Even the six-cylinder diesel can do 0 to 60 in around eight seconds. The supercharged V8 will do 0 to 60 in around five seconds.

This, then, is a massive improvement on the old car. It is a Range Rover, it does have a sporty feel, it does work off road and it is comfortable and well equipped. But it doesn't have a split, folding tailgate. That's why my eye is still on its more expensive bigger brother.

It seems unfair now to call it the proper model. Because the Sport's proper too. Gary has already ordered one.

9 June 2013

They only make one car. But it's a nice colour

Porsche Cayman S with PDK

Porsche's biggest problem is that it doesn't seem to employ any stylists. The 911 was created after someone accidentally sat on a clay model of the Volkswagen Beetle, and every single version that has come along since has looked exactly the same.

Enthusiasts of the breed point at new door handles and head-lamp clusters and say these subtle changes alter the whole appearance of the car. But that's like saying Tom Cruise's eye-brow wax makes him look completely different. It just doesn't. It makes him look like Tom Cruise with new eyebrows.

My colleagues on *Top Gear* always refer to the various 911 incar-nations by their model numbers and say the type 993 wasn't as good-looking as the 996, and the 997 doesn't have quite the right stance. To me it's like looking at a field full of babies. No matter where in the world they come from, they are all the same.

Porsche's designers had a stab at something new and different with the Boxster but, having created a front end, they were so exhausted by the effort that they simply fitted exactly the same thing to the back. Were it not for the colour of the rear lights, this car would look exactly the same going forwards as it does in reverse.

And then we have the Cayenne, the big off-roader. This has the nose of a 911 and the rest of it looks as if it's melted.

I suppose the trouble is that when you are a small company, and you can afford to launch a new car only every 300 years, you can't really employ a big styling division full of bright young things in polo neck jumpers and thin glasses. Because most of the time they'd have nothing to do. Except apply for a job with

General Motors. We see the same problem with Aston Martin, which designed one car back in the 14th century and is still making the same sort of thing today. At least that first car was pretty – which is emphatically not the case with Porsche's squashed Beetle.

All of this brings me on to the Cayman, which, in essence, is a Boxster coupé. The first effort looked, as you may have guessed, like a Boxster with a roof. It looked as though I'd designed it. And anyone who bought it was saying one thing very clearly, 'Hey. I can't afford a 911.'

But now there's a new Cayman and it no longer looks like a Boxster with a roof. It looks like a 911. Still, at least passers-by will no longer clock you as a man hanging to the bottom rung of the ladder. Only very keen Porsche enthusiasts will spot that it isn't a 911, and you don't want to be talking to those people anyway.

I have to, in a professional capacity, and what they're saying is interesting. They're saying that the latest 911 has lost some of the brand magic. That it's no longer as sporty or as involving as it should be. And that actually the much cheaper lookalike Cayman is much closer in spirit to the original dream.

I don't know about any of that, but what I can tell you is that my test car, an S version, really was the most lovely colour. It was a deep, rich metallic blue. It's the best colour I've seen on any car.

I can also tell you that, as a sports car, the Cayman S is simply spectacular. It seems to flow down the road the way honey would flow over the naked form of Cameron Diaz. Only faster.

The steering and the brakes and the feel through the seat of your pants are all exactly as they should be. And then there's the engine: a 3.4-litre flat six mounted just behind the cockpit. Not over the rear bumper, as it is in a 911.

This creates not just an innate sense of balance but also a surprisingly large amount of oomph. And it's all accompanied by a noise that's never intrusive or showy but is always there, in a deep, quiet, reassuring way. It's like driving along with Richard Burton in the boot, endlessly complimenting you on your clothes and your hair and your driving style.

We can wax lyrical as much as we like about the sporting prowess of cars such as the Jaguar F-type and the Lotus Elise and the Mitsubishi Evo. But all of them are left shivering in the cold, hard shadow of the Cayman's magnificence. It really is that good.

Which of course is all very well when you are on the road from Davos to Cortina and the sun's just coming up and everyone else in the world is in bed. But what about when you're in Rotherham and it's rush hour and you're knackered and you just want to get down the M1 as comfortably as possible?

Things begin well. The Cayman has two boots, which means it can swallow a surprisingly large amount of luggage. It has a roomy cockpit too, full of nice touches. The satellite navigation is easy to program, the air-conditioning works nicely and, joy of joys, there isn't a single button on the steering wheel. It all feels simple and unthreatening. As long as you stay away from the G-meter and the idiotic lap timer.

It also has surprisingly compliant suspension. Yes, it crashes a bit at low speeds on a badly maintained town centre high street. But once you're above 30, it's like a limo.

There's a lesson here for every car maker. If you make the chassis stiff, the suspension doesn't have to spend half its time masking deficiencies. It can concentrate on isolating occupants from potholes. Of which, in Rotherham, there are many.

Of course, you can ruin everything by engaging Sport mode, but if you leave that alone, and avoid the 20-inch wheel option, you'll be fine. And you'll be doing 30 mpg in a car that costs just £48,783.

However. I'm afraid I arrived back in London in some discomfort, which is the British way of saying 'screaming agony', because of the bloody seats. The shoulder bolsters are too close together, which means you get some idea of what it might be like to be a letter inside an envelope. Even this morning, after eight hours of deep sleep, my neck feels as though it's spasming.

There are other irritations too. The cupholder system is needlessly complicated and, no matter what you do, tins entrusted to the receptacles always rattle. And then there's the gearbox. I had the double-clutch system, which costs an extra £2,000, and for the most part it's very good. But at slow speeds, in Auto mode, you sense that a computer program is keeping you going, rather than neat mechanical design. In a sports car as pure as the Cayman, I think a manual is more in keeping.

That really is a tiny criticism, though. And I probably wouldn't have mentioned it at all if my neck didn't hurt so much.

Which brings me on to the final point. When I have finished writing this, I must drive the Cayman to Oxfordshire. With decent seats, that would be something I'd relish. But it doesn't have decent seats, so I'm rather dreading it.

16 June 2013

Say the magic word and the howling banshee turns sultry sorceress

McLaren 12C Spider

When I first reviewed the idiotically named McLaren MP4-12C, I said it was better in every measurable way than the Ferrari 458 Italia, but that it lacked sparkle, panache, zing. That it was too technical and too soulless. And that, given the choice, I'd take the Ferrari.

Other reviewers came to the same conclusions, and as a result McLaren acted fast to address the situation. The company made its car more noisy and tuned the exhaust to make it sound dirtier. It gave the car even more power. And it fitted door handles that actually worked. Very nice. And, most of all, it cut the roof off to create the more sensibly named 12C Spider.

The effect of this amputation has been dramatic. It's like one of those stern-looking girls you sometimes find in adult films who simply by letting their hair down are transformed into complete sex bombs.

Britain loves convertibles. We buy more of them than almost any other country in Europe, and it's easy to see why. Because the sun on these islands is a rare visitor, we don't want to waste the days when it's here by sitting under a metal roof. We want to savour it, because we know tomorrow it'll be gone.

I know the rules, of course. No man over the age of thirty-eight can drive a car with the roof down when he can be seen by other people, as it sends out all the wrong messages. You think you look good on the high street, sitting in the sun. You think you come across as suave yet carefree. But to other people it looks as if your gentleman sausage no longer works properly.

I don't care, though. I love to drive a car with the roof down.

I love the noise and the sense that it's just you hurtling through time and space; that you're not actually in a car. In fact, when you're in a convertible and the roof is down, the sensations are so vivid, it doesn't matter what the car's like at all. Worrying about handling when your hair is being torn out is like worrying about your ingrowing toenails when you are being attacked by a swarm of killer bees.

This is a good thing because as a general rule convertibles do not ride or handle or go anywhere near as well as cars that have roofs. That's because today the chassis of a car is its bodyshell. The front and the back ends are joined not by two huge rails, which used to be the case, but by the floor and the roof. Taking 50 per cent of that connection away means you have to add all sorts of strengthening beams that a) add weight and b) are never really a satisfactory substitute. Soft-top cars never feel stiff. Much like many of the people who drive them.

The McLaren, however, is different. Because the spine of the original was so rigid, no strengthening beams have been added at all. That means no extra weight – apart from the electric roof mechanism – and no compromises. As a result, this car feels exactly the same as the hard-top. Which is to say, it feels magical. As if it's being propelled by witchcraft.

No car in the world has better steering. It's very light, which suggests there's no feel. But in fact there is so much that when you run over a wasp, you can tell whether it was a male or a female. This means you can feel the precise moment when grip is about to be lost. Which means you always feel completely in control.

You're not, though. A computer is. You can turn it off, if you are a space shuttle commander and you have half an hour to kill, but there's no point. Because it will let you take diabolical liberties before it steps in, like a well-trained butler, with a gentle helping hand. It's the best traction control system I've yet encountered.

And it's almost never necessary. Fitting a car this well behaved with an electronic restraining bolt is like fitting the Archbishop of Canterbury with an ankle tag. It's pointless. Because the spine is so stiff, and because there are no anti-roll bars, there's no physical connection between any of the wheels – it's all done electronically – so the cornering speeds of this car are simply immense. Around a track, I know of no road car that could even get close.

If you have a Ferrari 458, do not attempt to keep up with a McLaren 12C. You will be either humiliated or killed.

And here's the clincher. When you have finished tearing up the laws of physics and your neck hurts from the cornering G-forces and it's time to go home, the 12C is as comfortable as a Rolls-Royce Phantom. Even though it will corner at Mach 3, the lack of anti-roll bars means it simply glides and floats over bumps and potholes. As I said, witchcraft.

As a piece of engineering, then, it's fabulous. Jaw-dropping. Mesmerizing. But as a car? Hmmm. There are one or two things that would drive you mad. For instance, every time you open the butterfly door to get in, the side window will take your eye out. I must also say that if you are tall, the cockpit is a little tight. And the satnav doesn't work. It always thinks it's where you were two hours earlier, but that's not really the end of the world, because with the roof down you can't see the screen anyway.

There's more too. Almost every feature is adjustable. You can even alter the volume of the wastegate chirrup. But only if you are six years old. Because when you go into the system menu, you'll find the typeface is in 2 pt and you won't be able to read it. Not without putting on a pair of reading glasses and peering into the binnacle, something that's not advisable when you are in a 600-plus brake horsepower soft-top and you're doing 200 mph.

These things may be enough to drive you in the direction of the Ferrari. But remember, that comes with a steering wheel that's unfathomable and electronic readouts that make the McLaren look as though it's been made by Playmobil. I'm afraid, then, that if you buy either of these cars, it will infuriate as often as it exhilarates. It was always thus in the world of the supercar, though.

It has been suggested by some that they are similar in other ways too and that choosing between them is difficult. But that's not so, actually. They may look the same, cost about the same and have the same basic design parameters. But they are completely different.

The McLaren is like a three-star restaurant. The food is immaculate, the service impeccable, the loos impressive and the temperature just so. Every detail is spot-on, and to give the place a bit of character, there's now a maître d' who has a twinkle in his eye.

The Ferrari, on the other hand, is a loud Italian joint full of shouting and massive pepper grinders. They're both restaurants, then. And they're both bloody good. But they are not remotely similar. As a result, I cannot tell you which is better. You have to choose what you want, and don't worry, because whichever way you go, I promise you this: you'll end up with a masterpiece.

23 June 2013

Take the doors off and put them back on? That'll be £24,000, sir

BMW M6 Gran Coupé

In the olden days it was jolly difficult to design a car. You had to use slide rules and pencils and guesswork. And you couldn't simply buy in parts from Lucas, because it was usually on strike, and even when it wasn't, the parts you'd bought didn't fit and wouldn't work anyway.

It would take years – and all the money the government had – to get your new model designed, and then you'd have to make all the tooling necessary to put it into production. This is why, when a car went on sale, it stayed on sale for 200 years. It's also why each car company made only a handful of models.

Today, though, you just fire up your laptop and ask it to design a car, and while you go for a chat at the water fountain, it comes up with the answer. An answer that'll be safe, economical and made from parts that will be delivered bang on time and that will work.

What's more, the finished product will be modular. Which means that all the expensive bits can be used on other models. It's for this reason that Volkswagen made just three models in 1960. And about four million today. Because while an Audi A3 looks different from a Golf, underneath it isn't.

This is good news, of course, but because it's now easy to make a new model, car makers are going a bit mad. Which brings me to the subject of this morning's column . . .

In the beginning there was the BMW M5. Then BMW made a two-door version of it called the M6. And now there's a four-door version of the two-door M6 that is called the M6 Gran Coupé.

It's going to be a tester for BMW's showroom salespeople, that's for sure. Because they will have to say to prospective customers, 'Yes, it has the same engine and running gear as the M5. And the same number of doors. But here's the thing, sir. It's £24,000 more expensive and there's less space inside.'

A car maker can get away with that when a coupé is dramatically and noticeably better-looking than the saloon on which it is based. People will always pay for style. But when the coupé isn't dramatically different? Hmmm. As I said. It's going to be tough for the hair-gel-and-Burton boys.

However, let us be in no doubt that the M6 Gran Coupé is extremely good-looking. It's better-looking, weirdly, than the two-door M6. And while there is a hefty price premium, it does come with some things the M5 doesn't have, such as a carbon-fibre roof for a lower centre of gravity.

There's more too. While I like the M5, it does come with a whiff of the enthusiast about it. Every one you see has been bought second-hand on the internet, fitted with private plates to disguise that fact and then polished to within an inch of its life. Then you have the driver, who always looks exactly like the sort of person you don't want to sit next to at a dinner party. The sort of person that refers to his car by its manufacturing code, not its name. With an M6 Gran Coupé you don't get that association. Yet.

Plus, I'm a sucker for pillarless doors and rear seats that are separated by an (optional) console full of knobs and dials. Sitting in the back of this thing is like sitting in a private jet, and no one's complained about that. Even though your knees are in your nipples, your head's on the ceiling, it's deafening and there's no lavatory.

So, yes, I will say that there is just enough in this car to warrant the price premium over both the M6 and the M5. Right now it's the M car to have. Provided that's what you want. But is it?

Well, not the first time you drive it, that's for sure. God, it's complicated, and there is an electric German on hand to stop you doing anything out of sequence. Or that feels natural. Or sensible. It won't let you do anything without ordering you to do something else first. This means that soon you will be screaming at it, 'I own your arse! And if I want to put you in Drive without pressing the switch first, I will!'

Sometimes, though, it asks you to do things that you can't do. Such as putting it in Park before getting out. Which is tricky because there is no Park button. 'Bong,' it says. And then 'bong' again. And then 'bong'. You feel like Dustin Hoffman in *Marathon Man*, constantly being asked if it's safe, and you panic because you don't know the answer.

Eventually, when you are mad and drooling, you will get out anyway, hoping the bloody thing does roll into a river. And this is the good bit. When you get out, it goes into Park all by itself. I wanted to kick it.

Some of the electronics, however, are very good. The satnav is huge and brilliant. The ability to choose settings for the suspension and the steering and the powertrain and then store your preferences for future reference is wonderful. And . . . I'm sounding like a stuck record. Because I said exactly the same thing when I reviewed the M5 last year.

There's another similarity too. A great sense of weight. When you push down on the accelerator, you sense that the 4.4-litre twin-turbo V8 is really having to gird all of its 552 loins to get the car rolling, and it's the same story when you turn the wheel. You feel as though you are asking the suspension to deal with something that's heavier than most monasteries.

And yet strangely it weighs less than two tons. It's not a lightweight, by any means, but by today's standards it's not a porker either. And, anyway, some people enjoy the sense of driving about in a hill. Rather than rolling down it while inside a balloon.

However, you won't be thinking about weight when you really mash the throttle into the carpet, because this car absolutely flies. It's really, properly fast, and, better still, it doesn't make much of a song and dance about it. There's a trend these days for fast cars to let you know they're fast by barking every time you go near the throttle. The BMW doesn't. It just gets on with its speed, efficiently and with no fuss.

Cornering? Yup, it does that too. And from memory it does it better than the two-door M6, which feels woolly and soft. Sadly, though, I'm not going to ring the man in BMW's suspension control department (electronics subdivision) and ask why this is so. Because undoubtedly he's the sort of chap who would enjoy telling me for hours.

It didn't take me long to work out that this car is special and unusual. An M5 with a hint more style. A genuinely nice place to sit. And, all things considered, it's not a bad price tag. Yes, its value will depreciate like a fat man falling off a tower block, but £97,490 in the showroom isn't bad. Not when you see how much Aston Martin wants for a Rapide.

Mercedes, of course, does the CLS 63 AMG, which is similar, and Audi has its RS 7 in the wings. But for now I think the BMW makes the most sense. If it had sensible controls and a Park button, I'd even consider giving it four stars, but it hasn't, so . . .

30 June 2013

Thunderbird and Mustang have gone, so what'll we call it, chaps?

Vauxhall Adam

This morning a man in a chunky-neck jumper and corduroy trousers is sitting down to his plate of kippers, blissfully unaware that he's the last person in Britain to have been christened Malcolm. It's much the same story with his wife, Brenda, and his friends from the lodge, Neville and Roger.

Who is Britain's youngest Simon? Is there a Clive aged under ten? Where is the last Derek? Do you live next door to the final Brian?

This cull of monikers doesn't happen in Iceland, because the government gives new parents a list of names from which to choose. But here the army of opinion-forming orange people have got it into their heads that they can call their poor little tyke pretty much anything that comes into their heads. And, frankly, why go for something traditional such as Edith or Gertrude when you can name your little girl after a sweet white wine, or a village where you had particularly enjoyable sex in Crete?

This, of course, brings me on to the naming of cars. By and large it's always been very simple. Expensive cars such as BMWs and Mercedes and Audis were given numbers and letters. Smaller, cheaper cars had names. And usually those names were absolutely terrible.

Fiat has always been especially hopeless. Over the years, it has had the Road and the One and the Point. But we can't forget Austin Rover, which named the car it said would save it from the dustbin after the Paris underground system. Can you imagine Renault calling its next little car the Tube? No. Neither can I.

Volkswagen isn't much better, but there's a reason for this. In the past it would give a shortlist of names to executives in the company, who were asked to rate them out of ten. Which meant the winner was invariably the name that was everyone's second or third favourite. How else could they have arrived at the Golf? That's like calling a car the Herpes.

Then we have Nissan, which for a long time kept alive traditional English names that Coleen and Wayne felt were beneath them. There was the Cedric, the Gloria and the Silvia.

Toyota, meanwhile, called the first car it tried to sell in America the Toyolet. Until the importers suggested that Toyopet might be a bit better. And then we had the Mitsubishi Starion. Which was supposed to have been the Stallion but there was a mix-up caused by the Japanese problem with the letter 'l'.

There have been some good names, though. The best by a mile – and I won't take any argument on this – is the Interceptor. The Pantera was pretty good as well but, really, for consistently good names you need to look to America, which has given us the Thunderbird and the Mustang, the Cougar and the Barracuda.

It's a confidence thing, I guess, the big, toothy ability to name an awful, slow car after a wild, ferocious animal: it's like calling your son Hercules, even if you have an inkling he'll grow up to be a six-stone weed with asthma and pipe cleaners for arms.

All of this brings me on to the new baby Vauxhall. The company has called it the Adam, which was the Christian name of the founder of Vauxhall's sister brand Opel, but the car maker says that's not why it chose the name. It says it chose Adam for the reason that UKTV changed the name of its G2 channel to Dave. Because it's a nice name. I think I agree.

The Adam is supposed to take Vauxhall into territory currently occupied by the Fiat 500 and the Mini. It's supposed to be a trendy car for young urbanites. But there's a small problem with that. The Fiat and the Mini hark back to cars people remember fondly, but what does the Adam hark back to? The Chevette? The Viva?

'The Prince Henry,' said a spokesman for General Motors, Vauxhall's owner. Well, it's true. The Prince Henry was indeed very special – the first performance car – but if you can remember that, I suspect you're not really in the market for a small car. Or indeed any car. Not since your final road journey in that hearse.

No. This new car cannot rely on people wanting to recapture a flavour of the Fifties and Sixties. It's going to have to stand up on its own four wheels. So does it?

There are three trim levels: Jam, Glam and Slam. But each is available with a bewildering array of options. There are, and I'm not making this up, billions and billions of permutations. And don't worry if you make a mistake and order 'Men in Brown' door mirrors – that's what they're called – because you can have them changed for the 'White My Fire' option in a jiffy.

In fact, when you become bored with the look of the interior you've selected, you can change it next month or next year for something completely different, for £70.

The upshot is that you cannot hate the way the Adam looks because you can make it look however you want. You can't really hate it as a town car either. There is space in the back for two people, provided their lower legs are no more than 3 inches thick, and there is a boot that's just about big enough for a mid-week shop.

Visibility is good, the clutch is light, the steering is nice and the ride comfort is exceptional. Take away all the connotations, and the fashion aspiration, view it as a town car only, and I have to say it's better than the Fiat and the Mini.

But as an all-round car, I'm not sure. The model I selected was a 1.4-litre Slam with a chessboard roof lining, yellow trim on the wheels and a billion other sporty features besides. This meant it looked like a hot hatch, and one thing's for sure: it wasn't.

The Adam is not at all fast. It doesn't handle with much enthusiasm and at 70 mph on the motorway it feels awfully busy – as if it's sort of surprised to be there.

There are more things too. It doesn't come with satellite navigation or a telephone, because it is designed to hook up to your smartphone and piggyback the features on that instead. In theory this is a properly good idea. I even asked a man from Vauxhall how it all worked, and in a matter of seconds, well, minutes – well, a quarter of an hour – he had the car talking to his phone.

But when I was left to fly solo, my phone treated the Adam in the way that a reluctant bitch treats a dog. There was no mating at all.

So. There are problems but overall it's a likeable and practical little car. The only thing that would stop me buying one if I were in the market for such a thing is its other name. Adam is fine. Vauxhall, though? They've still got some way to go with that.

7 July 2013

Ha! They'll never catch me now I'm the invisible man

VW Golf GTI 2.0 TSI Performance Pack

There are many wonderful cars on the market right now: the Ferrari 458 Italia, the McLaren 12C Spider, the Bentley Continental GT V8, the Mercedes SLS AMG, the Lexus LFA, the Aston Martin Vanquish and the BMW M6 Gran Coupé. All are fast, stylish and characterful and I'd happily own any one of them. But I can't, because driving around in a flash car is like driving around naked. You tend to get noticed. Which is not something I find very enjoyable.

When I'm out and about I'm asked constantly to pose for a photograph. 'It's for my sister who's going on a hockey tour,' they always say, while rummaging around for their cameraphone.

Several minutes later, after I've heard all about the hockey tour and how her boyfriend has a BMW M3, she has found the camera on her phone and is asking passers-by to take a shot. But they don't know which button to press so they end up taking a picture of their own nose. Or turning it off. And by the time they've had a lesson, someone else has arrived. 'Oh, my son would never forgive me if I didn't get a picture.'

Resisting the temptation to say, 'Well, don't tell him you saw me then,' I agree to a snap, only to discover he's a bit of a David Bailey and wants me to move into the shade because the shot's a bit too backlit. And soon my two-minute trip to the shop for a pint of milk has turned into a two-hour photoshoot.

This, of course, is a time-consuming by-product of appearing on the television. But when it happens on the road, it's actually pretty dangerous. People brake and swerve and cut across three lanes of traffic to get a shot of me. Once, a chap in a Mini was so busy videoing me he crashed into the car in front. A car that was being driven, amusingly, by a gorilla of a man.

One of these days someone is going to be killed and that's why I want my next car to be inconspicuous. And that's a problem, because every single car that offers the speed and excitement I crave comes with look-at-me styling. Except one: the Volkswagen Golf GTI.

So, I'm sorry. Every week I come here and review a car for your benefit, but this morning I'm reviewing a car for mine. Because the new GTI might just be the answer to all my prayers.

When Volkswagen first created the fast Golf thirty-seven years ago, it was truly classless. I knew housewives who scrimped and saved to buy one. And I knew someone who part-exchanged his Gordon-Keeble. It was also a car that was all things to all men. It could carry five people. It had a boot with rear seats that folded down. It had body panels that cost no more to repair than those on the normal Golf but, thanks to its 108-bhp engine, it was faster and more exciting than any of the sports cars that were kicking around at the time.

The Mk 2 GTI was pretty good as well, but since then the mojo has been slipping away. You had the impression that VW was making a GTI because it felt it had to, not because it was something that excited it in any way. But for the Mk 7 the company brought in the man who did the Porsche 911 GT3 RS. And from what I've been hearing, the original magic is back.

The engine is a 2-litre turbo that produces 217 bhp. But for an extra £980 you can have the performance pack, which takes the output up to 227 bhp. I tried that version and after a short time knew that in the real world this car could keep up with just about anything.

Quietly. There's no fuss with this GTI. No drama. No rorty exhaust noises. You see a gap. You put your foot down. The overtaking manoeuvre is completed. You reach a bend. The electronic front diff makes sure there's no understeer and no unseemly tugging at the steering wheel either. You come out on the other side. It is a machine built to make speedy progress. It is German.

There's more too. It costs just £195 more than the previous model and, thanks to a camera-based emergency braking system, it has fallen five insurance groups. And for a 150-mph-plus car it is also extremely economical.

What we have here then is a bundle of pure, undiluted common sense. Except for two things. First, you can't buy it with optional 19-inch wheels in combination with a sunroof – no idea why. And second, you can choose between Comfort, Normal and Sport settings for the front differential, the suspension, the gearbox and the steering, so I did a test. I took the car to *Top Gear*'s test track, put it in Normal and asked the Stig to do a lap. He did it in 1 minute 29.6 seconds. I then put it in Sport. This time he did a lap in 1 minute 29.6 seconds. So then I put it in Comfort, which softens everything up. He did it in 1 minute 29.5 seconds.

Adjustable suspension and gearboxes are fitted to many cars these days, and I've long harboured a suspicion they make no difference to how fast a car goes. And here's proof. Sport makes the ride uncomfortable but provides no benefit at all. In its Normal setting the GTI is tremendous. The sportiness is still there – the times prove that – but Comfort mode is sublime. It's phenomenal and brilliant.

Inside, it's as logical and as sensible as a German's knicker drawer and you have the impression that everything will still be working perfectly in ten years' time. Except for the radio, which broke after two days. And a bit of trim round the rear window, which fell off. But to be fair, I was testing a pre-production model. And the man responsible for these mistakes will have been shot by the time the lines start to roll for real.

Best of all, though, nobody took my picture as I drove along. I had a car that can rip holes in the physics books, that can scream to 62 mph in just over six seconds, that slices through the bends like a well-drilled monoskier and that is as comfortable as having a nice lie down. But nobody looked at it twice.

The only thing that annoyed me was the double-clutch flappy-paddle gearbox. It was impossible to set off from the lights smoothly, and by the fourth day I was being driven mad. By the seventh I was so angry my nose was beginning to itch. And then I discovered the 'auto hold' button.

Fitted to stop the car rolling backwards when you are doing a hill start – you can't ride the clutch with a flappy-paddle box and there's none of the in-built 'creep' you get from a traditional automatic – it applies the brake whenever you stop. And then, when you put your foot on the throttle, it takes the brake off again. But not fast enough. Hence the jerk.

I turned it off and all was well. Very well. For me this car is perfect. And if you're honest, it's perfect for you too.

14 July 2013

Coo! A baby thunderclap from Merc's OMG division

Mercedes-Benz A45 AMG

For a hundred years Mercedes was a byword for solid, sensible engineering. While the rest of the world let its hair down and listened to Jimi Hendrix, the company plodded on with its doleful recipe of longevity with just a sprinkling of toughness. The men of Stuttgart built no-frills cars that were made to last. They were tortoises to counter the hares from BMW.

If you want to drive across Africa next weekend, then by all means get yourself a Toyota Land Cruiser. But if you actually want to get there, you'd be better off with the standard Mercedes from the late 1970s and early 1980s. In a world of thongs and briefs and frilly bits of nothing, this was a sturdy pair of games knickers. It was a car that simply didn't know how to let you down. And it still doesn't today.

But then one day the company chiefs got bored with making games knickers, so they got together with a little-known tuning company called AMG and went berserk.

The cars that resulted are stupid. They are too big, too loud, too crazy, too brash, too sideways most of the time and too scary as a result. I like them a lot.

I like the way that a BMW or a quick Audi is designed to put a fast lap time on the board, whereas an AMG Mercedes is designed to put a smile on your face, and most of its rear tyres into the atmosphere. In a world where fuel economy is king and tall poppies are frowned upon, it's refreshing to find a range of cars built purely for blood-and-guts savagery. They are not sniper rifles. They are dirty great artillery pieces.

If I may liken all of the world's cars to weather, you have

many that are drizzle and some that are lovely sunny afternoons. You have those that are precise and fast, like lightning. And some that are just Tupperware grey as far as the eye can see. Then you have the AMG Mercs. They are cracks of thunder. They are V8 muscle cars. A blend of the American dream and German engineering. They are tremendous.

But then several months ago I drove an AMG-badged A-class Mercedes, and that wasn't a V8, or thunderous, or even very muscly. Its AMG badge was writing cheques the car simply couldn't cash. I gave it two stars and wondered what on earth Mercedes was thinking of. Putting that badge on that car was . . . well, it would be like calling a small river launch HMS *Ark Royal*.

And now the company has done it again with this car. It is called the A45 AMG, and, to be honest, I was expecting about 14 feet of solid, chewy disappointment. However . . . Let's start with the engine. It's a turbocharged 2-litre unit that meets emissions legislation that the EU hasn't even introduced yet. It's quite frugal too. Despite this, it's the most powerful four-cylinder engine in production.

The figures are fairly astonishing. You get 355 brake horsepower, which means you're getting almost 178 bhp a litre. To put that in perspective, the V8 in a Ferrari 458 Italia can manage only 125 bhp a litre. There is some very clever engineering in here.

And because the A45 is so clever and so potent, Mercedes decided it could not solely be front-wheel drive. Because asking the front wheels to do the steering while handling 355 rampaging German horses would be like asking a man who's on fire to solve a crossword puzzle. So it has a system that sends up to half the power to the rear wheels should those up front become a bit flustered.

On top of this, Mercedes fitted big brakes, lowered the suspension and slotted in a fast-acting, double-clutch, flappy-paddle gearbox. And the result is . . . quite boring.

It's all so planted and neutral and benign that you wonder whether you've climbed into the diesel version by mistake. And then you look at the speedometer and what it's saying is scarcely believable. Most of the time I was going almost exactly twice as fast as I'd guessed.

All previous AMG Mercs make you feel as if you're going faster than you really are. This one does the exact opposite.

That said, Mercedes has tried to give it some of its big brothers' traits. When you change up, the exhaust sounds like Rubeus Hagrid clearing his throat. And there's a feel of great heaviness. Probably because that's what the car is: heavy. But, whatever, you have to manhandle it through the bends as though you're trying to get a piano up a back staircase. You have to work for your rewards.

And, boy oh boy, are they there. The engine has an uncanny knack of delivering lots of meaty torque right the way up to 5000 rpm and then, just as you think it's game over and time for another gear, you get a frantic burst of power. And then you are going three times faster than you'd guessed. Happily, then, the brakes are immense and the handling is sublime. Obviously it won't stick its tail out and smoke like every other AMG product. But it doesn't understeer unduly either. It just goes round the corner in such a way that you get the impression it wasn't really trying.

I liked it enormously. It had more character than any hatchback since the Fiat Strada Abarth, the speed was immense and there's no getting away from the fact that it looks really rather handsome. Many Mercs are overstyled these days, but on this one the creases and the fussy little details seem to work.

However, there are a few issues. Let's start with the little ones. It's needlessly bumpy. This has been done simply to make it feel sporty, not for any handling benefits. The chassis is so stiff that Mercedes could easily have softened the suspension without affecting the performance at all. It was a mistake.

Then there's the width. Certainly you do not whizz through width restrictions the way you would in a 1-series BMW. You need to breathe in and grimace first.

Other stuff? The petrol tank is too small, which means you have to fill up every few minutes. And filling up with petrol is worse than trying on trousers.

Then there's the interior styling, which is completely over the top. Who, for instance, thought that it would be a good idea to make the air vents red? This is a Mercedes-Benz, for heaven's sake. Not a Gillette commercial. Red air vents are like red trousers. And there's a blog for people who wear those. (Google it.)

But the big sticking point for me is the price. It's £37,845. Of course, for an AMG Mercedes, this is extremely good value. But it's a shedload for what, when all is said and done, is a hot hatchback.

It's a whopping £9,000 more than you're asked to pay for a range-topping Volkswagen Golf GTI. Yes, the Mercedes is a better car. But £9,000 better? With that ride, that fuel tank and those stupid vents? No. I'm afraid not.

21 July 2013

From the nation that brought you Le Mans . . . A tent with wheels

Citroën DS3 cabrio D Sport

As I write, the sun is belting down with a fury we haven't seen for many years. Yet, after one of the coldest, most miserable springs on record, the countryside is still as green as an eco-mentalist's groin. It is truly beautiful out there and I am consumed by an overwhelming need to drive about in a sports car. But there's a problem with that. You can't actually buy such a thing these days.

The Mazda MX-5 comes close, but over the years it has swollen up and been given a bigger engine and a retractable metal roof, so that now, while it's still delightful, it's a bit too fast and a bit too sensible and a bit too grippy in the corners.

The Caterham 7 isn't bad either. But these days it's aimed mainly at the adenoidal track-day enthusiast rather than the chap who simply wants to slither about Oxfordshire in the sunshine. And it is extremely ugly.

The new Jaguar F-type is not ugly but it's too expensive and too powerful. Which brings me on to the BMW Z4, a machine that is neither of those things. But it is too smooth and too polished. It's lovely to behold and lovely to own but it's not a sports car.

An Alfa Romeo Spider. That was a sports car. So were the Fiat 124 Spider and the MG. The Triumph TR6 was a sports car, too, as was the Sunbeam Alpine. These cars were built for fun, for a laugh, ha-ha, ha-ha. You could keep them by the racehorse that the Aga Khan bought you for Christmas. Even by the standards of the day they were not especially fast, but they were pretty and they all came with simple canvas roofs. They were like tenting but without the dysentery. And I miss them all.

I miss the days when handling mattered and grip didn't. Today cars are built to go round a corner as quickly as possible. Which means you can't indulge in a big four-wheel drift at 20 mph. And they have to be safe, which means they have to be heavy. And the one thing a sports car cannot be is heavy.

Sports cars are for long, warm summer afternoons. And on a long, warm summer afternoon you want a light salad. Not a dirty great meat pie. That's why I was rather looking forward to the Citroën DS3 convertible.

I'm a big fan of the hard-top DS3, particularly the limited-edition Racing version. I'll be honest: it isn't much of a driver's car; the gearing is too weird for that. On a hot lap of the Monaco Grand Prix track I stuck it in second about two seconds after the start and didn't have cause to change gear at all until it was time to stop.

Then there is the suspension. That isn't very good either. But all of these little issues are smothered by an interesting body, lots of natty decals and a sense of fun.

There will be a convertible version of the Racing in the months to come, but the car I tested was simply a chopped-down version of the standard model. And that's fine by me, because who needs speed when the sun's out and there are wildflowers to look at?

There's more too. Before the car arrived, a chap at Citroën sent me a text saying that it was the world's only genuine five-seat convertible – apart from the woeful Jeep Wrangler – that its boot was almost twice as big as the boot in a Mini convertible and that the roof could be opened at up to 74 mph. It all sounded good.

But when the car arrived, I discovered that, actually, it isn't a convertible at all. It's a normal car with a big canvas sunroof. Back in the Seventies my grandfather had a Rover 3.5 that had a Tudor Webasto sunroof. And that wasn't a convertible either.

Then I discovered that if you push the sunroof button again, the back window flops down and the roof keeps on folding itself back. This was good news, until it stopped, completely obscuring the rear view. How can Citroën have thought this was a good idea?

Actually, scrub that. I know exactly how Citroën thought it was a good idea. Because its last attempt at making a convertible was the C3 Pluriel. And it came with a roof that detached all right, but only after half an hour of swearing and broken fingernails. And then, when it was off, there was nowhere to put it. You had to leave it where it was and hope it didn't rain while you were out. It was the stupidest piece of design since the Ronco Buttoneer.

I think the problem is that in all of automotive history the total number of sports cars made in France is, let's see ... um ... exactly none.

For a country where motoring is a demonstration of nonchalance and cheapness is king, it seems odd that they never combined the two things. And it's doubly odd when you look at their love of motor sport. Their engines dominate Formula One. A Peugeot has just smashed the record up Pikey's Peak in Colorado. They rule the roost in international rallying. And yet, despite all this, they have never made a sports car. And on the evidence this far, the DS3 convertible doesn't exactly change things.

But let us plough on. Let us treat it as a pretty hatchback that comes with a big sunroof. Then what? Well, it will cost you some money but not much. Because whatever price is quoted in the brochure, you can be assured that, being a Citroën, it will come with 0 per cent finance, £1,000 cashback, no VAT, an offer of an evening out with the dealer principal's daughter and £5,000 to spend on a holiday. You may need an incentive such as this because as a car it's not very good.

First of all there's the driving position. The steering wheel is mounted pretty much directly above the pedals, which means the only person who can get comfortable is someone whose arms and legs are the same length. To make matters worse, the seats were lined with a fabric that had the grip of KY Jelly.

That's why I can't tell you how this car handles. Because every time I tried to go round a corner with any gusto at all, I fell over. There were other issues too. There are no cupholders. And it comes with an entertainment system that couldn't even find Radio 2 half the time. The satellite navigation system, meanwhile, was unfathomable. And even if by some miracle I did manage to type in an address, it would take me on a route of its choosing to a destination that it plainly thought was near enough.

I suppose in the interests of fairness I should say that the 154-bhp engine is quite nice and that there is a decent amount of space in the boot. But, to be honest, that's like saying there's a decent amount of space in a postbox. There is, but you can't really get at it because the slot's too small.

It's odd. I haven't disliked driving a car as much as this for quite some time. It was so bad that even though it was half full of petrol paid for by someone else, and had a big sunroof, I spent the whole of last weekend driving around, in the sunshine, in my own hard-top car, using petrol I'd paid for myself.

For a Yorkshireman to do this? Well, it should tell you all you need to know.

28 July 2013

The fun begins once you've arm-wrestled Mary Poppins for control

Audi RS 5 cabriolet quattro 4.2 FSI

Over the past few years many companies and organizations have announced they are close to putting a driverless car into production. They all speak of a vehicle that will use satellite navigation to find a destination and radar to monitor its surroundings. A car that will pull up when the traffic stops and move off again when the road is clear. A car that will find a parking space, and reverse into it, all by itself.

But then what? That's what I've never understood. There's no point sending your car into town to buy a pint of milk because while it will be able to find the right shop and the nearest available parking bay, it will not be able to go inside and actually buy the milk.

Similarly, it would be able to find your office but then it would spend all day just sitting outside. It wouldn't be able to go to your desk and answer all the emails. A driverless car, then, is completely useless. You have to be inside or there's no point. And if you're inside it's not driverless.

That said, I'm sure there are plenty of people who would very much like to step into their car after a hard day at work, tell it to go home and then curl up in the back for a little sleep. But could you actually nod off? Would you trust the on-board systems to behave? Really? You'd put your life in the hands of the same people that built your laptop?

I flew to the Isle of Man last weekend on a brilliant little aeroplane. It had a whizz-bang glass cockpit and was completely up to date in every way. So the pilot set the autopilot and we sat back for a chat. Things were going very well until we descended out of the clouds near the island's airport to find that we were heading straight for a hill. If he'd been asleep we'd have hit it.

Which brings me to the Audi RS 5 cabriolet. This has something called active lane assist, a technology that's not new. But systems we've seen in the past simply vibrate the steering wheel if sensors think you're straying out of your lane on a motorway. The system in the Audi is different: if it thinks you're drifting out of your lane, it actually takes control of the steering and puts you back on the straight and narrow.

I tested it on the Westway in London and was amazed because it simply stayed in the outside lane, steering nicely round the long left-hander. I didn't have to do a thing, so I thought, Brilliant. I can get on with some texting . . .'

But no. Because after a short while a message flashed up on the dashboard saying I now had to take manual control of the steering. What's the point of that? Why build a car that is capable of steering itself but after a minute or so can't be bothered?

I'm afraid it gets worse. Because later, on the M40, I decided to pull into the middle lane, and the steering wheel wouldn't really let me. It was gently pushing the wheel to the right, thinking that I'd nodded off and that my life needed saving.

It was not a big push. It was like arm-wrestling a child. You can overcome the system, but unless you indicate – which tells the sensors you're moving on purpose – it argues every single time you cross a white line. This started to drive me mad. I therefore decided to turn it off. But I couldn't find out how to. And that made even my hair angry.

You spend an extra £370 on this system, hoping that one day it will save your life. Then it will drive you so mental that you'll want to drive into a wall and kill yourself. And it won't let you. My advice then is simple. There are many options you can have on the RS 5. But don't, whatever you do, buy this one.

Which brings me on to the nutty question. Should you be buying an RS 5 cabriolet at all? Well, let's start with the engine. It's a joy. To meet stringent emission regulations, most new engines are turbocharged. You can't really tell. There's no lag any more between putting your foot down and the commencement of acceleration. And yet . . .

When you put your foot down in the Audi and that naturally aspirated 4.2-litre engine is energized, it's just better. You can't put your finger on why this is so. But it is. This is a tremendous engine – one of the very best in production today.

And the rest of the car? Well, that depends. If you have the on-board drive control software set to Comfort, it's a fairly quiet, reasonably comfortable boulevardier. You don't drive it in this mode so much as promenade in it.

But if you engage the Dynamic mode, the feel of everything changes. The engine becomes more urgent, the gearbox snaps to attention. Even the noise is different. In Comfort mode the RS 5 is quiet, like a nuclear power station. You sense rather than hear the grunt being produced. But in Dynamic mode it's like a nuclear power station that's blown up. It's loud and a bit scary. I liked it.

It encourages you to drive a little more quickly, to explore the outer limits of the four-wheel-drive system. But sadly I can't report on how the handling stands up to right-foot brutality because the active lane assist system kept steering me where it wanted to go, rather than what was necessary.

It's remarkable, really, that I liked this car so much when it had such a deeply annoying feature. It would be like AA Gill enjoying his dinner, despite the large globule of chef phlegm that was clearly visible on the potatoes.

Perhaps it's because my week with the car coincided with what the government laughably called a 'level-three heatwave'. A convertible this good in weather that sensational was a joy. It made you question the need for a driverless car. It would be like having driverless sex. Why give over such pleasurable duties to a machine?

I had that canvas top up and down like a pair of whore's drawers and I learnt many things: that you can raise or lower it at speeds of up to 31 mph; that when it's up, it keeps the sound on the outside and the chill on the inside. And that when it's down, you get a burnt face.

I also decided that Audi does a pretty good interior these days. Everything is sensibly placed, intuitive and well screwed together.

There are a couple of things that grate, though. It's not the best-looking car in the world. It comes across as heavy, which it isn't especially. And it is bit pretentious. But worse than this is the price. It's not even adjacent to cheap. In fact, it's on the wrong side of bleeding expensive.

However, until BMW gets round to launching a convertible version of the forthcoming M4, this is as good as it gets. Just remember: do not fit the active lane assist. And spend the £370 you save on Red Bull. That way, you won't need it.

4 August 2013

Gliding gently into the parking slot reserved for losers

Peugeot 2008

My children are not even remotely interested in cars. My son has a Fiat Punto, simply because it's Italian and he quite likes pasta and Inter Milan. He has no clue about its engine and is really only bothered about fuel economy. Speed, he reckons, is dangerous and silly.

It's the same story with all his friends. They can tell you who plays centre-back for every football team in Europe, and how many left-footed goalkeepers there are in the Premier League, but could they tell the difference between an Audi and a Mercedes? Not in a million years. Would they be able to identify Kimi Räikkönen? Nope, not even if he was all alone in a shed wearing a name badge.

Then we have my friends. Of these, maybe three are what you might call interested in cars. The rest simply aren't bothered at all. Mostly they go out every couple of years and buy whatever Range Rover happens to be in the showroom that day.

I'm aware also that there are significant numbers of people who dislike cars in the same way they dislike soap and David Cameron. And all of this raises the question: who's watching *Top Gear*? Who's buying all the car magazines? Why has Ferrari just had a record year?

I've been pondering on this for quite some time and now I think I have the answer. It's the people who go shopping in London's gigantic Westfield centre.

The multistorey car park there is permanently packed with slammed Volkswagen Golfs, tarted-up Beemers and tricked-out Mercedes. It echoes constantly to the bellow of big-bore exhausts, the squeal of tortured 35-profile tyres and the boom of megawatt sound systems. It's a cathedral to the god of horsepower. A meeting point for the disciples of speed.

They've even worked out their own rules in there. The car park's owner has introduced a one-way system and various give-way points, but they've been replaced with a simpler system, which is: whoever has the most expensive car has the right of way. It works rather well.

Of course, you get people who cheat. Recently a chap in an AMG-badged Mercedes tried to nick the only remaining space outside Waitrose, but I'd already clocked that he was actually in an E 250 so gave no quarter at all.

However, last week it all went wrong, because I was driving a Peugeot 2008 Cross Dresser. And that, in the car park at Westfield, is the bottom of the food chain. It is the speck that insects eat. So you have to give way to absolutely everyone and then you must park in the loser lane, miles from the shops.

It's much the same story elsewhere. In Notting Hill two trendy-looking Dutch tourists stopped dead in their tracks, pointed at me and burst into peals of laughter. Then they took photographs and wandered off, laughing at those too. I think that if I'd just spent £19,145 on this car, and people laughed at me wherever I went, I'd be a bit disconcerted.

But I wouldn't have just spent £19,145 on this car because I simply don't understand the appeal. And it's the same story with all its rivals. The Mini Countryman, the Ford B-Max, the Vauxhall Whateveritis. And that dreary new Renault. You pay more than you would for a standard hatchback and all you get in return is the ability to drive while wearing a busby.

I think they're cars for people who've completely given up on life. They know they will never again have sex outside, or wake up with a hangover. Life has become a beige montage of comfortable shoes, nights in front of the television and excruciating anniversary dinners at the local Harvester. When you see someone go past in a car such as this, you know that he will be actively looking forward to the cold embrace of death.

But the fact is that many people do like cars of this type, in the same way that many people like *All Star Mr & Mrs* and marzipan. So it is my job to see how the Peugeot stacks up. And the truth is . . . it's really not bad at all.

It won't go round corners very fast, and it's about as exciting as being dead, but as a car for someone who sees no fun in anything at all, it makes a good deal of sense. Providing it doesn't go wrong. Which, because it's a Peugeot, it probably will.

But if it doesn't, it's surprisingly good. First of all, the suspension is so delightfully soft you don't feel potholes at all, or speed humps, or your neighbour's bicycle, or any of the other things that Peugeot drivers are prone to running over. It glides around like a hovercraft.

You can buy it with a petrol engine but why waste your money? You're not interested in revs or speed. You're just waiting for death, so save your pennies on fuel and have the diesel. That's the engine I tested, and again I was surprised. It pulls so well from low revs that you could stick it in top and leave it there all day. And it's not just torquey; it sips diesel in the same way that an old lady sips a sherry.

It's a nice place to sit in as well. There are natty materials in the cabin, with some nice stitching on the upholstery, and my model came with a glass roof, so it felt airy and pleasant. The satnav was dead easy to use, the car had many toys and even possessed a Range Rover-style traction system. You tell it what sort of surface you're driving on and it decides which wheel should get the power. The only slight niggle is that it's not actually four-wheel drive. So, really, it's just a knob for impressing your passenger. It doesn't really do anything at all.

I suppose, grudgingly, I will admit that because of the taller, boxier body you do get more space everywhere than you do in the standard hatchback. There really is room in the back for people, and the boot could handle three medium-sized dogs. Four, if you didn't like them very much.

So as a tool it must be said that this car ticks many boxes and does a lot of important things extremely well. But I hated it. I loathed the way it makes no attempt at all to be exciting or exhilarating. And I felt embarrassed to be in a Peugeot, which has become a badge of honour for the terminally uninterested.

In short, this car does absolutely nothing for me, and I don't blame my fellow disciples in the Westway shopping centre's car park for treating me – and it – with such disdain and derision.

But for the vast majority of people – and I mean, 97.3 per cent of the population – it makes a deal of sense. If – and it's a big if – it is reliable, then you really couldn't ask for more. The only trouble is, the sort of people I'm talking to have used this bit of the newspaper to line the budgie cage. So they'll never know.

10 August 2013

Where the hell did they hide the 'keeping up with Italians' button?

Jaguar F-type

I bet there'd be a hint of regret as well – a sadness that my life hadn't worked out quite as well as I'd hoped, and that the chap who'd just blasted by had more important things to do than sit behind my sorry arse all day long. That he was more important and more clever. And possibly fitted with a bigger gentleman sausage.

Well, last week I found out exactly what it is like to be passed by faster-moving traffic. I was driving the new Jaguar F-type; the one with the big engine. The V8 S. It can accelerate from 0 to 62 mph in less time than it takes someone with a mild stutter to say '62 mph' and it has a top speed of close to 190 mph. It is a very, very, very fast car. But I was in Italy, so absolutely everything else was even faster.

I pulled out of a restaurant one night, roof down, stars twinkling and Lake Como stretching away into the moonlight shadow of that glorious Alpine hinterland. With something zesty on the stereo, I engaged Dynamic mode and roared off into the night.

It was an epic evening to be driving, and an epic road, but after a short while I began to notice that a pair of headlights that had been some distance back were gaining awfully quickly. And, of course, not wanting to inconvenience a local by getting in his way, I increased to flank speed.

Now the Jag was really bellowing. That big 5-litre supercharged V8 was flexing its muscles after each corner and hurling me into the xenon glow that lay ahead. But it was no good. The headlights were right behind me by this stage, so I did what any decent human being would do: I pulled over in the next lay-by to let him past.

And guess what. He was driving a 1.25-litre Ford Fiesta. The old model with the Yamaha engine and the silly oval grille. That was not an inspiring car when it was new, and now, more than fifteen years later, I bet it's even worse. But I can report that with an Italian at the wheel it is still faster than a hot Jag.

And there's more. It is a fact that on a twisting mountain road a car is always going to be faster than even the fastest superbike. A bike simply doesn't have enough grip to corner anywhere near as quickly as a car . . . unless there's an Italian on board, in which case it somehow has all the grip in the world.

I love driving in Italy, I truly do, because there the car has nothing to do with environmentalism or politics. And to describe it as a means of transport is the same as describing a fresh sardine that's been grilled in a bit of butter and flour as a means of staying alive. To an Italian the car is an expression of your soul, your zest for life. Speed is not dangerous. It's necessary.

You may know the coastal motorway that heads from France towards Genoa. For an hour or so you are either in a tunnel or on a viaduct or going round a hairpin bend with a 1,000-foot drop on either side. If this were anywhere else in the world there would be a 30-mph speed limit, enforced by helicopter gunships. But it's Italy so it has the same 130-kph (81-mph) limit as all the other motorways.

And that, I presume, is a 130-kph minimum, because everyone – mums, nuns and the conker-brown, walnut-faced peasantry in their ancient Fiats – was doing more like 150 kph. Round blind bends where you could not possibly see if the road ahead was blocked. It was madness, and I loved it.

Twice, over the years, I have been pulled over by the Italian police while driving a Lamborghini. And on both occasions I was told very sternly that I wasn't driving fast enough. You have to love that. And on my most recent trip I was getting much the same sort of treatment from everyone. I really did get the distinct impression that many people were extremely annoyed with the lumpen, badly dressed Englishman in his enormous, slow-moving Jaguar.

I, meanwhile, was having a ball, because the F-type is one of those cars, and Italy is one of those places, where you stomp about all morning thinking up excuses to go for a drive.

It's actually quite a childish car. It makes a range of extremely childish noises, especially if you engage the Sports Exhaust setting, which makes the back end snort like a hippo when you change up, and bang and crackle when you lift off. I did this a lot in the tunnels.

The styling is quite childish as well. It's pretty in the same way as a child's fridge-door drawing of a princess is pretty. It's simple and clean, and I stand by my earlier claim that it is one of the best-looking cars yet made. Precisely because it isn't trying to be all grown-up and German.

But there are a lot of things that are not childish at all. The gearbox stands out. There are flappy paddles behind the wheel, but it isn't a boy-racer double-clutch affair that works well on a racetrack but falls to pieces in town. It's a conventional eight-speed automatic and it's a delight.

15 September 2013

Go and play with your flow chart, Comrade Killjoy, while I floor it

Audi RS 6 Avant

The Sunday-evening crawl back into London is enough to make most sentient beings wonder if they should pull onto the hard shoulder and shoot themselves in the head. The weekend is over. There is nothing but drudgery ahead. The kids are tired and crotchety. And the traffic is dreadful.

It was always thus. But now, on the M1, the government has found a way to make everything much, much worse. Because every few hundred yards there is an overhead gantry that informs motorists the speed limit has been lowered to, say, 50 mph. And that speed cameras are on hand to catch those who think that's stupid. This means that everyone drops down to the new limit. And there's a word for this: communism.

I don't doubt for a moment that many people with interesting hair and degrees in advanced mathematics have spent several weeks working with the principle of flow dynamics and have decided that when x number of cars are using the motorway, pi equals MC^2 and that the speed limit should be lowered to ensure a smooth passage for everyone. Certainly we know their arguments took in the former transport secretary, John Prescott, who announced that the slower you go, the faster you'll get there.

Unfortunately, human beings are not molecules. We cannot be likened to water flowing down a hosepipe, because we're all different. Some people are pushy and dynamic. Some are mice. It is a fact that if you gave everyone in the country £100 today, tomorrow some people would have £1,000 and some would have nothing. And that's what the mathematicians don't seem to understand.

When Russia experimented with the idea of making everyone the same, it wasn't long before it needed a secret police force to keep the system going. Gatso. KGB. Same thing, really.

I came down the M1 last Sunday evening, and I think I'm right in saying that I have never been in a situation on any road anywhere in the world that was quite so dangerous. Because all of a sudden the pushy, dynamic people were stuck, and the car in front could neither speed up, because it was being driven by a mouse, nor pull over, because everyone was doing 50, so all three lanes were clogged.

This sort of thing makes the alpha male mad, so he starts to tailgate and undertake, using gaps that aren't really there. And that causes the mice to panic-brake. Then you're in a world of squealing tyres and tortured metal, and pretty soon you have the headline: 'Dozens die in juggernaut dance of death'.

I suppose I should explain that by far the worst offender that night was me. This is because I was in a rage at the politicians who allowed this system to be implemented. I was in a rage at the lightly dented Ford Galaxy in front that would not pull over, even when its driver had the chance. I was in a rage at the mathematicians who were responsible for the 50-mph limit. But most of all I was in a rage because I was in an Audi. A big, twin-turbo RS 6 that was the colour of a dog's lipstick.

We all know that Audi drivers are by far the most aggressive you encounter, and I've often wondered which comes first: the temper or the car.

Well, now I have the answer. Most of the time I'm pretty calm behind the wheel. I occasionally mutter the odd profanity at another motorist's idiocy, but I don't tailgate, I don't shake my fist and I don't arrive at my destination with a face the colour of a plum and armpits like Lake Superior. And yet in that Audi I did all those things. I think the company put testosterone in the air-conditioning system.

Or maybe it's the small-man syndrome at work. We all know that people who can't reach things on high shelves (no names here, Richard) have a bad temper because they are not as tall as all their friends. Well, could it be that Audi drivers are in a permanent state of fury because they do not have a BMW or a Mercedes?

Either way, I was a menace that night, getting far too close to the car in front in a stupid and dangerous attempt to scare the f****** b****** into getting out of my f****** way.

And while engaging in this idiotic pursuit I noticed something strange. The Audi was fitted with a radar in its nose that warned you when you were travelling too close to the car in front. This is available in many cars these days, and normally it errs on the side of caution. Not in the Audi, it doesn't. It issues a red alert only when you are precisely 1 inch from the car ahead. And even in my deranged state I thought that was a bit silly.

And I suppose while we're looking at the negative points we should examine some of the other things that are wrong with the new RS 6.

No 1: it's not that nice to drive. You have a four-wheel-drive system that uses a mechanical centre differential to apportion power between the front and the back. You have adaptive air suspension. Then you have more diffs that send the power from side to side. And you have a steering system developed after more than a century of trial and error. But most of the time it's uninvolving, and then very occasionally, when you are really tanking along, it all gets overwhelmed by the torque and goes a bit wobbly. If you really do want a large estate car that feels like a Ferrari in wellies, an AMG Mercedes is better.

That said, the Audi's engine is a peach. The last RS 6 was propelled by a big 5-litre V10, but for this one the company has fitted the twin-turbo 4-litre V8 that Bentley is now using in the Continental GT.

There's less power than before but there is also less weight. A fifth of the car's body is now made from aluminium. The wiring is as thin as possible. The soundproofing is chosen for its similarity to helium. And as a result the performance is still somewhere between electric and mind-blowing. This is a car that will take two children back to boarding school after the summer holidays – and yet it will get you from 0 to 62 mph in less than four seconds. And it has a top speed of 155 mph.

It's not just brute force and ignorance, either, because when you are just pootling along, four of the eight cylinders shut themselves down. And to make sure the car doesn't shake itself to pieces as a result, the 'active engine mounts' are fitted with 'electromagnetic oscillation coil actuators [that] induce phase-offset counter-oscillations which largely cancel engine vibration'. You can tell it's German, can't you?

But this is exactly the sort of engineering that is missing from the Jaguar F-type V8 S, which I wrote about last week. The sort of stuff that makes you go, 'Huh?' The entire RS 6 is riddled with it. Clever solutions to problems you simply didn't know existed. Some of it has to do with weight. Some with delivering music from the entertainment system. You sense all the time that you are driving not so much a car as an engineer's homework.

Maybe that's why it feels a bit detached. A bit uninvolving. Because, unlike the Jag, it wasn't built with passion; it was built with maths. And maths, as we know from the Stalinist cameras on the M1, doesn't always work.

22 September 2013

Who lent Scrooge the ninja costume?

Lexus IS 300h F Sport

Over the millennia, man has been consumed by a need for speed. In the Stone Age the fastest runners would catch the best food, and that made them the kings of the hill. Then came the horse, and it was the same story here. Genghis Khan was successful because his cavalry soldiers wore silk armour, and that made them faster.

In the days of steam, engine drivers would compete to see who could wrest the best times out of their locomotives, and at sea, liners would stage races across the Atlantic. Then, in the Cold War, whoever had the fastest jets was deemed to be winning.

When I started driving, it was all my friends and I talked about. Which one of us had the fastest car? I would spend hours scouring the auto-porn magazines for evidence that my Volkswagen Scirocco GLi was faster than Andy Scott's Vauxhall Chevette HS. And when we were out and about he would do everything in his power to demonstrate that it was not.

But now something strange has happened. Speed no longer seems to matter. Concorde has been replaced by the fuel-efficient Boeing 787 Dreamliner. HS2 is being questioned because of the cost and the impact it will have on 'communities'. And on the roads everything possible is being done to slow us down. Not that long ago Frank Beard, the drummer with ZZ Top, said he had a Ferrari because that way 'I can leave for the party later, get there first, stay longer and still be in bed with someone before anyone else'.

Today, though, I listen to teenage boys discussing their first cars, and all they ever seem to talk about is fuel consumption. My son is extremely proud of his Fiat Punto TwinAir, not because of the snazzy wheels or the turbocharger, but because it can do more than 60 mpg. Which means he gets to the party after everyone else but has more money to spend on beer. Which means he can't drive home afterwards and has to sleep on the floor.

These, then, are strange times, which brings us to a strange car. The new Lexus IS 300h F Sport. It looks extremely aggressive. There are fat alloys, sharp daytime running lights, a lean-forward stance and a grille so big I'm surprised it doesn't have its own moon. This is a car that trumpets a very clear message to the rear-view mirror of drivers in front: 'Get out of my way.'

It's the same story on the inside. The dash is a direct copy of the cockpit in the brilliant Lexus LFA supercar. You have dials that move about, information you didn't know you needed, a mouse to operate the command and control system and a device that coughs discreetly, like a butler, when you are approaching a speed camera. You can turn this off. Not sure why you'd want to, though. The seats are body-hugging and superb. There is space for many, and you get a large, sensible boot into which you can put things.

So, what we have here appears to be an interesting alternative to the BMW 3-series. A front-engined, rear-wheel-drive sports saloon car with the added benefit of Japanese electronics and Lexus quality. However . . . it takes about four seconds for you to realize that this is not a sports saloon at all. Instead it is a car tailored for today and our new-found desire to save money. This is a car built for one thing: economy. This is a hybrid.

The four-cylinder petrol engine is designed not to produce as much power as possible but as little friction. Then you have the electric motor, which cuts in and out seamlessly. And then there's the electronic gearbox . . .

It simply doesn't feel or sound like any car you've driven. The revs rise and fall instantly. There are no gear changes as such. And because the noise it makes has nothing to do with the speed you're going, you do tend to arrive at corners either far too quickly or nowhere near fast enough.

Meanwhile, on the dash you get read-outs telling you all sorts of things that you don't really understand. You can learn, for instance, which motor is driving the wheels at any given moment and when momentum is being used to make electricity. Drive lightly and you are told you are being economical. Mash your foot into the carpet and you are told you are using energy. I know that already. My foot's halfway through the bloody firewall.

They say it will get from standstill to 62 mph in 8.4 seconds, which is respectable enough, but at no time does it feel even remotely sprightly. You put your foot down on the motorway and it's as though something is broken. There's more noise but no more speed. Not until you're going past Penrith, at least.

You can put it in Sport S+ mode, if you like, which brings up a rev counter but precious little else. So although the front end is barking orders at the car ahead, you'd better hope it doesn't pull over, because you sure as hell aren't going past. The IS 300h feels like a car. It looks like a car. But it doesn't behave like one, and I'm afraid I just found it annoying.

But I'm being a dinosaur, aren't I? I'm judging the baby Lexus on speed, which these days is bit like judging a dog on its ability to write poetry. I care about speed. Frank Beard cares about it. But everyone else? No. Not really.

Which brings us on to the important news. If you go for the non-F Sport model on skinny tyres you get an output of just 99 carbon dioxides. A meaningless figure to the likes of me, but if you're a higher-rate-tax-paying company-car driver, that low, low figure is going to save you a fair amount.

Then there's the question of fuel consumption. Well, officially, you're going to get about 60 mpg, which sounds almost unbelievable. And that's because in the real world it is. In town, where the hybrid system really works for a living, the Lexus will be massively more economical than all its rivals. But on the motorway or the open road you will have to work that throttle hard to keep up, and that will bring the economy figures way down.

Which raises a question. Why did Lexus not do what every other car maker does and fit a diesel? Why go to all the bother of fitting two motors? Why have all that extra weight? Why make it all feel so different? Simple answer, apparently. Lexus doesn't do diesels. Doesn't know how.

I make no bones about this at all. If I were in the market for a mid-size executive saloon car and I had one eye on the fuel bills, I'd buy a BMW 3-series with a diesel engine. It would be torquey, fast, cheap to run, smooth and conventional.

The Lexus is the future: of that we can be fairly sure. But I'm not sure we're ready for it yet, because the dinosaurs have their petrol engines and the new youth have their diesels. Which means hybrids are catering for a market that doesn't yet exist.

29 September 2013

Crikey, the Terminator has joined the *Carry On* team

Mercedes-Benz SLS AMG Black Series

At present your car's annual tax bill is based on how much carbon dioxide is emitted from its rear. And not since William III's window tax have we seen anything quite so stupid. You might as well levy people on how many armpit hairs they have.

My problem with taxing a gas is that to cut down emissions of it, cars are being ruined. Hopeless electric power steering is now replacing the 'feelsome' hydraulic systems of old because it is less of a drain on the engine. Double-clutch gearboxes are replacing smooth slushmatics because without a torque converter the economy is better. Which means less CO_2.

It's probable that fairly soon the last V8 will roll off a production line somewhere in the world. I like V8s. They are inherently unbalanced, which is what makes them sound all gruff and rumbly. But each cylinder has to be fed with fuel and why feed eight when technology means you can now get as much power from feeding six?

This means the turbocharger is back with a vengeance. And while many of these blown engines are incredibly good, and remarkably free of noticeable lag, you know as you sit there that the throttle response has to be dulled. Which is the same as giving a connoisseur of fine food a plate of Smash. It's nearly mashed potato, and yet it just isn't.

And it's all going to get worse. Because every year the madmen in charge insist on less and less carbon dioxide, and the only way to achieve that is for cars to burn less and less fuel. Which, to start with, will mean more hybrids, and then as the lunatics keep on going, cars that are purely electric.

I have nothing against electric power at all, except for the total impracticality and the fact the emissions are simply being made at power stations rather than under the bonnet, but I do suspect that when we are all humming around the place in near silence we shall miss the good old days of crackling exhausts and instant responses and limitless range.

And that's why I've been thinking: is there another way of taxing cars that keeps both the ecomentalists and the petrolheads happy? And I believe there is – tax weight instead.

Weight is the enemy of everyone except for the gun-toting, attack-dog enthusiast in a few Southern states of pick-up-truck America. But despite this, cars keep on getting heavier and heavier. It's our fault. We demand more space on the inside, more luxury equipment and more rigid safety cells, all of which makes a car fatter.

But if engineers can make an engine produce 130 bhp per litre of capacity – and Mercedes has done just that – then surely they can build a safe, big, well-equipped car that needs mooring ropes to stop it floating away at the lights.

Dragging extra pounds around means spending more pounds at the pumps. And that means more emissions, which is bad news – if you believe that sort of thing – for Johnny Polar Bear. So tax it. I certainly won't complain because weight also blunts a car's performance – not just its acceleration but its ability to go round corners. A heavy car will never be as much fun to drive as a light car.

I am particularly keen to have a go in the new Alfa Romeo 4C, which on the face of it sounds a bit hopeless. It costs around £45,000 yet it only comes with a four-cylinder 1742 cc engine. That's white-collar money for blue-collar power. And yet this is a car that tips the scales at just 895 kg – about half what the vehicle on your drive weighs.

It therefore doesn't need a big engine: 237 bhp – the stuff of hatchbacks – will give it a power-to-weight ratio of 268 bhp per tonne. And that's the stuff of full-blooded supercars. Along with more than 40 mpg, which you're lucky to get from a Toyota Prius. Frankly, if I were in charge, the 4C would be tax-free.

To get the weight this far down, Alfa Romeo has gone the extra mile and then it's gone round the corner and kept right on going. The wiring, for instance, is made as thin as possible. And the chassis is a carbon-fibre tub that weighs about the same as a loaf of bread. It's going to be good, this car. I can feel it in my bones.

And now I'm going to unpick every single thing I've just said by reviewing a car I have driven. The Mercedes SLS AMG Black Series. A lightweight car that isn't quite as good as its heavier brother.

To recap. AMG-badged cars are semi-lunatic versions of ordinary Mercs. Black Series cars are semi-lunatic versions of the AMGs. I have one, a CLK. It's bonkers.

But bonkers in a good way. Because it's not really built to go round a track as fast as the laws of physics will allow. It's not a Porsche or a Ferrari. Yes, it's lighter and more powerful than the standard AMG car, but these modifications have only been made to increase my smiles per hour. It's built to be a laugh.

It's much the same story with the standard SLS AMG. Oh, sure, it has a carbon-fibre prop shaft that weighs only 4 kg and an engine that can read Latin. But you try going round a corner quickly. The tail will swing wide and pretty soon you'll be making more smoke than a second world war destroyer. You'll also be giggling like an infant.

With the Black Series, though, Mercedes has put its sense of humour back in the box and gone all sensible. Odd that. It is normally so carefree. But whatever, the SLS AMG Black Series now has a Ferrari-style electronic differential that tames the rear end. It also uses exactly the same gearbox that Ferrari puts in the F12berlinetta. Though in the Mercedes it's tuned to last.

Oh, and try this for size. While the 6.2-litre V8 develops more horsepower than the standard unit, it delivers 11 fewer torques. That means less fire and brimstone when you put your foot down. And then, finally, various bits and bobs are now made from carbon fibre, which means less weight . . .

It should be good. And on a track it is. Very good indeed. Way faster than the standard SLS. But if you're going on a track, why use a pantomime horse that's been converted? Why not get a car that was built to be quick in the first place? A much cheaper Porsche 911 GT3, for example.

And on the road? Well, it still has all the creature comforts and the ride's not bad, so it feels quite similar to its heavier, slower brother. But it now comes with lots of showy spoilers and flaps. Imagine Kenneth Williams pretending to be the Terminator and you're sort of there.

I still love the standard SLS. I like the shape, and the noise and the hysterical muscle-car handling. It's one of my favourite cars. It makes me happy just thinking about it.

The Black Series doesn't. It's trying to be something it's not. If I wanted a serious car I'd wait to try the new featherweight Alfa 4C. I'm doing just that tomorrow. And that makes me happy as well.

6 October 2013

Grab her lead and forget all about the mess on the floor

Alfa Romeo 4C

My coffee machine is a complete and utter pain in the backside. It's a wall-mounted Gaggia and I cannot recall a single occasion when, after pushing the button, I have taken delivery of a cup of actual coffee.

It always wants water, and after you've filled up its bowl, it says, 'Empty trays.' So you empty them, and then it says they aren't emptied properly. So you empty them again and then again, and then you scrub them until they shine like a furnace worker's face. And then you put them back and it says, 'Trays missing.' So you put them in again more firmly, several times, until it says, 'Empty trays.'

Eventually, of course, you resort to extreme brute force, whereupon it becomes Italian and changes tack. 'Add beans,' it says. So you open another tin of £900 Illy coffee beans and, being careful not to upset the trays in any way, you pour them into – as I write, I can hear it doing things in the kitchen, but I don't know what – the bean drawer. And then it says, 'Clean unit.' So you have to go against every male instinct and find the instruction book, which tells you to hold clamp A while squeezing nozzle B for about a couple of hours, and then when you put it all back together it says it wants decalcifying.

Usually I don't get my morning coffee until it's time for afternoon tea. But, of course, it's worth persevering, because when the moment finally arrives the result tastes a whole lot nicer than the instant alternative.

It's the same story with your choice of pet. A dog requires almost constant attention. It raids your bin, gets the bones it's nicked stuck in its throat, bites the postman, eats the milk lady, poos on the carpet, wants a walk when it's raining, barks in the night for no reason and gets ill on Christmas Day, when the vet is too drunk to come over. But despite all this it's so much more satisfying than a feed-and-forget cat.

Which naturally brings me on to Alfa Romeo, an experience that's subtly different. I had one once, a GTV6, and it was like a coffee machine – that had been designed by a dog. At night it would let all the air escape from its tyres, its clutch would weld itself to the flywheel and once it dropped its gear linkage onto the prop shaft, causing an extremely loud noise to happen, followed by the rear wheels locking up. It was a constant nightmare.

But here's the thing: even when it was a sunny day, and it wasn't being premenstrual, it was a pretty horrible car to drive. The steering was too heavy, the driving position was tailored for an ape, second gear was impossible to find and it handled as though it was running on heroin.

It's not alone, either. At present, the Giulietta is ho-hum and the MiTo is ghastly. And if we plunge into the pages of recent history, we find the 8C, which wasn't quite as good as it looked, and the SZ, which was the other way round. But only because it looked as if it had been designed by a madman. The 33, the 75, the 156, the 159 and the 164? There's not a great car there. Just many puddles of oil on your garage floor.

And yet Alfa Romeo is still my favourite car maker. I still believe you can't really call yourself a petrolhead until you've owned one. So why is this?

It's no good going back to the Sixties and saying, 'It's because of the GTA.' Yes, it was fabulous, but it was one car in a torrent of rubbish. Judging Alfa on this one achievement would be the same as ignoring all of Mussolini's crimes simply because he once bought his mother some flowers.

I've had a good, long think and reckon that in all its history Alfa has made only four or five really good cars. Memorable cars. And that in the past thirty years it hasn't made one.

Yet the love remains, and I think it's because we all sort of know what Alfa could and should be making. We have in our minds a mini Ferrari. A supercar on a shoestring. Pretty as hell, lithe as a greyhound, cheap as chips and built for fun. We have in our minds the 4C. It is utterly gorgeous. Spoilt, some say, by the headlamps. Yes, maybe, in the way Cindy Crawford is spoilt by her mole – that is, not spoilt at all.

But it's not the looks that impress most with the 4C. It's how it's made. Before this, if you wanted a car with an all-carbon-fibre tub, you had a choice: you bought a machine such as a McLaren MP4-12C or you bought a Formula One racer. It's expensive to make a car this way, but that's what Alfa has done.

The benefit is lightness, and that's a theme it has continued throughout. So, if you're after luxury and soundproofing and lots of standard equipment, forget it. There's no satellite navigation. You don't even get power steering.

The result is a car that tips the scales, fat with fluids, at well under a ton. Which means it doesn't need a big engine. Instead, mounted in the middle of the car, is a 1742 cc turbo unit that itself is made to be so light it has to be bolted in place to stop it floating away.

Disappointed that it only has the four-cylinder engine from a motorized pencil sharpener? Well, don't be. Because, thanks to the lightness, you can get to 62 mph in 4.5 seconds and onwards past 160 mph. Way past, I found. Oh, and 40 mpg-plus is on the cards as well.

I shall make no bones about it. I loved this car. It's like being at the controls of a housefly. You can brake later than you think possible into corners, knowing that there's barely any weight to transfer. And it has so much grip. Then there's the noise. Or rather noises. It makes thousands. All loud. All mad.

Yes, the interior trim is shocking, but if you want that lightness, it's the price you pay. And you do want it. Because lightness is coming. It has to. It makes both the polar bear and the petrolhead happy. And in the Alfa it made me very happy indeed. I drove the car round Lake Como on a sunny evening and there was almost a tear in my eye. I kept thinking that life didn't really get any better.

Now the boring stuff. I fitted easily. The boot is big. The dash readout is clever and clear so you don't need spectacles to see how fast you're going. And you can choose how you want your car to feel. Really. Just put it in Dynamic mode. And leave it there.

There are only a couple of drawbacks. The gearbox is a bit dim-witted and the steering isn't quite as sharp as I had been expecting. Also, it's wider than a Mercedes SLS AMG, which means it's wider than Utah. And it costs around £45,000. That, for a carbon fibre-tubbed mini-supercar, is not bad at all. But it does put it in the same price bracket as a Porsche Cayman.

Of course, the Cayman is more in tune with where we are now. It feels sturdy, and well made and luxurious. But that sort of thing will have to stop. We will have to go down Alfa's route, which means, in fact, the 4C feels like the future.

It also feels like the Alfa that the company made only in your dreams. It feels wonderful. I'm sure, naturally, that it will be like my coffee machine to own. But, unlike with any other Alfa in living memory, the rewards will make all the effort worth it.

13 October 2013

Goodbye, Dino. It's the age of the mosquito

McLaren P1

Let us be in no doubt about this. The Toyota Prius is a stupid car for sanctimonious people. It has two power sources and is made from rare materials that have to be shipped all over the globe before the finished product is finally delivered with the morning muesli and a copy of the *Guardian* to some malfunctioning eco-house in some trendy part of town where the coffee shops sell stuff that no one understands.

Remember how people used to sew CND badges to their parkas in the 1960s? This simple act didn't actually stop the SS-9 missiles rolling off the Soviet production lines, but it did tell everyone that you were interested in nuclear disarmament. Well, that's what the Prius is: a badge. A full metal jacket that tells other people you are interested in sandals as well. It's a knowing wink, a friendly nod. And I hate it.

However, it will be viewed by historians as one of the most important cars to have seen the light of day. A genuine game-changer. Because versions of its hybrid drive system will eventually be fitted to every single car on the market. McLaren is already there. Its new 903-bhp P1 uses a 727-bhp 3.8-litre twin-turbo V8 that works in tandem with a 176-bhp electric motor. This has not been done to save the polar bear, but to produce more speed. A lot more. Yes, you can turn the V8 off and use the electric motor to drive you silently around town, but mostly it's used to fill in the performance hole while the turbos spool up, and to fire rev-generating backwards torque at the petrol engine during gear changes.

I asked a McLaren engineer if the P1 would have been even

faster if it weren't fitted with 324 very heavy laptop-style batteries and the extra complexity of the electric motor, and he was most emphatic. 'No,' he said. 'Really, no.'

So, in other words, McLaren has taken Toyota's concept and turned it into something else entirely. You can think of this car as Viagra. Designed originally as a drug to mend a patient's broken heart, it is now sold to keep you going harder and faster for longer and longer.

And McLaren is not alone. Ferrari is working on a car called, weirdly, the Ferrari the Ferrari, which uses much the same technology. Porsche is nearly there with a hybrid called the 918 Spyder. Already it has lapped Germany's Nürburgring in six minutes and fifty-seven seconds. That's faster than any road car has gone before. Mercedes is working on a hybrid S-class.

They're not statements. They're not cars for eco-lunatics. They are cars for people who want the speed they have now – and in some cases even more – but not the petrol bills. What we've done, then, is taken a technology intended for the greens . . . and hijacked it. We've weaponized the muesli.

There's more, too, because we are about to see a shift in the way cars are made. For years they've all been built along pretty much the same lines. The body is a sort of frame onto which the engine, the suspension, the outer panels and the interior fixtures and fittings are bolted.

This is fine, but as the demand for more luxury and more safety grows stronger, the penalty is weight. Twenty years ago a Vauxhall Nova weighed about 800 kg. Its modern-day equivalent is more than 1,000 kg. Many larger cars tip the scales at more than 2 tons. And weight blunts performance, ruins handling and costs you at the pumps. Try playing tennis with a dead dog on your back and you'll soon see the problem.

Happily, there is a solution. It's called the carbon-fibre tub and it's been the basis of all Formula One cars for years. It really is just a tub, which is used instead of the frame. And because it's made from carbon fibre it weighs less than Richard Hammond.

Seriously. But it is much stronger. Ferrari uses a similar thing in its road cars. So does McLaren. And now it's starting to filter down the food chain. The Alfa Romeo 4C has a tub. Maybe one day the Ford Fiesta will too.

Preposterous? Not really. I remember when the video recorder first went on sale. The Panasonic model was £800 and was viewed by the bitter and mealy-mouthed as being another example of life being all right for some. And here we are today with DVD players being available on benefits.

For about forty years cars have inched along, getting a little more refined and a little easier to use with each generation. They have been evolving at about the same rate as the trees in your garden. But, in part because of the law makers in Brussels and the need to meet tough rules on what comes out of the tailpipe, we are about to witness a seismic shift. The meteorite has landed, and if the species is to survive, it needs to change.

I look at all the cars out there now and all the cars in this supplement and I get the impression they are all dinosaurs, roaming about in the fields, chewing grass and bumping into one another, blissfully unaware that the dust cloud is coming.

Some of them have V12 engines. And they're not going to survive the storm. Nor will V8s. And that'll be sad. We'll all miss the rumble. In the same way, I'm sure, as when the last apatosaurus keeled over, the species that were left may have shed a bit of a tear.

But look at it this way. It's argued by some that dinosaurs actually evolved into birds. The velociraptor became the white tern. The Tyrannosaurus rex became the peregrine falcon. And cars will have to do the same thing. It's already happening, in fact. And it's not necessarily a bad thing. Ford has squeezed 124 bhp out of the 1-litre three-cylinder engine it fits into the Fiesta. And I challenge anyone to get out after a drive in that thing without wearing a grin the size of Jupiter's third moon. It's a riot and yet it can do more than 60 mpg.

That Alfa Romeo 4C is a pointer as well. It's light, so it needs only a little petrol-sipping 1742-cc engine to reach 160 mph. But imagine if it were a hybrid, if it had a small electric motor firing gobs of instant torque at the rear wheels while the petrol engine was waking up, and adding horsepower to the mix when the road ahead opened up. It'd be like driving a mosquito that had somehow mated with a water boatman.

I have enjoyed my time with the dinosaurs. I shall look back at the Mercedes SLS AMG and the Ferrari 458 Italia and the Aston Martin Vanquish with a teary eye. And I shall always keep a picture of the wondrous Lexus LFA in my wallet. But that chapter is closing now. We're about to start a new one, and from the snippets I've seen so far, it looks rather good.

20 October 2013

Watch out, pedestrians, I'm packing lasers

Mercedes-Benz S 500 L AMG Line

Because I spend pretty much all of my life at airports, I've learnt a great deal about the human spirit. And what I've learnt most of all is that a man is genetically programmed to go into a branch of Dixons.

You watch him with his little-wheeled hand luggage and his laptop bag, wandering past all the shops selling perfume, and all the other ones selling Chinese bears in Beefeater suits. He drifts past Smythson like a trout in a slow-moving river and looks neither left nor right as he meanders past the art gallery selling massive horses. He doesn't even register it – never even stops for a minute to think, How would you get an actual life-sized ceramic horse in the overhead bins?

But then, carried by the current of his tiny mind, and by impulses over which he has no control, he will slither into Dixons to have a look at all the new machines that beep when you push their buttons. It doesn't matter if the passenger is late and doing that half-run businessman thing. He will still go to Dixons. Nor does it matter if he's naked and plainly in need of some new trousers. He will still consider a quick browse in gadget central to be more important. Hungry? Thirsty? Minutes to live? None of these things will get between a man and his need to examine the latest GoPro camera.

Which brings me on to the new Mercedes S-class. Over the years this flagship has been the pad from which most of the important motoring innovations have been launched. Crumple zones. Collapsible steering columns. Airbags. That sort of stuff. If it matters, we saw it first on an S-class.

So what manner of new stuff is to be found on the new model, I hear you ask. Well, stand by and roll the drums, because . . . it comes with the option of having a choice of fragrances in the air-conditioning system. Don't mock. In thirty years' time, when the S-class is a minicab, you will welcome anything that masks the overpowering aroma of the driver's armpits.

I haven't finished with the air-conditioning system either, because you are also able to go into the on-board computer and alter the level of ionization in the air being delivered. And what is ionization? Good question. Glad you asked. Because it's the process by which an atom or a molecule acquires a positive or a negative charge. So, in my book, that means Mercedes has developed a car that can fire lightning out of the air vents. It says that this makes the interior more relaxing. If I were ever to use an exclamation mark after a sentence, I'd have used one then.

Anyway, sticking with the air-conditioning system, I can tell you that its innards are filled with dried coconut shells that absorb not just some of the world's more unpleasant gases but, with the help of the ionization, many of the world's viruses. This, then, is a car in which you cannot catch smallpox, cholera or even ebola. Which, as a safety feature, beats the crap out of a collapsible steering column, if you ask me.

Before we move away from the air-conditioning system, I should explain that different levels can be set for each part of the car. So if you have a passenger whose company you do not enjoy, you can make him hot, fire lightning into his face and give him the bubonic plague.

I'm aware I have now spent a long time covering the air-conditioning, so let's take a deep breath and move on to the seats. As you would expect, they can all deliver a 'hot stone' massage, the armrests are heated and in the long-wheelbase model those in the back can recline until they are pretty much flat. Pillows are provided, naturally, and they are not just heated but also air-conditioned. This . . . (That's enough air-conditioning stuff. Ed.)

There's more, of course, but we must move on at this point to the optional thermal-imaging camera. When engaged at night, it projects a greyscale image of the road ahead to a screen on the dash, and – get this – when it detects a pedestrian, the person in question is highlighted in red.

This is fantastic. Because when you drive down Baker Street you feel as if you are commanding an Apache gunship and that all the people are targets.

They are as well. Because if one of them does something that appears to be threatening, such as, say, stepping into the road, the Mercedes fires a laser into his eyes to warn him you're coming. I know you think I'm making this all up. But I promise I'm not.

Mercedes says the equipment is so sophisticated, it can tell the difference between a person and an animal. But this isn't so. Because when I reached my London flat late last Sunday night, the camera detected what it thought was a human hiding in the bushes, and a little red square highlighted his exact position. I could see nothing with the naked eye, so I drove over to find it was a paparazzo. Not a human at all.

Sadly I was not able to run him over because the Mercedes has a system that applies the brakes even if you don't. The same system allows you to engage the cruise control in stop-start traffic and sit back while the car quite literally drives itself, maintaining a safe distance between itself and the car in front.

But that is nothing compared with the cameras that guide what's called the Magic Body Control. I kid you not. When they see a speed hump or a pothole approaching, they don't just soften the suspension to minimize the jolt, they actually lift the wheel clear. So there's absolutely no jolt at all.

Oh, heavens. I've forgotten the interior lighting. You can choose from a vast array of colours and then choose how bright you'd like it all to be. And then, with a swivel of the knob, you can turn on the wi-fi. You then input the seventeen-digit code into your phone – don't worry about drifting onto the other side of the road while you do this because the Mercedes knows when you're about to cross the white line – and, if you haven't indicated, it will steer you back again.

Did I mention the fridge? Or the larder? Or the availability of TV screens that fold out of the centre armrest? Nope? Well, what about the button that allows you to choose just how high you'd like the electric boot lid to rise?

Some are saying that all of this stuff is ridiculous and the S-class is no longer sitting at the prow of motoring innovation. Others say – and I have some sympathy with this argument – that true luxury is achieved with a clever use of space, light and silence, and that a billion gadgets is no match for the sheer opulence you find in a Rolls-Royce.

And yet. My inner man loved foraging about in the Merc, finding solutions to problems that no one knew existed. It also comes with an engine.

27 October 2013

I can see the mankini peeking out over your waistband

BMW 435i M Sport coupé

When I first moved to London, during the war – with Argentina, that is – Knightsbridge was a quite genteel place full of old ladies and sausage dogs. It was an oasis of calm in the centre of that magnificent 1980s whirlwind. It isn't any more. Now it's one of the noisiest places on earth.

This is because it has been bought, pretty much completely, by gentlemen from the Middle East, all of whom drive extremely loud supercars. You can hear them start up from three streets away, and you can hear them leaving the lights from back in Qatar. A Ferrari drove past me last night fitted with exhausts like 120-mm guns being played through the Grateful Dead's sound system. And then there was a Lamborghini with a bark so loud it could frighten an old lady's sausage dog to death.

There are rules on how much noise a car can make, but because they are rules they can be broken. Many supercar makers have worked out how. They have noted that the EU noise inspectors, who test a car before it is allowed to go on sale, take their reading when the engine is turning at 3000 rpm. So the manufacturer fits a valve in the exhaust system that opens at 3001 rpm. This means all is lovely and quiet for the men with clipboards. But then, just after they've smiled and put ticks in the right boxes, all hell breaks loose.

Today most fast cars make a racket. Jaguar even fits a discreet little button that enables you to turn the silencer into a trumpet. So when an F-type goes down your street, it feels as if the Royal Artillery has just opened up with everything it's got. I shall be honest. I like a car to make a noise. I loved the muscle-car rumble you got from AMG Mercedeses before the turbocharg-ers came along. I love the melancholy howl of a Ferrari F12. And the shriek from the Lexus LFA was up there with Roger Daltrey's scream towards the end of 'Won't Get Fooled Again'.

So I was surprised and, yes, a little bit disappointed when I put my foot down for the first time in BMW's new 435i coupé. This is the sporty two-door version of the 3-series. In the past it would have been called the 3-series coupé but BMW has decided to give it a name of its own, which has allowed the stylists to have a freer hand. A much freer hand, as it turns out, because the only panel this car shares with its four-door stablemate is the bonnet.

The rest is all different and all new and nowhere near as dra-matic as I'd been expecting. Yes, it's lower and wider than the saloon – the rear wheels are about 3 inches further apart – but it lacks visual presence. And on the face of it, that doesn't sound such a good idea.

When someone buys the two-door version of a four-door car, they are spending more money and getting less practicality. And the only reason they would want to do this is: they want more style. And with the 4-series I'm not sure they're getting it.

Which brings me back to the noise. Floor the throttle and all you get is a gentle hum, the sound of an engine that is doing a spot of gardening or maybe popping down the road for a pint of milk. It doesn't really sound as though it's making much of an effort at all.

Maybe it isn't. Because even though it's a 3-litre turbocharged straight six, it's producing only 302 brake horsepower. That's 14 bhp less than you get from the 3-litre turbocharged straight six in the smaller BMW M135i. What we have, then, is a car that is more expensive than the more practical 3-series and slower than the 1-series. A bad start.

But here's the thing. While it is extremely enjoyable to put your foot down in an F-type Jag and listen to all those pops and bangs as you lift it off again, I have a sneaking suspicion that after a while it might become wearisome. Certainly if I'm arriving at an adult's house in my AMG Mercedes I do everything in my power to stop the engine from sounding as if a yobbo is pulling up outside.

So, in the long run, a car that hums rather than shouts might be a more rewarding companion. And a car that is quietly stylish without being tartan might be burgled and vandalized less often. We see this a lot with BMW these days. There was a time when they were brash and driven an inch from your tailgate by men with inadequate sexual organs. But not any more. BMWs have become . . . gentle.

With the 435i you still have the near-perfect weight distribution and an extremely low centre of gravity. The engineers have worked their traditional magic to make everything balanced and just so. It may not be the fastest car in the world but it is extremely rewarding to feel it turn one way as you go round a roundabout and then the other as you leave. Some cars wobble about; some lurch. A BMW just does as it's told.

A BMW also has extremely impressive antilock braking. Many good cars these days are fitted with a system that cuts in too early. It thinks you're panicking and about to crash when you are not. A BMW waits for you to be an inch from the tree before it says, 'Excuse me, sir. Can I be of any assistance in these troubling times?'

A BMW accepts that you are not a nincompoop. It accepts that you may be a very good driver, and that you may want to have some fun before the electronic nanny tells you to come inside and wash your hands before dinner.

Let me put it this way. If you have an Audi or a Mercedes or a Jaguar, you are telling the world that life is treating you well. If you have a BMW these days you're not really saying much of anything at all. You're like the quiet, grey man on the bus. Nobody notices you. And certainly nobody would guess that under your dignified, grown-up clothes, you're wearing a lime-green mankini.

In the 435i you really are, thanks to a little button that changes the characteristics of everything. Most of the time it'll be set to Comfort, but you can go to Sport+, which sharpens up all the important stuff. It's nice. I loved driving this car. I loved being in it as well.

As with all BMWs, there's no unnecessary detailing, no silly fuss and no superfluous gadgetry. You have a little readout at the bottom of the rev counter that is unintelligible and has something to do with polar bears, but everything else is A* common sense. I even found myself thinking that the optional head-up display was a good idea.

Let's be in no doubt. The 435i is not for everyone. It is not good value and it is no use for showing off. But if you miss the way Knightsbridge was. Or if you want a getaway car that no one will notice. Or if you are a grown-up. I can't really think of any car that is better.

17 November 2013

The crisp-baked crust hides a splodge of soggy dough

Kia Pro_Cee'd GT Tech

Peter Mandelson has been responsible for many important initiatives over the years. Though for the life of me I can't actually recall off the top of my head what any of them were. Er . . . he had a moustache, and is reputed to have once mistaken mushy peas in a Hartlepool chip shop for guacamole. Oh, and without him the snazzy car I'm going to discuss might never have happened.

Mandelson was business secretary in the aftermath of the global economic crash and decided that what the British car industry needed to help it through the ensuing recession was a bit of government support. He therefore came up with the scrappage scheme.

The idea was simple. If you bought a new car, you'd get £2,000 for your old one, no matter how asthmatic and rusty it had become – well, as long as it had an MoT. The effect was instant. Sales of new cars rocketed by 30 per cent, airfields filled up with old bangers that would go to the crusher and everyone was very happy.

Those of a green disposition were pleased because old cars produce a lot more carbon dioxides than new ones. Shiny-suited car salesmen were happy because they didn't have to spend all day at work playing solitaire. You were happy because you got £2,000 for a car that was worth just shy of £7.50, and even the taxpayer was happy because, thanks to VAT, the government was earning more than it was giving away in subsidies. But the people wearing the biggest smiles of all were the Korean car makers: Kia and Hyundai.

You see, people were not exchanging their old cars for BMWs or Audis or Range Rovers. No. They were going for cheap runabouts. The scrappage scheme – not just here but in Germany and France as well – took Korean car dealerships from small lots on industrial estates in towns you've never heard of to centre stage in the cities. Let me put it this way: 98.8 per cent of all the cars Kia sold under the scrappage scheme were to people who had never owned a Kia before. Frankly, Mandelson should be given the freedom of Seoul. Certainly he should be adopted as a bonnet mascot for Kia. Especially as the name is derived – and I'm not making this up – from the Korean word 'to come out'.

Kia did not begin making cars properly until 1986. And even then only in very small numbers. In fact, it made just twenty-six. But the following year its numbers were up to more than 95,000. And all of them were extremely horrible. They continued to be extremely horrible even as the 21st century dawned.

I still maintain the worst car I have driven is the Kia Rooney. Or was it Rio? And the second-worst is the equally Korean Hyundai Accent with a three-cylinder diesel. Both still make me shiver. I'd rather sit in a bucket of vomit.

Had people bought one of these under the scrappage scheme, they'd have been breaking into the airfields and the scrapyards to get their old bangers back.

Today, though, things are very different. The Kia Sportage SUV is charismatic. The latest Kia Cee'd is verging on excellent, as is the coupé version, the Pro_Cee'd, and then there's the hot version of that. A hot Kia. A vehicle to rival the Volkswagen Golf GTI and the Ford Focus ST. Who'd have thought it?

First of all, the Pro_Cee'd GT is extremely good-looking. Sporty without being brash. It hints at its potential with discreet nudges, such as the red brake callipers, the deep sills and the little red stripe across the radiator grille. The lines are exquisite.

Then you get inside and it's much the same story. Yes, legroom in the back is tight, and after you've tilted and slid the front seats forwards to let someone in, they return to a position that would suit only Richard Hammond. But we see this idiocy in many cars these days. What we don't usually see is quite so many buttons.

I was driving the high-spec Tech version and it was festooned with switches. And equipment. You get a tremendous satnav system, and a button that changes the look of the instrument panel. You get voice activation and iPod connectivity. You get far more than you would get on a Golf GTI, that's for sure.

After you set off, it continues to impress. It's quiet and comfortable, and the driving position is perfect. There are lots of places to put your stuff, and nothing is rattling around in the enormous boot because a cargo net is holding everything in place. I especially liked the position of the gear lever in relation to the wheel, and was deeply impressed by the tiny gate through which it moves. First, third and fifth are no more than a few millimetres apart. It's as if Kia has taken the best bit of every hot hatch that's yet been made and put them all in the Pro_Cee'd GT.

Then the company has given it a seven-year warranty and a price tag of £22,495 – about £3,600 less than VW charges for an entry-level GTI. So that's all great and tremendous . . .

Unfortunately after about half an hour you realize that the whole car is a sham. A matrix. A veneer of excellence draped over a lot of pile-'em-high-and-sell-'em-cheap rubbish.

And it isn't hard to see why. Mothers Pride is a brand of sliced bread, and if all you want is a base on which you can serve a helping of baked beans, that's absolutely fine. But what Kia has done with the Pro_Cee'd GT is wrap a slice of what my grandfather used to call 'wet vest bread' in a faux crusty exterior with bits of nut sprinkled here and there.

The electric power steering and the brakes feel cheap. So does the gear-change action, and so especially does the 1.6-litre turbo engine. It's so lacking in torque that you often stall when trying to dribble away in second. And at the top end it sounds like a cement mixer full of gravel. Speed? Well, there's some, but nowhere near as much as the red brake callipers and all those buttons would have you believe.

Then there's the weight. If Kia had been serious about making a proper hot hatch, the GT wouldn't weigh more than the Pacific Ocean. And it's a heaviness you can sense when you are cornering, accelerating and braking.

Kia needs to understand that the real hot hatches from Ford and Volkswagen are designed by engineers who care, and signed off by accountants who really wish they didn't. The Pro_Cee'd GT? It's just a half-witted attempt to pull the wool over our eyes. It's not cheap just because it costs less than a Golf GTI. It is, in fact, expensive, because it costs so much more than it should.

In the Seventies various long-forgotten electronics companies made music systems that appeared to offer the same level of performance as Wharfedale, Marsden Hall and Garrard. They had many bells and whistles and they were cheap, but to anyone with ears they sounded dreadful. Well, that's exactly what's going on with the hot Kia. It's a good-looking, well-equipped bag of Virgin Cola. I hated it.

24 November 2013

A menace to cyclists, cars, even low-flying aircraft

Audi SQ5 3.0 BiTDI quattro

Every lunchtime on Radio 2 Jeremy Vine hosts a topical news and discussion show in which the 'motorist' is always portrayed as a swivel-eyed, testosterone-fuelled speed freak with the social conscience of a tiger and a total disregard for the wellbeing of others.

This always strikes me as odd because just about everyone over the age of seventeen is a motorist. Which leads us to the conclusion that in Vineworld all adults are men, and we are all mad or murderers or a worrying mix of the two.

There was a debate recently on the show about pelican crossings and how elderly people are not given enough time to reach the other side of the road before the lights go green. I know, I know. It was a slow news day. Apart from the tornados in America, the typhoon in the Philippines and the floods in Sardinia.

Anyway, Vine said that when an elderly lady is marooned in the middle of the road and the lights go green for traffic, motorists start to rev their engines. Really? What motorists do this? I have been driving for thirty-six years and not once have I ever been tempted to rev my engine to encourage an old woman to get a bloody move on. What's more, I've never heard anyone else do it either. The idea that an adult would do such a thing is preposterous.

But, of course, you can't be bothered to telephone the show and say that, because you would be faced with someone who says it happens all the time. And then you'd be in a does/doesn't argument until it was time for 'Mandy' by Barry Manilow.

This meant the counter-argument was put by a lunatic from

a 'motoring' organization who said that if the lights at pelicans were retuned to give old people time to cross the road, it would be bad for the economy. At that point I switched over to Radio 4.

Later in the show they were going to be discussing bicycles and why, in London alone, in the past month seven million cyclists have been killed by motorists on purpose. I couldn't bring myself to listen to that because at no point would anyone say, 'If you're going to put thousands of bicycles on the streets of London it is inevitable that some of them are going to be squished.' That would be the voice of reason. And that isn't allowed in Vineworld.

There are other issues, too, that are always held aloft as shining examples of the motorist's stupidity. We all drive with our rear fog lights on, apparently, even when the weather is dry and clear. Really? I ask only because I haven't seen anyone do that for twenty years or more.

We all hog the middle lane as well. This, of course, is true, but usually because the inside lane is crammed full of lorries. So technically we're not hogging it. We're just using it. We also block yellow junctions. Nope. You're confusing us with bus drivers.

Then we have young motorists who tear about at breakneck speed. This is a given. A fact. There is no arguing with it. Even though it simply isn't true. Most young people I know drive extremely slow cars very carefully because they can't afford the petrol that breakneck speed requires.

Yes, in the late 1980s and early 1990s there was a problem with twockers, and kids on the Blackbird Leys estate in Oxford tearing hither and thither in other people's hot hatchbacks. But that doesn't happen any more. So complaining about it is like complaining about BT giving people party lines. And the quality of the recordings on Dial-a-Disc. And French 101 lavatories.

There is, however, one Vine discussion topic that is worth the time of day. The new-found fondness people have for SUVs. Naturally in Vineland they're called Chelsea tractors and they're all driven by silly rich women and they all have bull bars. And pretty soon the producer will put a caller through from the Labour party, who will say, 'They were designed to go off road but all they ever do is put a wheel on the pavement.' And then I switch over to Radio 4 again.

The fact is this. There are two types of off-road car. There's an off-road car that is designed to go off road. A Range Rover, for instance. And then you have off-road cars that are not designed to go off road. These are called SUVs and they annoy me.

I look at everyone in their Honda CR-Vs and their BMW X3s and their Audi Q3s and I think, Are you all mad? An ordinary estate or hatchback costs less to buy and less to run and is nicer to drive, more comfortable and just as practical. But it doesn't take up so much bloody space.

I parked yesterday between two of the damn things in a London square, and because they were so wide I couldn't open my door, which meant I was stuck inside, being forced to listen to Vine's callers phoning up to moan about secondary picketing.

Now, though, things are getting completely out of hand because Audi has decided that what the world really needs is another fast SUV. And so welcome to the SQ5, the fastest-accelerating diesel SUV of them all.

First things first: it's not fast. If Audi had really wanted it to blister tarmac and earn its own slot on Jeremy Vine, the company would have given it a big petrol V8. But instead it has a twin-turbo diesel unit that is made to sound fast by the fitting of a speaker to the exhaust system.

Furthermore, if Audi had actually been serious about making it a high-riding modern-day take on the old quattro, it would have entrusted the suspension alterations to its in-house performance division. But it didn't. It simply added some fat tyres and lowered the suspension and left it at that.

You read that right. It lowered the suspension. So Audi made a car that was jacked up to suit the weird new trend. And then to capitalize still further on that trend, it lowered it again.

Oh, it's not completely horrid to drive. It zooms along with a fair degree of urgency, and I have to say the compromise between ride and handling isn't bad at all. Even though it's not as good as it would have been if the roof weren't a menace to much of Europe's air traffic.

Inside? Well, the back bench slithers backwards and forwards – a nice touch – but you don't get satnav as standard, which seems a bit mean. The worst thing, however, is the visibility. The pillars, the headrests and the door mirrors all seem to conspire to make everything outside disappear. You could easily run over a cyclist in this vehicle and simply not know it had happened.

Which brings us to an inevitable conclusion. No. Motorists get a bad-enough press as it is, without driving around in cars such as this. I drove it for one day. And then went to Yalta, on the Crimean peninsula, to get away from it. I'll come back when it's gone.

1 December 2013

I'm sorry, Comrade. No Iron Curtain, no deal

Dacia Sandero Access 1.2

It's strange. Today there are far fewer car makers than there were thirty years ago. And yet choosing what sort of car you would like next has never been more difficult. This is because thirty years ago only one thing mattered: the letter at the beginning of the numberplate. That's what told your neighbours you had a new car.

The idea of identifying a vehicle's age by a letter on the numberplate started in 1963. But quite quickly the car makers noticed that it was creating a massive problem. Because the letter denoting age changed on January 1, everyone wanted to take delivery of their new vehicle on New Year's Day. This meant car salesmen had to sit about dusting the pot plants for eleven months and then work like mad ants over the Christmas holidays. It made life tough in the car factories as well and created a cash-flow headache seen previously only in the nation's turkey industry.

And so in 1967 the changeover date became August 1. This, it was felt, would create two spikes. One at the beginning of the year, when people could take delivery of a 1968 model. And one in August, when the new letter became available. But it didn't work. That letter meant more than the endeavours of Pope Gregory.

That letter trumped everything. It said you were doing well. That life was being kind. It was critical. Nobody cared what sort of car they bought just as long as other road users knew it was new. And strangely the people this helped most of all were the comrades behind the Iron Curtain.

Cars made in the Soviet bloc were cheap. They were therefore the easiest way of getting the right letter on your driveway. People would see the H-registration plate and say, 'Have you seen the Joneses at number forty-seven have a new car?' and simply not notice that it was a Moskvich. Which wasn't really a car, so much as a collection of pig iron fashioned into a rough approximation of a car.

Or the FSO Polonez. Made in Poland by people who didn't care, from steel that was both heavy and see-through, it was utterly dreadful. The steering wheel was connected to the front wheels by cement, and when you pressed the accelerator, it felt as though you had sent a signal to an overweight and sweaty man in a vest, who rose in a disgruntled manner from his seat in the boot to put some more coal on the fire. Eventually this would cause you to go 1 mph faster.

Braking? Yes. It had that. Though really it was like trying to stop an overloaded wheelbarrow on a steep, muddy hill. Certainly in both cases you tended to end up with brown trousers. But despite all this the FSO sold in respectable numbers because it was available at no extra cost with a V on its numberplate.

Then there was the Lada Riva. It was originally designed by Fiat when Ben-Hur was still the star attraction in Rome – I mean the actual Ben-Hur, not Charlton Heston – and the design rights were sold to Lada, which promptly didn't develop it at all. Why should they? There was a thirty-year waiting list at home, there was no competition and there were plenty of people in Britain who'd buy one because of its numberplate.

Oh, the company had other reasons. It would argue that because the Lada was designed and built to handle Russian roads, it was tough. This was untrue. It was actually designed to handle Italian roads and it had the crash protection of a paper bag. The pillars supporting the roof had the strength of drinking straws, which meant that if you rolled during an accident your head was going to end up adjacent to your heart.

Skoda, bless it, tried its hardest with some interesting designs. But back at home it had no yardstick against which these designs could be measured. So often they didn't work. Though when I say 'often', I mean 'always'.

Unusual rear-suspension design on rear-engined cars meant that if you tried to take any corner at any speed, the rear wheel would fold up, the car would spin and you'd hit a tree and die screaming in a terrifying fireball. And at your funeral they'd say how sad it was because you'd just bought a new car.

Of course it was inevitable that one day the alphabet would run out of new letters for car registration plates, and so someone came up with the system we have today. You can still tell a car's age from its registration plate, but only if you have a calculator and the brain of an elephant.

Some say it was Ronald Reagan's proposed Star Wars technology that finished the Cold War and brought down the Berlin Wall. Others reckon it was the accident at Chernobyl that caused Mikhail Gorbachev to come to do the business with Margaret Thatcher. But actually it was the registration-plate change. Because that one single thing ended the demand for cheap-at-any-price new cars.

Which brings me, after quite a long run-up, to the Dacia Sandero. Prices start at £5,995, which on the face of it is astonishing value for money. Yes, it's built in Romania, which has exactly the same car building history as Ghana, but it's actually based on the one-before-last Renault Clio. So. This is a car that is based on a 2007 Renault, that does more than 55 mpg and that is yours for less than the price of most holidays.

There is probably an inconsequential issue with its name. Even though it's spelt Dacia, which in English rhymes with fascia, its maker insists that actually it should be pronounced 'datcha'. Which means you could end up with a Russian country house.

Using the Romanian pronunciation is silly. It'd be like the Florentine marketing board urging British people to visit Firenze. We wouldn't know where to go.

But what of the Sandero itself? Drawbacks? Yes. Plenty. It

looks as if it's been styled by someone who's never actually seen a car before. And it is a bit spartan. And a bit cramped in the back. And a bit slow. And a bit roly-poly in the corners. And, compared with some of its rivals, it produces quite a few carbon dioxides, which means you have to pay £125 a year in vehicle tax.

In the olden days none of these things would have mattered, because for less than £6,000 you could have had the latest registration prefix. But now you just get a thirteen or a sixty-three and nobody really knows what any of that means.

So the Sandero must be judged as a car, and I'm sorry, but for £5,995 you can do quite a lot better by trawling through driving. co.uk, *The Sunday Times*' second-hand-car website. This, then, is why buying a cheap new car is so much more difficult than it was. Because without anything that identifies it as new, you may as well plunge into the vastly more complex world of pre-owned.

8 December 2013

You're off by a country mile with this soggy pudding, Subaru

Subaru Forester 2.0 Lineartronic XT

When I was growing up, in the days before either health or safety had been invented, commercial breaks on the television were often filled with important public information films. They were designed to open our eyes to all of life's hidden perils, and some of them were jolly frightening.

In one we were told not to put a rug on a recently polished floor. In another we were warned about the dangers of fishing while under electricity cables. And in my favourite we saw a pretty young woman in a short skirt running down the pavement. Sadly she wasn't really concentrating, and as she rounded the corner she crashed head first into a large pane of glass being carried by two workmen. 'Don't run,' said the voiceover sternly. It's a piece of advice I've followed ever since.

Unfortunately far fewer of these films are made today, partly because some of the television presenters who fronted them are currently troubling the Operation Yewtree investigation. But mostly because after you've spent the day in a high-visibility jacket and a hard hat, filling in risk assessment forms, the last thing you need is for your evening's viewing pleasure to be interrupted with yet more reminders to stay safe.

However, while we are no longer told to learn to swim and wear a seatbelt and think once and then twice about motor-cycles, there remains one safety drum the government is still banging: we are still being told not to drive when we are tired.

Is this really the most important safety message it can come up with? What about driving when you are under the influence of marijuana or Vera Lynn? What about a message telling us not to swerve for badgers? Or cats? Or how about a simple film that explains to those recently arrived from eastern Europe about how a roundabout works?

Or maybe it's just me because I'm only ever tired about two hours after I get into bed and turn out the lights. During the day, sleep for me is impossible. (Except after the first corner in a grand prix. Then I can nod off no problem at all.)

However, last Friday night I set off up the M1. It was dark and the middle of rush hour but, unusually, traffic was flowing quite well. In the outside lane everyone was doing 60 mph, Simon Mayo was on the radio with his 'all request' Friday and I was going to have dinner with my boy.

It was all very warm and safe and pleasant and the engine was moaning out its one long song and I started to feel the same sensation I get after a lovely Sunday lunch and Sebastian Vettel has just taken the lead in the second corner. My eyelids became heavy. My head began to nod. And way off in the distance I noticed the brake lights were coming on . . .

Ordinarily this would cause me to slow a little and to cover the brake pedal. But I simply couldn't be bothered. It would have meant moving my leg and I was just too warm and cosy for that. Much easier, I reckoned, to plough into the back of the car in front.

Naturally the traffic wasn't actually stopping. It was just a moron in a Peugeot braking for no reason and causing everyone behind to brake as well. So I didn't have the accident. But I did for the first time pull over at the next services for a little walk in the fresh air and a cup of coffee. Weirdly I didn't need a government film to tell me to do this. It was just common sense.

Once I was back on the road, with some matches in my eyes and a drawing pin on the seat – that works well, by the way – I began to wonder what on earth had brought about this drowsiness. Yes, I have just finished a relentless spell of travelling and, yes, there have been a few late nights. But that's nothing new.

Which led me to the conclusion that I was being sent to sleep by the car I was driving – a Subaru Forester XT.

I've never felt drowsy in a Subaru before. This is because the cars are built for slightly over-the-limit rural types who wear extremely heavy shoes from Countrywide and have little interest in comfort.

This applies to all the models it has sold here. You had the original pick-up truck. Sold through agricultural suppliers and farm shops, it had a corrugated iron cover over the back, some sheep in the passenger seat and at the wheel a slightly over-the-limit driver with heavy shoes who'd never been to London.

Then you had the much-talked-about and greatly missed Impreza. Available in many stages of tune over the years, it came with a bonnet scoop the size of the Sydney Opera House and a turbocharger that was even larger than the driver's shoes. Imprezas made their mark in international rallying, a sport that's very popular with rural types who went to school with the local bobby and have no need to worry about the breathalyser kit in the back of his panda car.

And then there was the no-nonsense, go-anywhere Forester. It had no styling at all but it was extremely well made, a feature much prized in the shires.

In recent years, though, Subaru has been having a tough time. Sales in Britain have plummeted, and many have said this is because the strong Japanese yen made the cars expensive. That is rubbish. Sales were falling because of Tony Blair's crusade to make the countryside illegal.

So now it seems Subaru is fighting back by going all skinny latte and arugula metrosexual. The new Forester has been styled and the bonnet scoop has gone and it is big and well equipped and pricy. Which means that it's just another stupid sports utility crossover vehicle to make people in Surrey feel as if they live in the countryside.

Happily, unlike many other vehicles of this type, it does at least have some off-road credentials. A device that stops it running away on steep slopes, for example. And a boxer engine that a) makes a nice noise and b) gives you a lower centre of gravity. But sadly the gearbox is now what Subaru calls Lineartronic.

In essence, it's a continuously variable transmission affair, and CVT gearboxes don't work, even if they are fitted with eight artificial steps. A CVT gearbox detaches you from the sensation of driving, or being in control. Couple this to the electric power steering and a strangely mushy-feeling brake pedal and the sense of isolation is complete. As I discovered, you don't feel as if you're driving this car. It's just somewhere warm to sit as the world drones by.

Yes, it is comfortable and quiet and it does still appear to be well made, but there are now too many frills, none of which will be of the slightest interest to country types. I mean, an electric tailgate? Do me a favour.

I used to like the rugged, no-nonsense, rural nature of Subarus. But this one? I dunno. It feels as if Barbour has tried to make a dinner jacket. And failed.

15 December 2013

You can't play bumper cars, but the bouncy castle's brilliant

Volvo V40 T5 R-Design Lux

Many years ago, when *Top Gear* was made in the Midlands, I was sent some promotional bumf that said, 'You are invited to the opening of Birmingham's biggest restaurant.'

This confused me because at night people tend to say they want to go out for an Indian or a pizza, or to somewhere warm and cosy. Certainly I have never heard anyone say, 'You know what I fancy tonight? I fancy going somewhere really big.' Enormousness just isn't a selling point.

It was much the same story with cars. Back then, if you wanted something sporty you bought a BMW. If you wanted something reliable you bought a Volkswagen. If you wanted something durable you bought a Mercedes-Benz and if you wanted something safe you bought a Volvo.

And this was a problem for all the other car makers, because those four brands had all the important bases covered. I remember an adman at Audi sitting with his head in his hands, explaining that there were no other reasons for choosing one car over another.

People could see a clever ad about beating a German to the beach, or read a pithy review in *Autocar*, or talk to friends in the pub, but when push came to shove they wanted only one thing: safety, durability, sportiness or reliability. And those were already bagged. I suggested he go for 'Germany's biggest car'. But he said this would be silly and went instead for Vorsprung durch Technik.

Things have changed since then because Mercedes started to make little hatchbacks, BMW moved into diesels, Volvo went touring-car racing and today, if you want a reliable vehicle, you don't have to buy a Volkswagen. Anything will do. Except a Citroën. Or a Peugeot.

The motor industry is one big blur, with all the manufacturers offering something to suit everyone. Just about the sportiest car made today is a Nissan, and the least sporty is a BMW. The most durable car I know is a Toyota, and the least, probably a Mercedes-Benz from ten years ago. And yet in the midst of all this we have Volvo, which is sitting at the back with its hand up, still claiming that it's the one-stop shop for those who want to be safe. Indeed its engineers recently announced they were working on a range of developments that would mean soon no one would ever die while in one of Volvo's products.

As a general rule I hate safety. It makes me nervous because when I feel safe I have a nagging doubt in the back of my mind that I can't really be having much fun. As a general rule, the two things are mutually exclusive.

And, anyway, there's no point trying to be safe because things can often conspire to prove you aren't. For example, the most rigorously tested and inspected item *Top Gear* filmed was the jet drag-racing car Richard Hammond drove several years ago, and we all know what happened there. Whereas the least tested and inspected was the 'Hovervan', in which I found myself loose in a lock with a rampaging van full of blades. And I was not hurt in any way.

I laugh openly at the Royal Society for the Prevention of Accidents because here we have a body of worthies whose aim is to prevent something that, by its very nature, cannot be prevented.

And I'm afraid I scoff at Volvo's claim that soon no one will die in one of their cars because what if you are driving along in your shiny new V70 and a giant meteorite crashes into the roof? What if there's an earthquake? What if you are an arms dealer and a rival puts six tons of plastic explosive in your seat? Has Volvo considered all these possibilities? Quite.

Mind you, it does seem to have thought about pretty much everything else. Especially the business of protecting those in less fortunate surroundings. Because the car I've just been driving – a V40 T5 R-Design – is designed to make sure that you cannot run anyone down. And that if by some miracle you do, they will walk away from the impact thanking you very much for giving them such a good giggle.

In short, there are sensors that scan the road ahead, looking for people you might be about to hit. Warnings are sounded, and if you ignore them, the car will brake itself. And if this doesn't work and you crash into the poor unfortunate soul, the front of the car will turn into a giant bouncy castle, ensuring he or she has not just a soft landing but a fun one too.

I'm afraid it is impossible to test these claims in the real world so I cannot report objectively on whether they work. But I can ponder awhile on it: all this technology costs money. Which means you are forking out cash to pay for the wellbeing of other people.

In a darkened room, when nobody is listening, you may wonder about that. You may even decide to buy a Golf GTI instead. And you may use the money you've saved on a luxury holiday for your family in Barbados. This would make you very pleased.

And you'd stay pleased right up to the time when, through no fault of your own, you ran over a small boy and killed him. Then you'd wonder as you faced a life of shame and regret if perhaps the Volvo hadn't been the more sensible choice.

This, of course, is the trouble with safety. You don't want it in your life right up to the moment when you do. And, anyway, the Volvo doesn't just look after other people. It's claimed that it does a pretty good job of looking after you and your no-claims bonus as well. For example, when you are reversing out of a side turning into a main road, you are warned if the car detects oncoming traffic. Again, this was something I couldn't easily test.

But I did have a go with the automatic braking system. At

speeds of up to 31 mph the car will stop if it thinks you are about to crash into something. And certainly it works a whole lot better than the company's website, which doesn't work at all. It doesn't even seem to be sure that the front-wheel-drive five-cylinder T5 exists.

There are countless other touches too. Such as the key. You pop it into a slot high up on the dash and then push a button. This is annoying. But if you do somehow have a crash there isn't a bit of metal poking out of the steering column, waiting to rearrange your right knee. That's how my dad lost one of his kneecaps. He lost the other many years later while exiting a Ford Anglia through the windscreen.

And, oh dear, I seem to have reached pretty much the end of this week's missive without talking too much about the actual car. Which is fine. Because there's not much to say. It's very good-looking, quite nice to drive, reasonably fast, fairly comfortable and decently spacious. It is also a lovely place to sit, even if some of the controls are unfathomable. However, it is fantastically expensive.

So go ahead. Buy the Golf GTI. It's much better value. And a better car. But you will have to drive it with your fingers crossed.

22 December 2013

Drives on water and raises Lazarus in 4.1 seconds

Aston Martin Vanquish Volante

It was the week before Christmas. Rush hour. Central London. And the weather was every American's idea of what it's always like in Britain. Awful. The wind was coming in great shuddering lumps and the rain was a collection of stair rods. It was a night for being in.

But I wasn't in. I was out in the new Aston Martin Vanquish Volante, trying to find a parking space in St James's. Nobody's temper was even on that frightful night. The bus drivers had given up trying to run down lone cyclists and had just decided to kill everyone. The taxi drivers were hampered by steamed-up windows. Pedestrians were blind behind their inside-out umbrellas and, even with my wipers whizzing back and forth like a drowning man's arms, the whole scene was streaked with neon, headlamps and Christmas decorations. It was like driving on an acid trip, into a kaleidoscope. It was like having all of the headaches I'd ever had, at once.

At a time such as this you want to be in a car only because it's dry. You certainly don't want to be in a £199,995 Aston Martin with bone-hard suspension and a roof that has seemingly been designed specifically to make everything abaft your head invisible. At an oblique junction the only way you can pull out safely is by having a deep and fervent belief in God.

The next morning I was down at the *Top Gear* test track and it was the sort of day we dream about. Crisp and cold and bleached. The sun was pale. And the air was as clear as a lake of gin. What's more, the track was quiet, empty and beckoning. But even though the Aston has a 565 brake horsepower V12 engine, I didn't bother taking it out there and opening the taps of that mountain of muscle. Because I've done some track work in its hard-top sister, so I know what it's like.

Although it's largely made from carbon fibre, it's a heavy car, and it gets all bolshie and uninterested when you push it hard. The tyres don't last very well either. After three laps they lose their bite and you end up with 300 yards of dreary understeer. And the gearbox, a smooth-changing automatic, doesn't much like to be hurried. Taking this car on a track? It's as wrong as playing rugby in a dinner jacket.

Later that night I had to go to Oxfordshire on the M40, something I did at exactly 65 mph. The Vanquish will go a lot faster than this – 118 mph faster, to be exact – but, well, er, the last time I drove a Vanquish on a motorway, I ended up having a little chat with some policemen and women policemen. And afterwards they took away my driving licence for two months.

So here we have a car that is deeply unhappy on a wet night in town, that doesn't much care for track work and that fills me with a teeth-itchingly morbid fear of being stopped by the police. Oh, and it had been decorated by someone who had a mental age of four.

They'd gone, as pre-school kids often do, for a very garish teal colour, and then for no reason at all had decided to paint the brake callipers yellow. Somehow pleased with the effect, they had decided it should be mirrored on the inside, so, yep, that meant teal seats with yellow flashings and, yes, wow, yellow tips on the paddle-shift levers. I've seen less gaudy birds of paradise.

I think I know what Aston is playing at. It is hoping that by going for extreme colours, it would stop me noticing that the interior of this supposedly brand-new car is a bit old-fashioned.

Which, of course, it is. As I said when I reviewed the hard-top version, Aston is a small company with limited resources. It simply doesn't have the £500 million you need to design a new air-conditioning system, or £200 million for a new instrument binnacle. So it keeps having to fit the same stuff it used in the previous car. The satnav is new(ish), and while it's better than the original setup, the screen does look a bit like the sort of drawing that proud parents put on a fridge door.

And I think that's enough now. I could give you a thousand reasons for not buying this car, even before we got to the whopping price tag. I could tell you that a Ferrari 458 Italia is better, and that this isn't even the best Aston. The Vantage S holds that crown. But I'm afraid there's no getting round the fact that I loved it. And the main reason I loved it is: you loved it even more.

Normally when I drive an obviously expensive car, people hate it and me. It turns their mouths to meal, and at petrol stations they sneer. 'Bet you don't get many miles to the gallon out of that,' they say. At road junctions they will not let me out. And at night they like to run coins down the side. Expensive cars make people cross. Porsches especially.

But the Aston has exactly the opposite effect. It makes everyone happy. One distinguished-looking man walked up to me in a traffic jam, clutched my forearm and said, 'That really does make the most glorious noise, old chap.' Later I came out of a shop in Notting Hill to find a young man staring at it. 'That's just . . .' – he paused for a long time, searching for the right word – 'beautiful.'

I got some idea of what it might have been like to be Jesus. One young woman – and I sincerely hope she's reading this – was so busy looking at the car that she tripped over the kerb and went flying. Hand on heart, I have never, in thirty years of writing about cars, driven anything that engenders such affection.

So who cares if it's expensive, or not as fast as it should be? Who cares that the instruments are now a bit old-fashioned and that you can't see out of the back? Why worry about fuel consumption or how the gearbox works or why there's understeer? This is a car that makes people like you. And that raises an interesting question.

At present, *Daily Maily* bits of Britain insist that MPs must spend no money at all. If there's even a whiff of a salary or an expenses claim or a new pair of shoes, they are hounded into a stammering, stuttering apology that makes them look weak and hopeless.

Naturally they feel they have to campaign on foot or on a bicycle, and that if they have to use a car it must be some form of hybrid. They think this makes them look 'real'. But actually it makes them look daft. Because we can see it's all phoney.

So I wonder what would happen if one of them decided that for the next election he should campaign from behind the wheel of a Vanquish Volante. Could a Tory take Rochdale this way? Could a socialist win the hearts and minds of the people in Stow-on-the-Wold? You know what? The car's allure is so powerful, I reckon he probably could.

29 December 2013